Kyoodoz™ GEOMETRY SOLUTIONS BOOK

Yvonne Low

For Vicky, mom and dad who have always believe the best in me.

I'd like to thank my husband, daughter and sisters for their support and love.

First Edition: April, 2006

Published by:
Kyoodoz
P.O. Box 60231
Sunnyvale, CA 94088–0231
www.kyoodoz.com

All the characters and adventures portrayed in this book are fictitious.

Library of Congress Control Number: 2006921591
ISBN-13: 978-0-9771172-1-5
ISBN-10: 0-9771172-1-9

10 9 8 7 6 5 4 3 2 1

Printed in the United States of America

Preface

Welcome to Kyoodoz™ Geometry Solutions Book, a highly focus and topic driven quality questions and answers book. I am truly grateful for the success and positive reviews for Kyoodoz™ Algebra Solutions Book, and for this book, I have continued with my endeavor to help you learn math through a creative and fun environment. Just like all other Kyoodoz™ books, Kyoodoz™ Geometry Solutions Book has been created to cater for teachers and students from the beginners, to the intermediate, and up to the advance levels. The purpose of this book is to help you build a solid foundation in geometry, through constant practice and exposure to various questions and solutions using a fun and interesting back drop that will not only aide in understanding, but also giving you a new experience in math.

If you are wondering what sets Kyoodoz™ Geometry Solutions Book apart? Firstly, there are no lengthy explanations, just pure questions and solutions presented in a clear and neat format that is easy for students to learn, practice, and reference. Secondly, I have researched and developed a list of quality questions that has been designed to help students understand and master the concepts behind the various geometry topics. My aim is to help you appreciate and learn the topics inside out, so you can grasp any concepts that come your way. Thirdly, I have provided complete step-by-step solutions for every question to guide you through the workings and calculations (yes I know that should be the way in a good solutions book!). The solutions are given right after every question to facilitate easy reference. So, no more flipping to the back for answers, and no more wondering how answers are derived! How cool is that!

For this book, I have included a few surprises, that I am sure will enhance your practice sessions. I have incorporated a storyline to the cartoons, so you can get acquainted with the cartoon characters, Wiki and Zoe as they embark on an adventure in search of a legendary missing dinosaur. I hope our heroes and their friends will keep you going with their misadventures in the Land of Geometrica while teaching and reminding you of useful math tips, and formulas. Of course, if you wish to skip to a specific chapter, you will have no trouble as every chapter unfolds a new adventure on its own, as such you will still benefit from the story as the essence of the cartoons and storyline are mainly to highlight formulas, tips and tricks in a more interesting and interactive manner . In addition, I have included a summary of essential concepts at the beginning of every topic for easy revision and reference. Also, to help you prepare for high school geometry as well as the new SAT requirements for math I have taken care to include all core topics in geometry including an extensive coverage on planes, transformations, and coordinate geometry.

Math should be enjoyable and easy to understand, and that is what this book endeavor to share with you. And with practice, you'll sharpen your skills and strengthen your confidence. So put away your fear of math and prepare yourself for a fun learning experience, your journey begins here! Good Luck!

Yours truly,
Yvonne Low

How to use this book

If you are studying alone:

1. This book is most useful when use in tandem with a textbook.
2. Refresh your understanding with the chapter's preceding essential notes (found in the first page of every chapter).
3. Practice each question. The questions have been designed to show you how every concept can be applied and tested in different forms.
4. Check your answers and workings with the accompanying step-by-step solutions.
5. The cartoon panels provide you with tips, tricks, shortcuts and reminders. Common pitfalls and mistakes are also pointed in the cartoons.
6. Remember continuous practice and exposure to quality questions will sharpen and enhance your geometry skills and ability.

If you are mentoring or tutoring:

1. Use Kyoodoz™ Geometry Solutions Book together with a textbook.
2. Refresh and revise your students' understanding with the chapter's preceding essential notes (found in the first page of every chapter).
3. Go through each question with your students and remind them to check their workings and answers against the step-by-step solutions that are provided. Learning the correct steps in calculation will set and strengthen the foundation for more advance calculation in mathematics.
4. The questions in this book are also ideal for quizzes, tests and assignment homework as they will test students' understanding and appreciation of the various theories and concepts.

There is no easy way to success but this book will definitely help you in your pursuit for excellence!

Contents

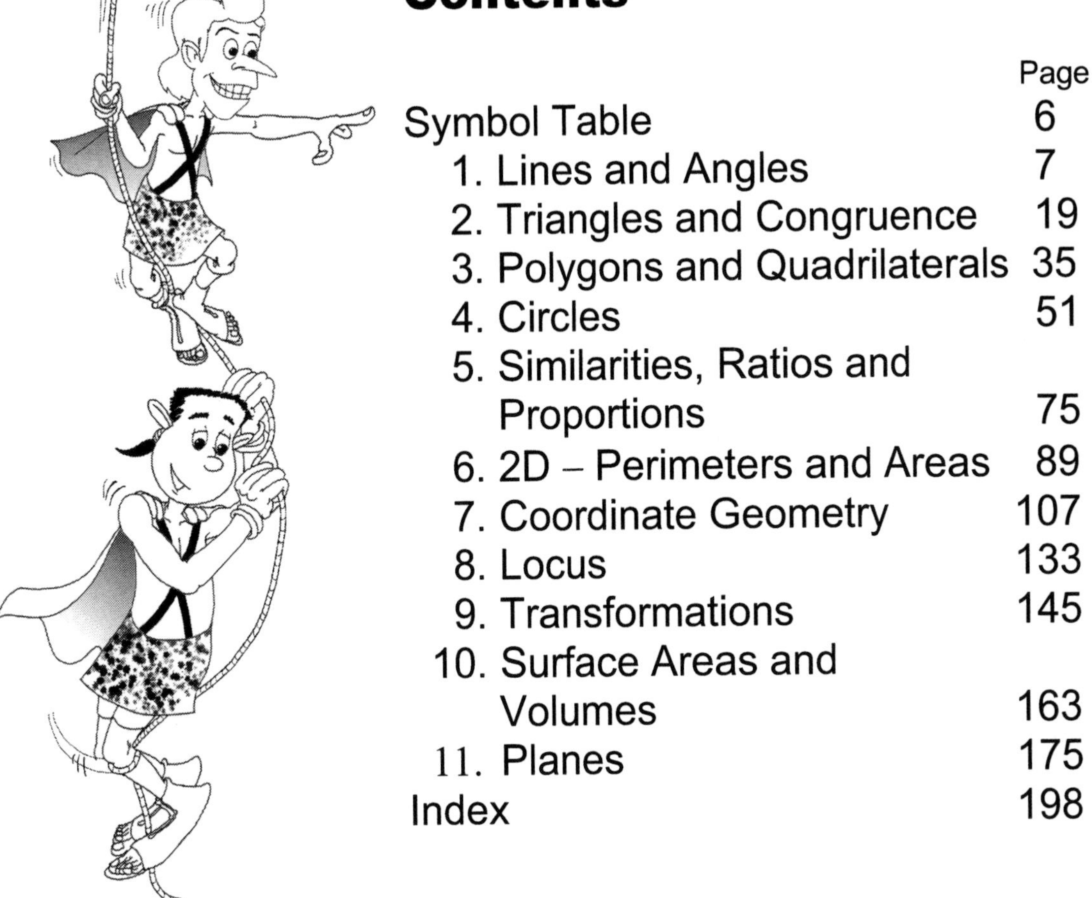

Symbol Table

Symbol	Meaning
$\overleftrightarrow{AB}$	Line AB that extends indefinitely
$\overline{AB}$	Line segment with endpoints A and B
$\overrightarrow{AB}$	Ray with endpoint A and extends indefinitely in the direction of B
$\overset{\frown}{AB}$	Arc AB
$m\overset{\frown}{AB}$	measure of arc AB
$\cong$	Congruent to
$=$	Equal to
$\triangle$	Triangle
$\angle$	Angle
$^\circ$	Degree
$\perp$	Perpendicular to
α	Alpha
β	Beta
θ	Theta
$\therefore$	Therefore
π	Pi, ratio of the circumference of a circle to its diameter
$\pm$	Plus or minus
$\parallel$	Parallel to
$+$	Plus or positive
$-$	Minus or negative
$\times$	Multiply
$\div$	Divide
$=>$	As a result / Can be deduced
$\approx$	Approximately
$\underset{\sim}{n}$	Normal vector
sin	Sine
cos	Cosine
tan	Tangent
$\otimes$	Points where loci intersect
rad	Radian
$\sqrt{a}$	Square root of 'a'
V	Volume
//	Final answer
$\lvert a \rvert$	Absolute value of 'a'
″ or in	Inch (plural Inches)
′ or ft	Foot (plural Feet)
cm	Centimeter
yd	Yard
mm	Millimeter

Chapter 1
Lines and Angles

$\overline{AB}$, **Line segment** – a section of a line between 2 endpoints A and B

$\angle$, **Angle** – formed by two intersecting lines

Straight line angles = 180°:

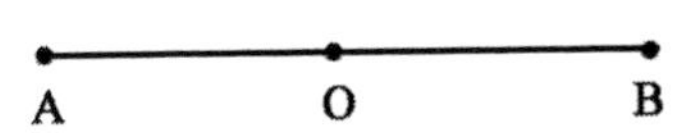

Sum of **supplementary angles** = 180°:

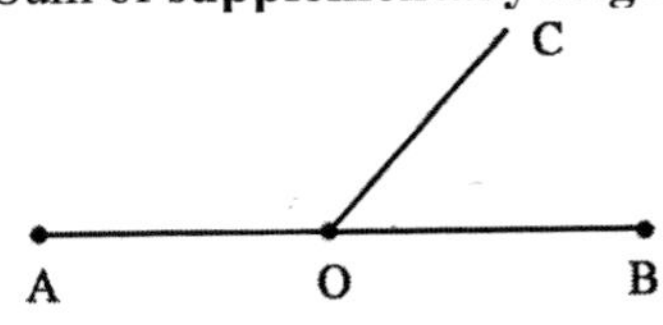

$\angle AOC$ and $\angle BOC$ are supplementary angles

$\perp$, **Perpendicular lines** $\overline{AO} \perp \overline{OB}$ **:**

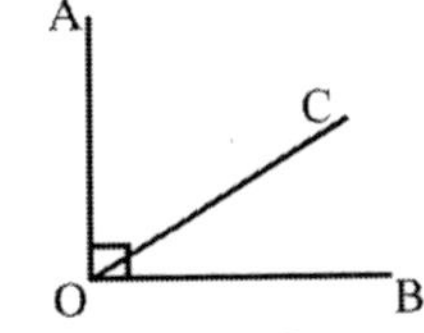

$\angle AOC$ and $\angle BOC$ are **complementary angles**, i.e. $\angle AOC + \angle BOC = 90°$

Adjacent angles:

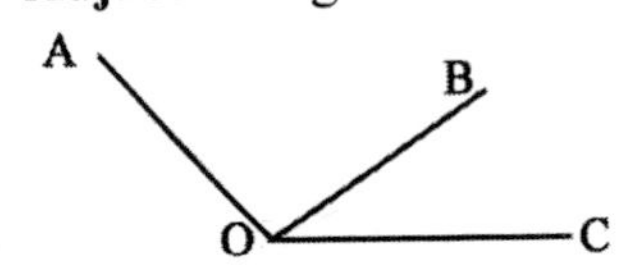

$\angle AOB$ & $\angle BOC$ are adjacent angles

Vertical angles:

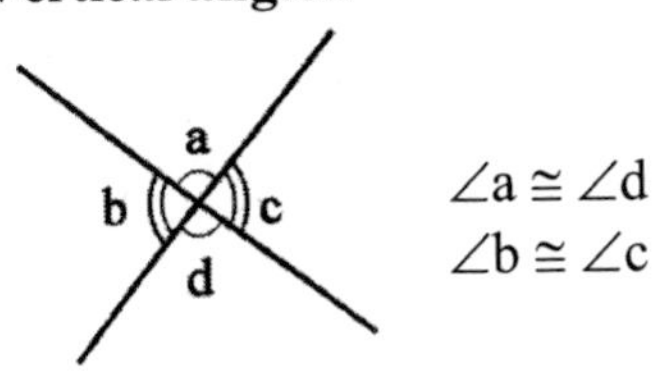

$\angle a \cong \angle d$

$\angle b \cong \angle c$

Acute angle: less than 90°

Obtuse angle, $\angle x$: $90° < x < 180°$

Reflex angle, $\angle x$: $180° < x < 360°$

Bisector – divides an angle into 2 equal angles

Transversal – line that intersects 2 or more lines

$\cong$, **Congruent** – same shape and same size

Corresponding angles are congruent:

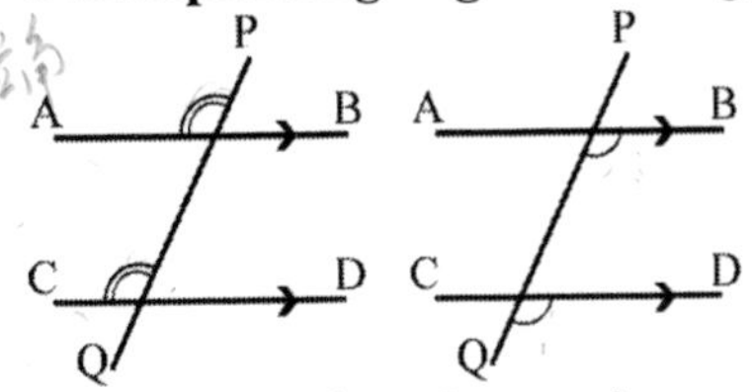

Consecutive interior angles are supplementary (= 180°):

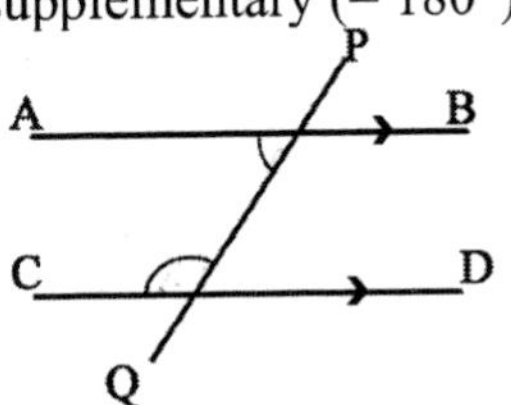

Alternate interior angles are congruent:

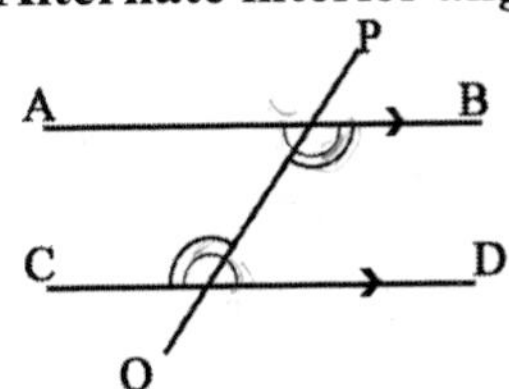

Alternate exterior angles are congruent:

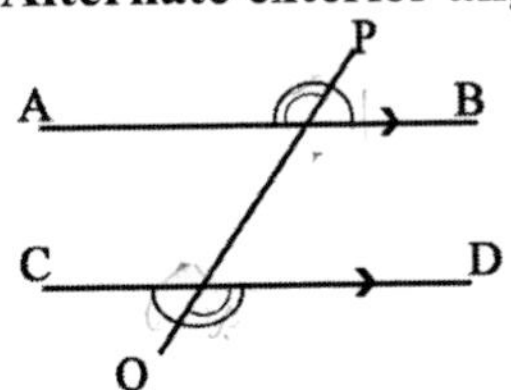

||, **Parallel lines** – lines in the same plane that never meet

Reflexive property of equality: AB = BA

Collinear – on the same line

Chapter 1 Lines & Angles

1. Find the value of θ in Figure 1.1 below.

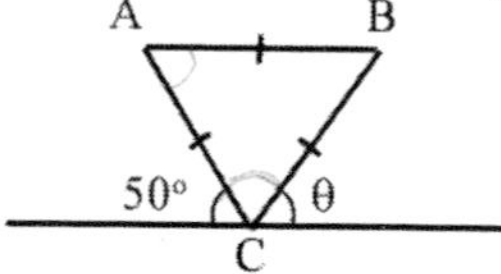

Figure 1. 1

Answer:
Given $\triangle ABC$ = equilateral triangle:
$\angle ACB = 180° \div 3 = 60°$ ...*
Straight line angles = 180°
$50° + \angle ACB + \theta = 180°$
$50° + 60° + \theta = 180°$ ⇦ From *
$\theta = 180° - 60° - 50°$
$\therefore \theta = 70°$ //

2. In Figure 1.2, $\overleftrightarrow{AB}$ and $\overleftrightarrow{CD}$ are parallel lines. Find the values of α, β, and θ.

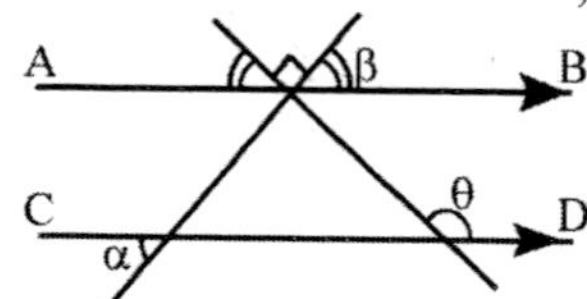

Figure 1. 2

Answers:
Given $\overleftrightarrow{AB} \parallel \overleftrightarrow{CD}$
Sum of straight line angles = 180°
$\beta + \beta + 90° = 180°$
$2\beta = 90°$
$\therefore \beta = 45°$ //

Alternate exterior angles are $\cong$:
$\alpha \cong \beta$
$\therefore \alpha = 45°$ //

Vertical angles are $\cong$:

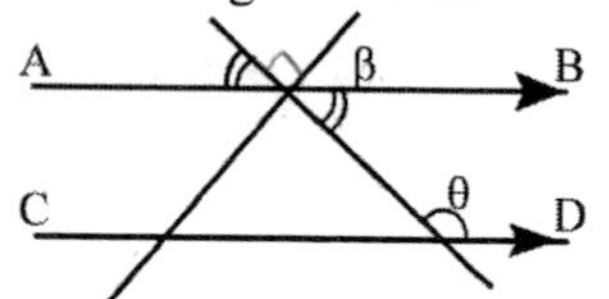

Consecutive angles are supplementary:
$\beta + \theta = 180°$
$\theta = 180° - 45°$
$\therefore \theta = 135°$ //

3. In Figure 1.3, $\overline{BE}$ and $\overline{CF}$ are parallel lines while $\overline{CE}$ and $\overline{DF}$ are parallel lines. If $\overrightarrow{DA}$ is a straight line, find the values of x and y.

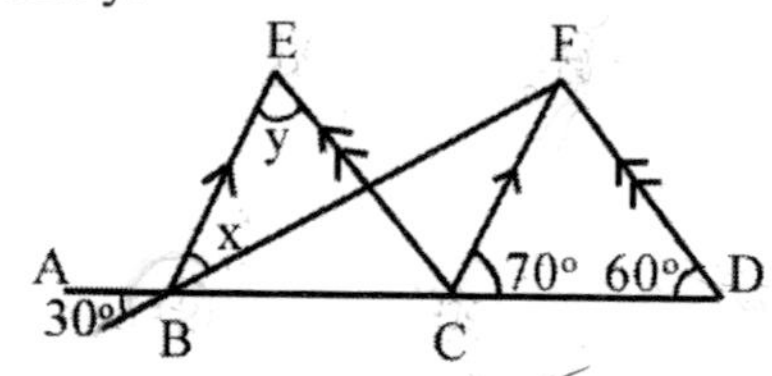

Figure 1. 3

Answers:
Vertical angles are $\cong$:
=> $\angle DBF \cong 30°$
Given $\overline{BE} \parallel \overline{CF}$, corresponding angles are $\cong$:
$\angle DBE = \angle DCF$
$\angle DBF + \angle EBF = 70°$
$30° + x = 70°$
$\therefore x = 40°$ //
Given $\overline{CE} \parallel \overline{DF}$:
Corresponding angles are $\cong$:
$\angle BCE \cong \angle BDF = 60°$
Since $\overline{BE} \parallel \overline{CF}$:

Alternate interior angles are ≅:
$\angle BEC = \angle ECF = y$
Sum of straight line angles = 180°
$\angle ECF + \angle BCE + \angle DCF = 180°$
$y + 60° + 70° = 180°$
$\therefore y = 50°$ //

4. A theorem is a statement that can be proven true. True or false?

Answer:
True. A theorem is a statement that can be proven true. A *postulate* is a statement that is accepted as true without proof. //

5. a) What angle is its own complement?
b) Hence what angle is its own supplement?

Answers:
a) 45°. Sum of angles that are complement must equal 90°, angle with measure 45° is therefore its own complement. //

b) 90°. Sum of the angles that are supplementary must equal 180°. Thus 90° + 90° = 180°, therefore 90° is its own supplement. //

6. In Figure 1.4, $\overleftrightarrow{K}$ and $\overleftrightarrow{L}$ are parallel lines while $\overleftrightarrow{M}$ is a transversal. Find the sum of x, y and z.

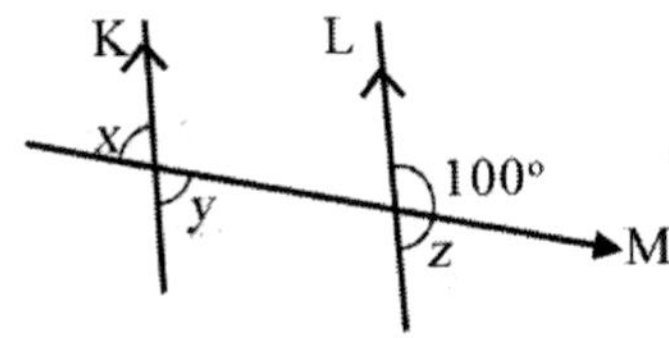

Figure 1. 4
Answer:
Sum of straight line angles = 180°
$100° + z = 180°$
$\therefore z = 80°$
Given $\overleftrightarrow{K}$ and $\overleftrightarrow{L}$ are parallel lines:
Corresponding angles are ≅:
$y \cong z$
$\therefore y = 80°$
Alternate exterior angles are ≅:
$x \cong z$
$\therefore x = 80°$
Thus,
$x + y + z = 80° + 80° + 80°$
$= 240°$ //

7. In Figure 1.5, A, B and C are the intersection points of three lines. Express the sum of x and y in terms of m.

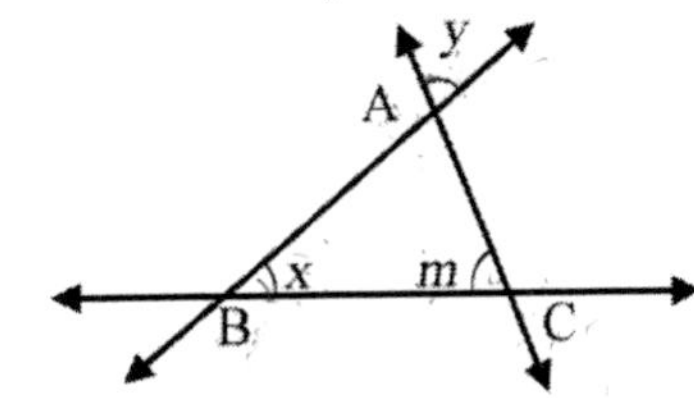

Figure 1. 5

Answer:
Vertical angles are ≅:
$\angle BAC \cong y$
Sum of angles in a triangle = 180°
$\angle BAC + \angle ABC + \angle ACB = 180°$
$y + x + m = 180°$
$\therefore y + x = 180° - m$ //

8. In Figure 1.6, $\overleftrightarrow{AB}$ and $\overleftrightarrow{CD}$ are parallel lines. Find the value of x.

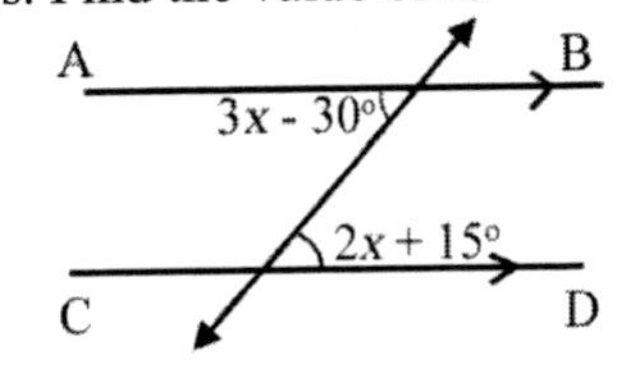

Figure 1. 6

Answer:
Given $\overleftrightarrow{AB} \parallel \overleftrightarrow{CD}$
Alternate interior angles are ≅:
$3x - 30° = 2x + 15°$
$\therefore x = 45°$ //

9. In Figure 1.7, determine the exterior angle to the interior angle, a.

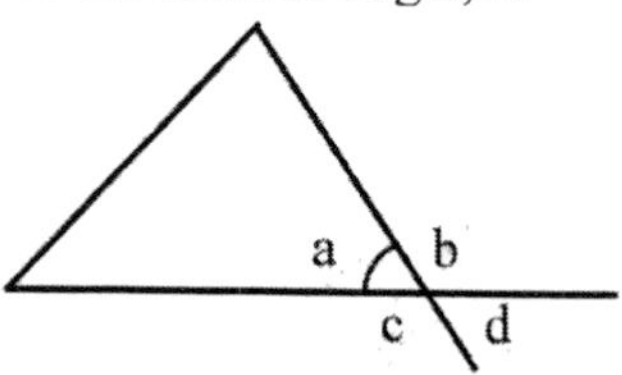

Figure 1. 7

Answer:
∠b or ∠c.
Exterior angle refers to the angle that is *adjacent* and *supplementary* (sum of the two angles equals 180°) to the interior angle. //

10. Find the exterior angle to x in Figure 1.8.

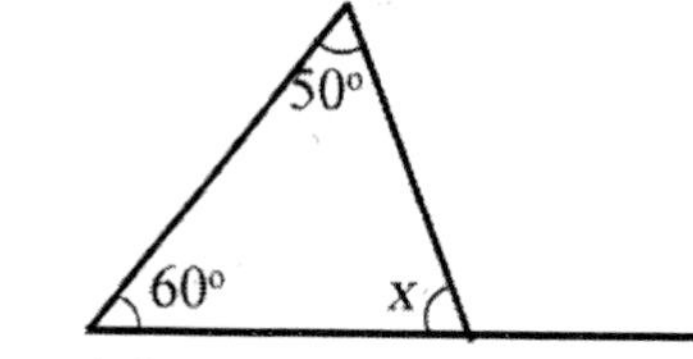

Figure 1. 8

Answer:
Let y = exterior angle of x
Reproduce Figure 1.8:

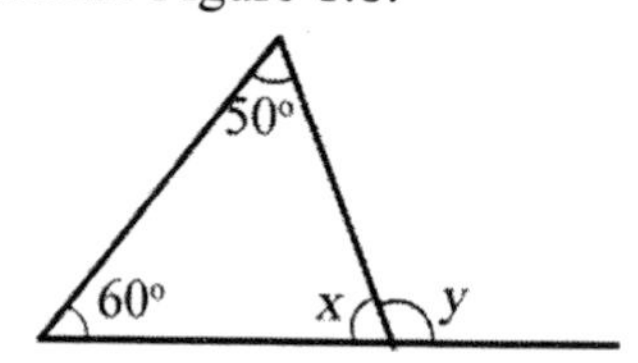

Sum of internal angles in a triangle = 180°
$60° + 50° + x = 180°$
$x = 70°$
Sum of straight line angles = 180°
$x + y = 180°$
$y = 180° - 70°$
$\therefore y = 110°$ //

Alternatively:
$y = a + b$
$y = 50° + 60°$
$\therefore y = 110°$ //

11. Given $\overrightarrow{AE}$ and $\overrightarrow{CH}$ are parallel lines (see Figure 1.9) and $\overline{DFG}$ and $\overline{CF}$ are straight lines. Find the value of x.

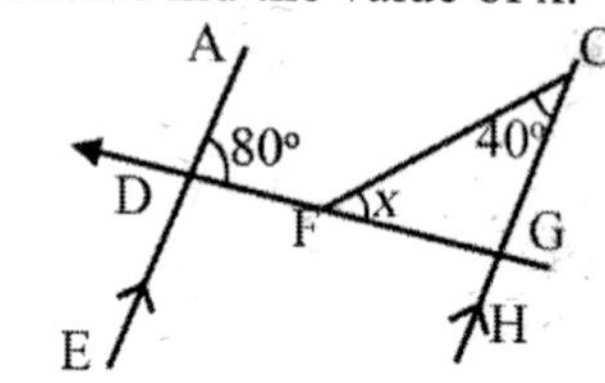

Figure 1. 9

Answer:
Given $\overleftrightarrow{AE} \parallel \overleftrightarrow{CH}$:
Alternate interior angles are ≅:
∠HGD ≅ ∠ADG
∠HGD = 80°
Exterior angle = sum of 2 nonadjacent angles in a triangle:
∠HGD = ∠CFG + ∠HCF
80° = x + 40°
∴ x = 40° //

Alternatively:
Consecutive interior angles = 180°:
∠CGD = 180° – ∠ADG
= 180° – 80°
= 100°
Sum of internal angles in triangle = 180°
∠CFG + ∠FCH + ∠CGD = 180°
x + 40° + 100° = 180°
∴ x = 40° //

12. $\overleftrightarrow{L}$ and $\overleftrightarrow{M}$ are parallel lines (see Figure 1.10) on the same plane. Find the values of x and y.

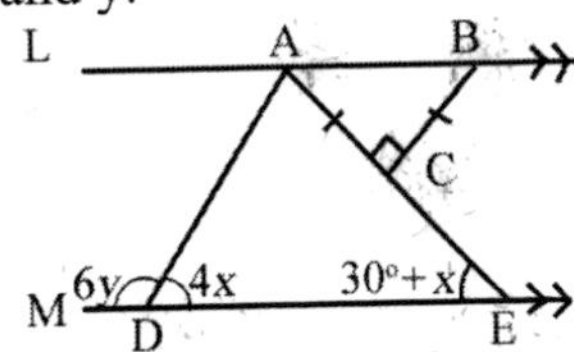

Figure 1. 10

Answers:
Given $\overline{AC} = \overline{BC}$
=> △ABC is an isosceles triangle
=> ∠CAB = ∠CBA ...*
∠CAB + ∠CBA + ∠ACB = 180°
∠CAB + ∠CAB + 90° = 180° ⇦ From *
2∠CAB = 90°
∠CAB = 45°
Also given line L || line M:
Alternate interior angles are ≅:
∠AED ≅ ∠CAB
30° + x = 45°
∴ x = 15° //
Sum of straight line angles = 180°
∠ADM + ∠ADE = 180°
6y + 4x = 180°
6y + 4(15°) = 180°
6y = 180° – 60°
6y = 120°
∴ y = 20° //

13. In Figure 1.11, $\overline{PQ}$ and $\overline{SR}$ are parallel lines and $\overline{PR}$ and $\overline{QS}$ are perpendicular lines. Find the value of x.

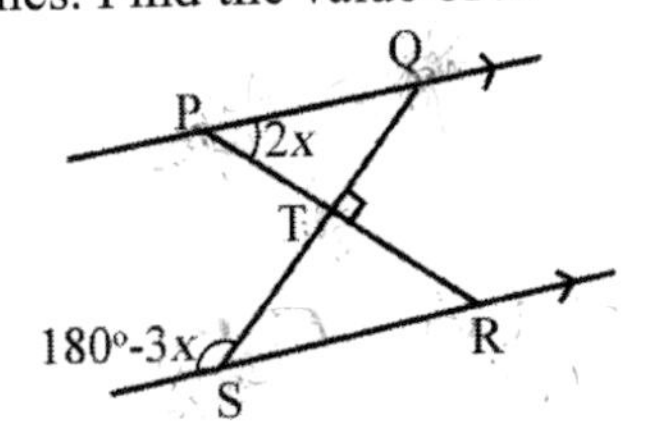

Figure 1. 11

Answer:
Given $\overline{QS} \perp \overline{PR}$
=> ∠PTQ = ∠QTR = 90°
Sum of straight line angles = 180°
∠QSR = 180° – (180° – 3x)
= 3x
Also given $\overline{PQ} \parallel \overline{SR}$
Alternate interior angles are ≅:
∠PQS ≅ ∠QSR
∠PQS = 3x
Sum of interior angles in △PQT = 180°
∠QPR + ∠PQS + ∠PTQ = 180°
2x + 3x + 90° = 180°
5x = 90°
∴ x = 18° //

14. In Figure 1.12, $\overrightarrow{TP}$ and $\overrightarrow{VQ}$ are parallel lines while $\triangle RST$ is a scalene triangle. Find the values of:
a) w
b) x
c) y + z.

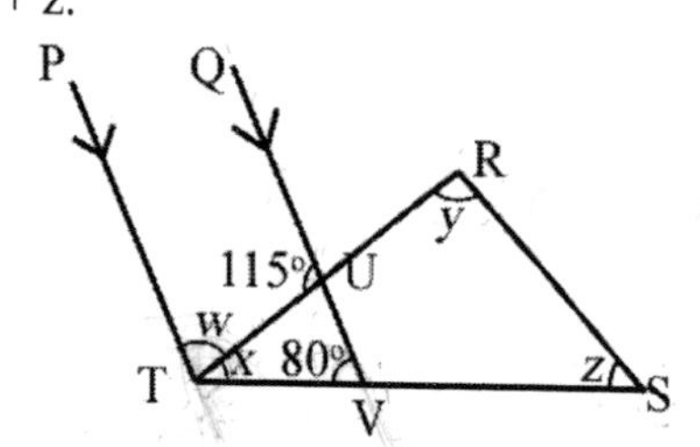

Figure 1. 12

Answers:
Given $\overrightarrow{TP} \parallel \overrightarrow{VQ}$
a) Consecutive angles are supplementary:
$\angle PTR + \angle QUT = 180°$
$w + 115° = 180°$
$\therefore w = 65°$ //

b) Consecutive angles are supplementary:
$\angle PTS + \angle QVT = 180°$
$\angle PTR + \angle RTS + 80° = 180°$
$w + x = 100°$
$x = 100° - 65°$
$\therefore x = 35°$ //

c) Sum of internal angles in triangle = 180°
$\angle RTS + \angle SRT + \angle TSR = 180°$
$x + y + z = 180°$
$y + z = 180° - 35°$
$\therefore y + z = 145°$ //

15. In Figure 1.13, $\overline{CA}$ is a straight line. If $\overrightarrow{CD}$ is parallel to $\overrightarrow{BG}$ and $\overrightarrow{CE}$ is parallel to $\overrightarrow{BF}$, find the value of x.

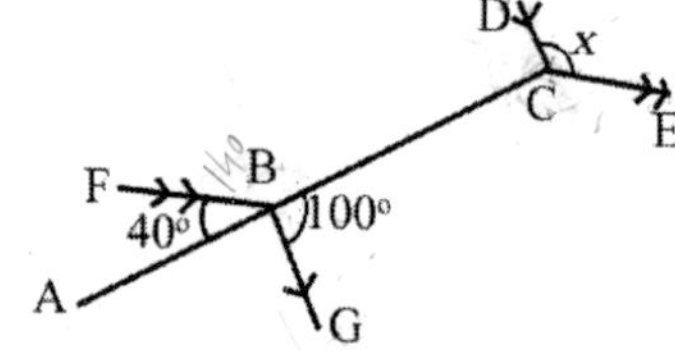

Figure 1. 13

Answer:
Sum of straight line angles = 180°
$\angle ABF + \angle CBF = 180°$
$\angle CBF = 180° - 40° = 140°$
Given $\overrightarrow{CE} \parallel \overrightarrow{BF}$:
Alternate interior angles are $\cong$:
$\angle CBF \cong \angle ECA$
$\angle ECA = 140°$
Given $\overrightarrow{BG} \parallel \overrightarrow{CD}$:
Alternate interior angles are $\cong$:
$\angle ACD \cong \angle CBG$
$\angle ACD = 100°$
Sum of angles around a point = 360°
$\angle DCE + \angle ECA + \angle ACD = 360°$
$x + 140° + 100° = 360°$
$\therefore x = 120°$ //

16. In Figure 1.14, $\overline{ABC}$ is a straight line while $\overrightarrow{BF}$ and $\overrightarrow{CE}$ are parallel lines. If CE = CD, solve for x + y.

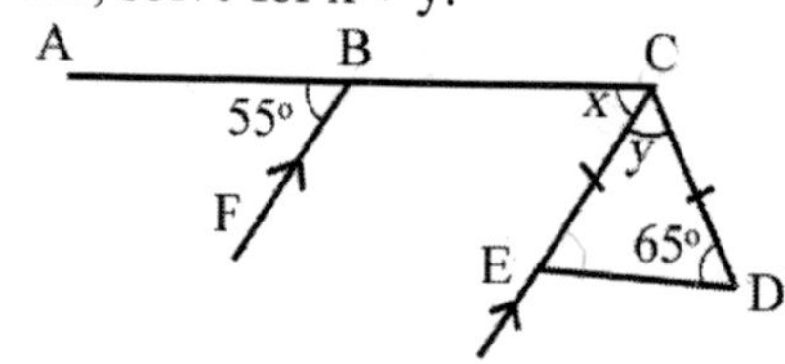

Figure 1. 14

Answer:
Given $\overrightarrow{BF} \parallel \overrightarrow{CE}$:
Corresponding angles are $\cong$:
$\angle ACE \cong \angle ABF$
$x = 55°$
Given CE = CD
=> $\triangle CDE$ = isosceles $\triangle$

$\Rightarrow \angle CDE \cong \angle CED = 65°$ ...*
Sum of internal angles in a triangle = 180°
$\angle ECD + \angle CDE + \angle CED = 180°$
$y + 65° + 65° = 180°$ ⇦ From *
$y = 50°$
Thus,
$x + y = 55° + 50°$
$\therefore x + y = 105°$ //

17. $\overline{AB}$ and $\overrightarrow{EF}$ are parallel lines in Figure 1.15. If it is further known that $\overline{AB}$ and $\overline{AC}$ are equal in length, find the value of x.

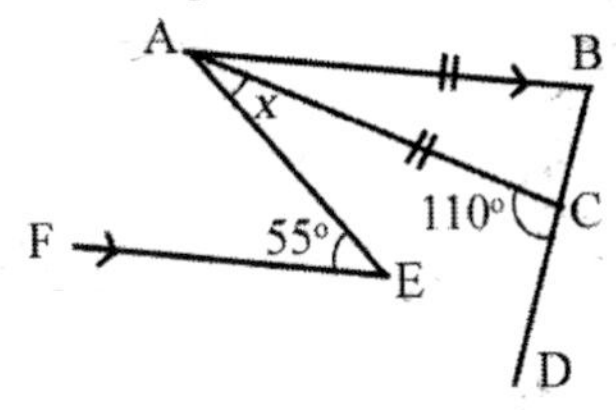

Figure 1. 15

Answer:
Sum of straight line angles = 180°
$\angle ACD + \angle ACB = 180°$
$\angle ACB = 180° - 110°$
$\angle ACB = 70°$

Given AB = AC
$\Rightarrow \triangle ABC$ = isosceles $\triangle$
$\Rightarrow \angle ACB \cong \angle ABD = 70°$ ...(1)
Sum of internal angles in a triangle = 180°
$\angle ACB + \angle ABD + \angle BAC = 180°$
$70° + 70° + \angle BAC = 180°$ ⇦ From (1)
$\angle BAC = 180° - 70° - 70° = 40°$ ...(2)

Also given $\overline{AB} \parallel \overrightarrow{EF}$:
Alternate interior angles are ≅:
$\angle BAE \cong \angle AEF$
$\angle BAC + \angle CAE = 55°$ ⇦ From (2)
$40° + x = 55°$
$\therefore x = 15°$ //

18. In Figure 1.16, $\overrightarrow{BA}$ and $\overrightarrow{GE}$ are parallel lines while $\overrightarrow{FC}$ and $\overrightarrow{GD}$ are parallel lines. Determine the values of x and y.

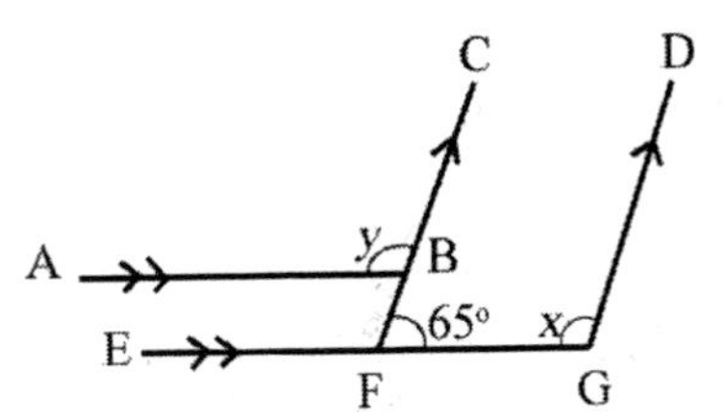

Figure 1. 16

Answers:
Given $\overrightarrow{FC} \parallel \overrightarrow{GD}$:
Consecutive angles are supplementary:
$\angle DGE + \angle CFG = 180°$
$x + 65° = 180°$
$\therefore x = 115°$ //

Sum of straight line angles = 180°
$\angle CFE + \angle CFG = 180°$
$\angle CFE + 65° = 180°$
$\angle CFE = 115°$
Also given $\overrightarrow{BA} \parallel \overrightarrow{GE}$:
Corresponding angles are ≅:
$\angle ABC \cong \angle CFE$
$\therefore y = 115°$ //

19. In Figure 1.17, $\overleftrightarrow{AD}$ is a straight line. It is further given that $\overrightarrow{AE}$ and $\overrightarrow{GB}$ and $\overrightarrow{CF}$ are parallel lines while $\overline{AH}$ and $\overline{GC}$ are parallel lines. Find the values of
a) x
b) y
c) z

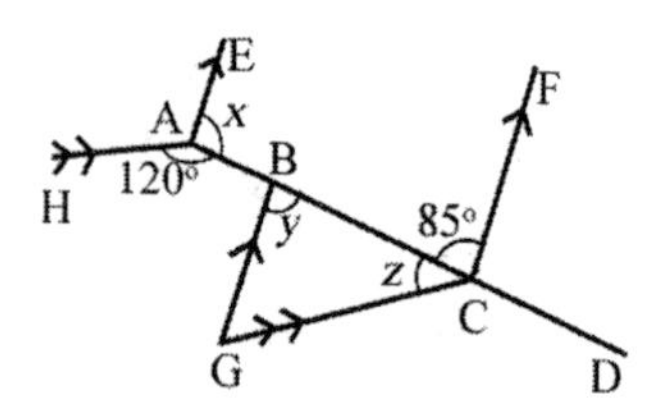

Figure 1. 17

Answers:

a) Given $\overleftrightarrow{AE} \parallel \overleftrightarrow{CF}$
Consecutive angles are supplementary:
$\angle EAD + \angle ACF = 180°$
$x + 85° = 180°$
$\therefore x = 95°$ //

b) Given $\overline{GB} \parallel \overleftrightarrow{CF}$
Alternate interior angles are ≅:
$\angle GBD \cong \angle ACF$
$\therefore y = 85°$ //

c) Given $\overleftrightarrow{AH} \parallel \overline{GC}$
Consecutive angles are supplementary:
$\angle GCA + \angle HAD = 180°$
$z + 120° = 180°$
$\therefore z = 60°$ //

20. In Figure 1.18, $\overline{PQ}$ and $\overline{ST}$ are parallel lines and $\overline{SQ}$ and $\overleftrightarrow{TU}$ are parallel lines. If $\overline{PRT}$ is a straight line, find the value of x.

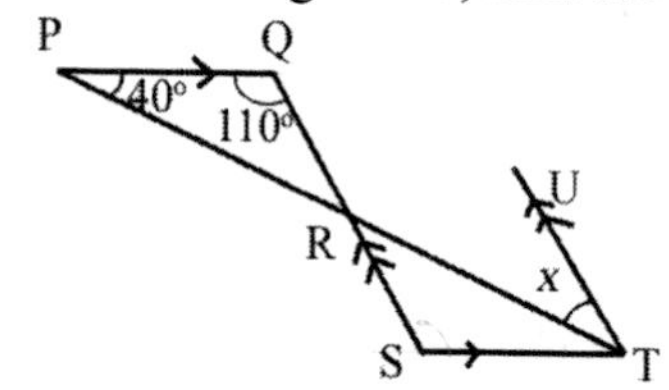

Figure 1. 18

Answer:
Given $\overline{PQ} \parallel \overline{ST}$
Alternate interior angles are ≅:
$\angle QST \cong \angle PQS$
$\angle QST = 110°$ ...(1)
Also,
$\angle PTS \cong \angle QPT$
$\angle PTS = 40°$ ...(2)
Since $\overline{SQ} \parallel \overleftrightarrow{TU}$,
Consecutive angles are supplementary:
$\angle QST + \angle UTS = 180°$
$\angle QST + \angle PTS + \angle UTP = 180°$
Substitute (1) & (2):
$110° + 40° + x = 180°$
$\therefore x = 30°$ //

Alternatively:
Given $\triangle PQR$
Sum of interior angles in a triangle = 180°
$\angle QRP + \angle PQS + \angle QPT = 180°$
$\angle QRP + 110° + 40° = 180°$
$\angle QRP = 30°$
Since $\overline{SQ} \parallel \overleftrightarrow{TU}$
Corresponding angles are ≅:
$\angle UTP \cong \angle QRP$
$\therefore x = 30°$ //

21. In Figure 1.19, $\overleftrightarrow{BG}$ and $\overleftrightarrow{CF}$ are parallel lines. Find the values of x and y.

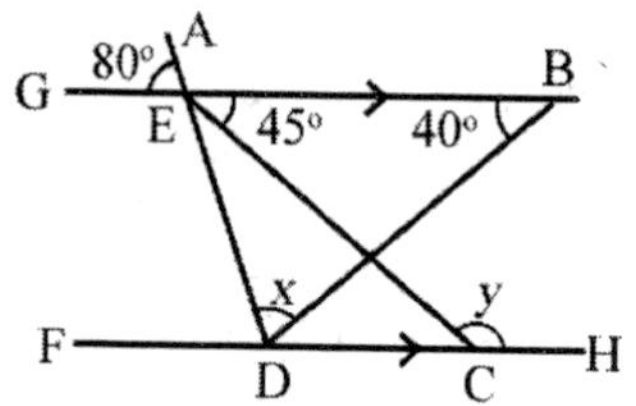

Figure 1. 19

Answers:
Given $\overline{BG} \parallel \overline{CF}$
Corresponding angles are ≅:
$\angle ADF \cong \angle AEG$
$\angle ADF = 80°$
Alternate interior angles are ≅:

$\angle BDC \cong \angle DBG$
$\angle BDC = 40°$
Sum of straight line angles = 180°
$\angle ADB + \angle ADF + \angle BDC = 180°$
$x + 80° + 40° = 180°$
$\therefore x = 60°$ //
Alternate interior angles are ≅:
$\angle ECF \cong \angle BEC$
$\angle ECF = 45°$
Sum of straight line angles = 180°
$\angle ECH + \angle ECF = 180°$
$y + 45° = 180°$
$\therefore y = 135°$ //

22. In Figure 1.20, $\overline{AC}$ and $\overline{FE}$ are parallel lines, while $\overline{FA}$ and $\overline{EB}$ are parallel lines. If $\overline{CDF}$ is a straight line, find the values of x and y.

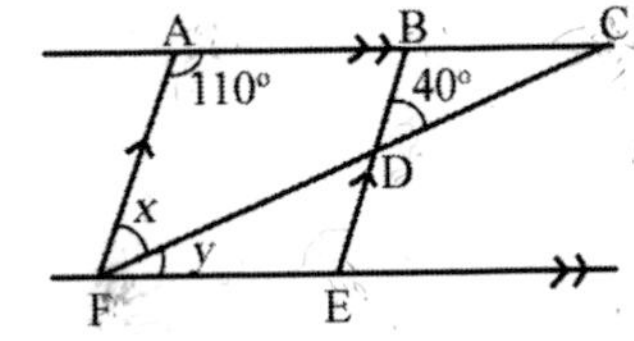

Figure 1. 20

Answers:
Given $\overline{FA} \parallel \overline{EB}$
Corresponding angles are ≅:
$\angle AFC \cong \angle BDC$
$\therefore x = 40°$ //
Given $\overline{AC} \parallel \overline{FE}$
Consecutive angles are supplementary:
$\angle EFA + \angle CAF = 180°$
$\angle AFC + \angle CFE + \angle CAF = 180°$
$x + y + 110° = 180°$
$40° + y = 70°$
$\therefore y = 30°$ //

23. Given in Figure 1.21, $\overleftrightarrow{FG}$ and $\overleftrightarrow{DE}$ are parallel lines. If $\overleftrightarrow{AC}$ and $\overline{BD}$ are straight lines, find the values of x and y.

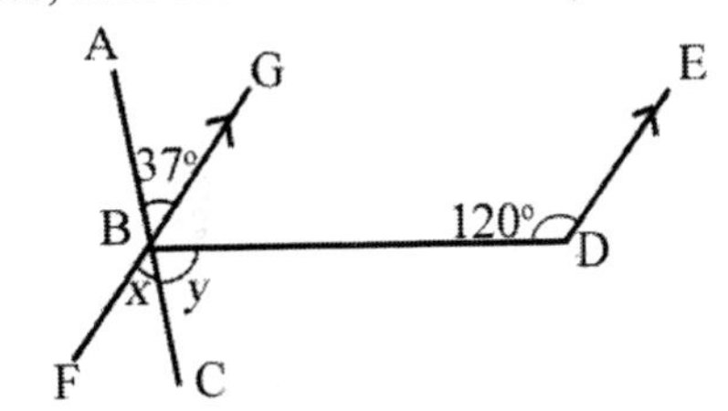

Figure 1. 21

Answers:
Vertical angles are ≅:
$\angle FBC \cong \angle ABG$
$\therefore x = 37°$ //
Given $\overleftrightarrow{FG} \parallel \overleftrightarrow{DE}$
Alternate interior angles are ≅:
$\angle DBF \cong \angle BDE$
$\angle FBC + \angle CBD = 120°$
$x + y = 120°$
$37° + y = 120°$
$\therefore y = 83°$ //

24. Given $\overrightarrow{BA}$ parallel to $\overrightarrow{CG}$ and $\overline{CF}$ parallel to $\overline{DE}$ (see Figure 1.22). If $\overline{DB}$ is a straight line, find the values of θ and α.

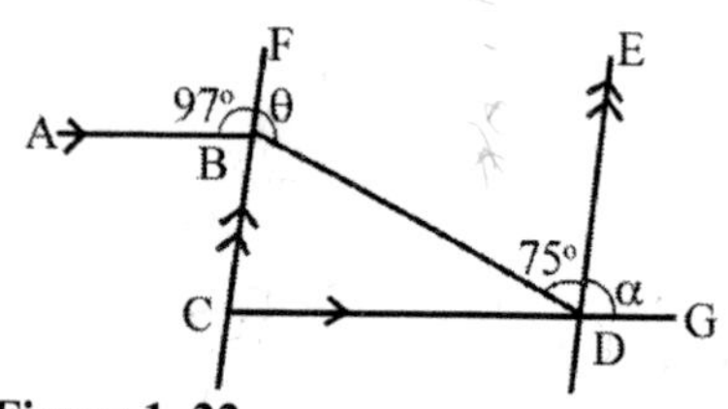

Figure 1. 22

Answers:
Given $\overrightarrow{CF} \parallel \overrightarrow{DE}$
Alternate interior angles are ≅:
$\angle DBC \cong \angle BDE$
$\angle DBC = 75°$
Sum of straight line angles = 180°
$\angle DBF + \angle DBC = 180°$
$\theta + 75° = 180°$
$\therefore \theta = 105°$ //

Sum of straight line angles = 180°
$\angle ABC = 180° - \angle ABF$
$= 180° - 97° = 83°$
Given $\overrightarrow{BA} \parallel \overrightarrow{CG}$
Alternate interior angles are ≅:
$\angle FCG \cong \angle ABC = 83°$
Since $\overrightarrow{CF} \parallel \overrightarrow{DE}$
Corresponding angles are ≅:
$\angle EDG \cong \angle FCG$
$\alpha = 83°$ //

25. In Figure 1.23, it is given $\overleftrightarrow{AH}$, $\overline{DC}$ and $\overline{GF}$ are parallel lines. If $\overline{BG}$ and $\overline{DF}$ are transversals, find the values of x, y and z.

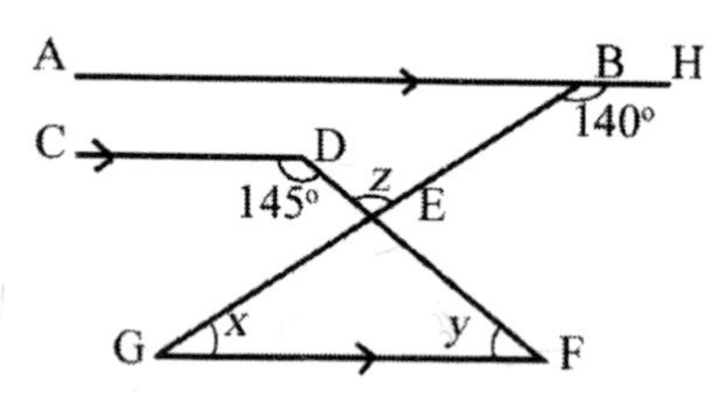

Figure 1. 23

Answers:
Given $\overleftrightarrow{AH} \parallel \overline{GF}$
Consecutive angles are supplementary:
$\angle BGF + \angle GBH = 180°$
$x + 140° = 180°$
$\therefore x = 40°$ //
Given $\overline{DC} \parallel \overline{GF}$
Consecutive angles are supplementary:
$\angle DFG + \angle CDF = 180°$
$y + 145° = 180°$
$\therefore y = 35°$ //
Sum of interior angles in a triangle = 180°
$\angle FEG + \angle BGF + \angle DFG = 180°$
$\angle FEG + x + y = 180°$
$\angle FEG + 40° + 35° = 180°$
$\angle FEG = 105°$
Vertical angles are ≅:
$\angle FEG \cong \angle BED$
$\therefore z = 105°$ //

26. In Figure 1.24, $\overleftrightarrow{AB}$ and $\overline{DC}$ are parallel lines. Find the value of y – x.

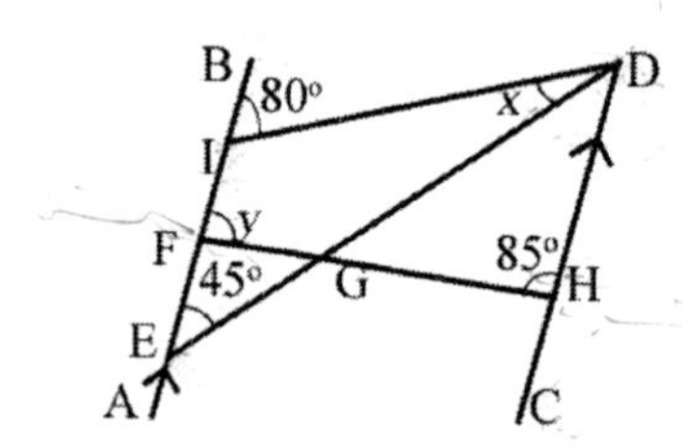

Figure 1. 24

Answer:
Given $\overleftrightarrow{AB} \parallel \overline{DC}$
Alternate interior angles are ≅:
$\angle CDE \cong \angle BED$
$\angle CDE = 45°$
Also
$\angle CDI \cong \angle BID$
$\angle IDE + \angle CDE \cong \angle BID$

$x + 45° = 80°$
$x = 35°$ ...(1)
Consecutive angles are supplementary:
$\angle BFH + \angle DHF = 180°$
$y + 85° = 180°$
$\therefore y = 95°$ ...(2)
From (1) and (2):
$y - x = 95° - 35°$
$= 60°$ //

27. In Figure 1.25, $\overrightarrow{DE}$ and $\overrightarrow{CF}$ are parallel lines while $\overrightarrow{CA}$ and $\overrightarrow{GH}$ are parallel lines. If $\overrightarrow{BI}$ is a straight line, determine the values of x, y and x + y.

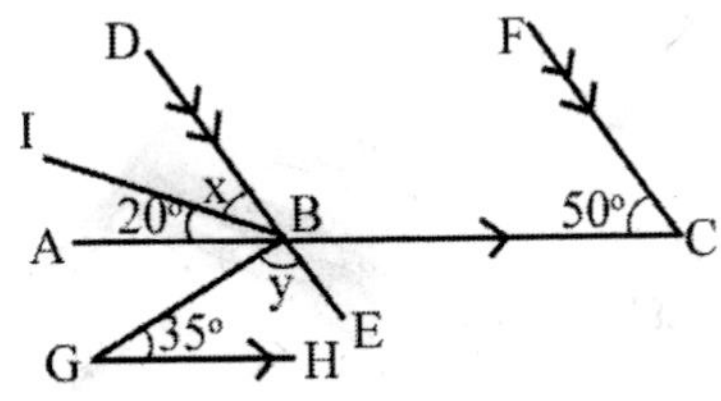

Figure 1. 25

Answers:
Given $\overleftrightarrow{DE} \parallel \overleftrightarrow{CF}$
Corresponding angles are ≅:
$\angle DBA \cong \angle FCA$
$\angle DBI + \angle ABI \cong \angle FCA$
$x + 20° = 50°$
$x = 30°$ //
Given $\overleftrightarrow{CA} \parallel \overleftrightarrow{GH}$
Alternate interior angles are ≅:
$\angle ABG \cong \angle BGH = 35°$
Sum of straight line angles = 180°
$\angle EBG + \angle ABG + \angle ABI + \angle DBI = 180°$
$y + 35° + 20° + x = 180°$
$y + 35° + 20° + 30° = 180°$
$\therefore y = 95°$ //
Thus,
$x + y = 30° + 95°$
$= 125°$ //

28. In Figure 1.26, $\overrightarrow{DA}$ and $\overleftrightarrow{EH}$ are parallel lines while $\overline{BL}$ and $\overline{CF}$ are parallel lines. If $\overline{DJ}$ is a straight line, find the values of x and y.

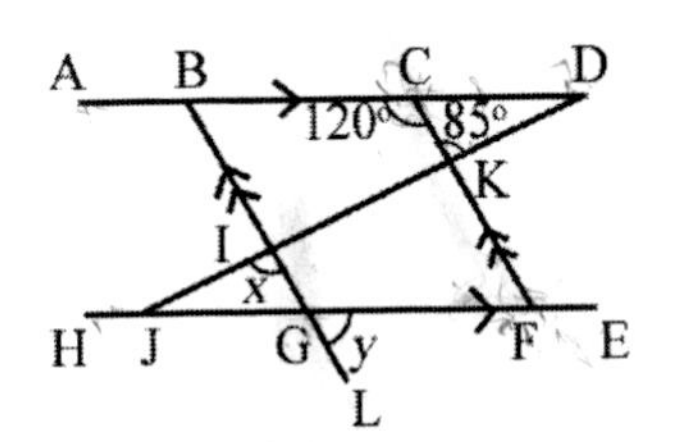

Figure 1. 26

Answers:
Given $\overrightarrow{BL} \parallel \overline{CF}$
Alternate exterior angles are ≅:
$\angle JIL \cong \angle CKD$
$\therefore x = 85°$ //
Given $\overrightarrow{DA} \parallel \overrightarrow{EH}$
Consecutive angles are supplementary:
$\angle CFH + \angle ACF = 180°$
$\angle CFH + 120° = 180°$
$\angle CFH = 60°$
Since $\overrightarrow{BL} \parallel \overline{CF}$
Alternate interior angles are ≅:
$\angle LGE \cong \angle CFH$
$\therefore y = 60°$ //

29. In Figure 1.27, $\overrightarrow{AB}$ and $\overleftrightarrow{CE}$ are parallel lines, while $\overrightarrow{AF}$ and $\overleftrightarrow{GH}$ are parallel lines. Find the values of x and y.

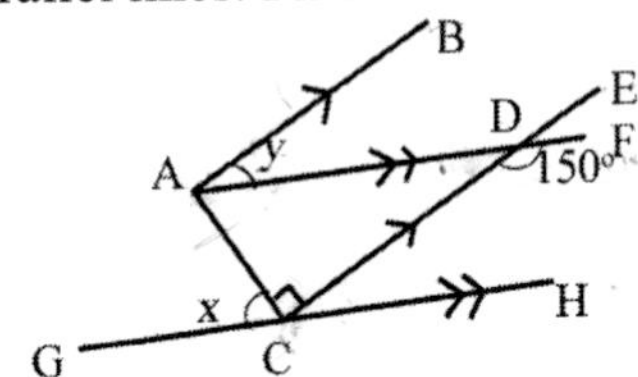

Figure 1. 27

Answers:
Sum of straight line angles = 180°
$\angle ADC + \angle FDC = 180°$
$\angle ADC + 150° = 180°$
$\angle ADC = 30°$
Given $\overrightarrow{AB} \parallel \overrightarrow{CE}$
Alternate interior angles are ≅:
$\angle BAD \cong \angle ADC$
$\therefore y = 30°$ //
Given $\overrightarrow{AF} \parallel \overleftrightarrow{GH}$
Alternate interior angles are ≅:
$\angle ECH \cong \angle ADC$
$\angle ECH = 30°$

Also given, $\angle ACE = 90°$
Sum of straight line angles = 180°
$\angle ACG + \angle ACE + \angle ECH = 180°$
$x + 90° + 30° = 180°$
$\therefore x = 60°$ //

30. In Figure 1.28, $\overrightarrow{QP}$ and $\overrightarrow{SR}$ are parallel lines. $\overrightarrow{AF}$, $\overrightarrow{BE}$, and $\overrightarrow{AB}$ are straight lines. Find the values of x, y, and z.

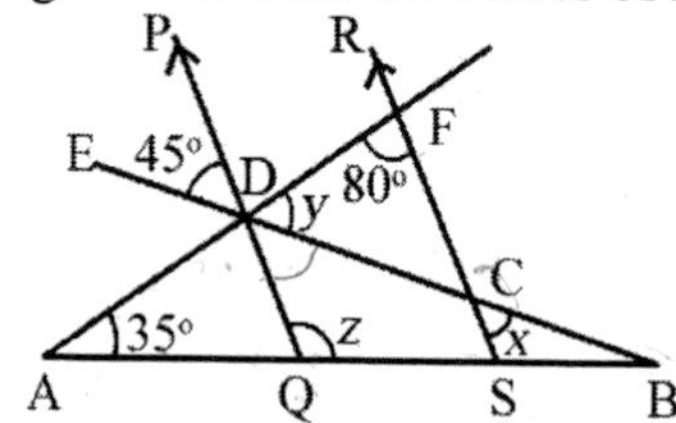

Figure 1. 28

Answers:
Given $\overrightarrow{QP} \parallel \overrightarrow{SR}$
Alternate exterior angles are $\cong$:
$\angle BCS \cong \angle EDP$
$\therefore x = 45°$ //
Vertical angles are $\cong$:
$\angle BDQ \cong \angle EDP = 45°$
Consecutive angles are supplementary:
$\angle QDF + \angle AFS = 180°$
$\angle BDF + \angle BDQ + \angle AFS = 180°$
$y + 45° + 80° = 180°$
$\therefore y = 55°$ //
Sum of straight line angles = 180°
$\angle ADQ + \angle BDQ + \angle BDF = 180°$
$\angle ADQ + 45° + 55° = 180°$
$\angle ADQ = 80°$
Exterior angle of $\triangle$ = sum of 2 nonadjacent interior angles of $\triangle$
$\angle BQP = \angle BAF + \angle ADQ$
$z = 35° + 80°$
$\therefore z = 115°$ //

31. In Figure 1.29, it is given $\overrightarrow{BA}$ is parallel to $\overleftrightarrow{CD}$. If $\overrightarrow{EI}$ and $\overline{BE}$ are straight lines, find the value of x.

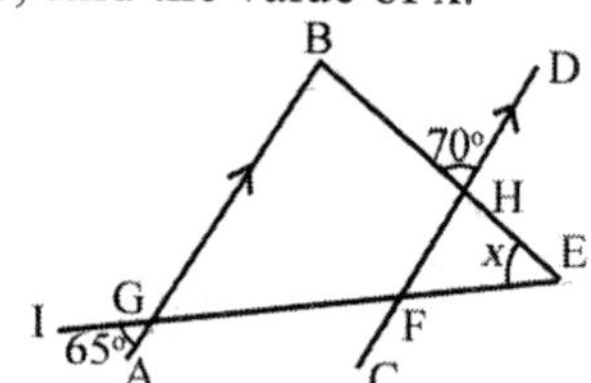

Figure 1. 29

Answer:
Given $\overrightarrow{BA} \parallel \overleftrightarrow{CD}$
Alternate exterior angles are $\cong$:
$\angle DFE \cong \angle AGI = 65°$
Vertical angles are $\cong$:
$\angle CHE \cong \angle BHD = 70°$
Sum of interior angles of $\triangle EFH$:
$\angle BEG + \angle CHE + \angle DFE = 180°$
$x + 70° + 65° = 180°$
$\therefore x = 45°$ //

Alternatively:
Vertical angles are $\cong$:
$\angle BGE \cong \angle AGI = 65°$
Alternate interior angles are $\cong$:
$\angle EBG \cong \angle BHD = 70°$
Sum of internal angles in $\triangle BEG = 180°$
$\angle BEG + \angle BGE + \angle EBG = 180°$
$x + 65° + 70° = 180°$
$\therefore x = 45°$ //

Chapter 2
Triangles & Congruence

△, Triangle – figure with 3 sides & angles
Vertex – angular points of a triangle
Classifying triangles by sides:
Equilateral triangle– all sides are equal in length and equiangular

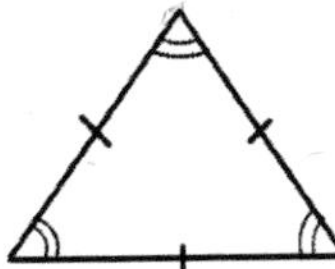

Isosceles triangle – 2 sides are equal in length and angle measure

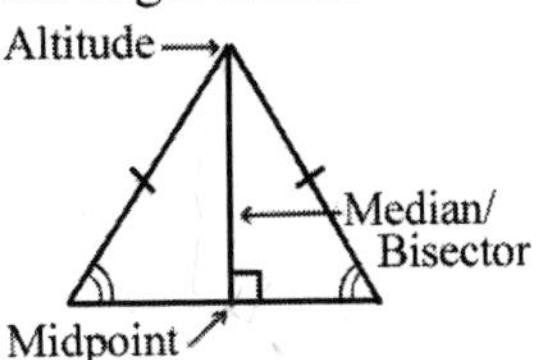

Scalene triangle – all sides are different in length

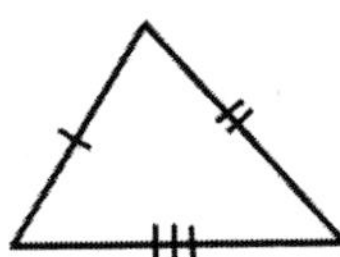

Classifying triangles by angles:
Right triangle – one of the angles = 90°

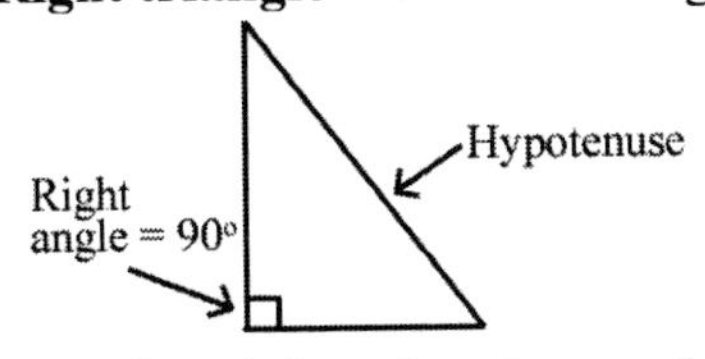

Isosceles right triangle – angles are 90°, 45° and 45°

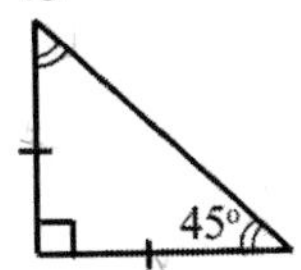

Acute triangle – every angle less than 90°

Obtuse triangle – 1 angle is more than 90°

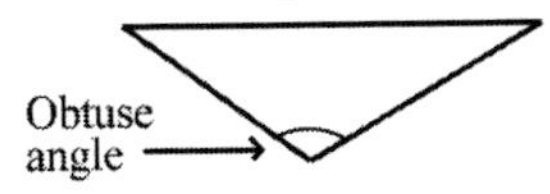

Pythagorean Theorem: (right triangle)
Hypotenuse, c:
$c^2 = a^2 + b^2$

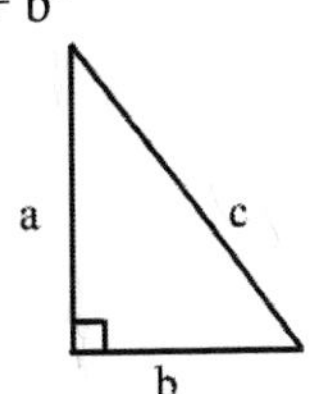

Prove of **congruence**:

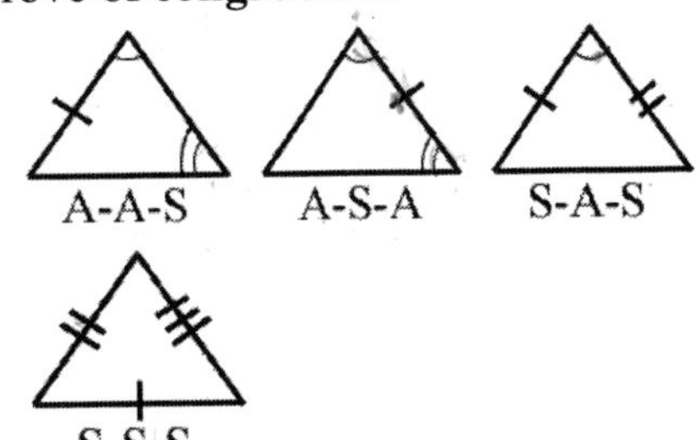

A–A–S: angle, angle, side theorem
A–S–A: angle, side, angle theorem
S–A–S: side, angle, side theorem
S–S–S: side, side, side theorem
≅, **congruent** – same shape and same size
Important principle: Corresponding parts of congruent triangles are congruent.

1. In Figure 2.1, it is known that $\alpha = \beta$. Deduce that $\triangle ABC$ is an isosceles triangle.

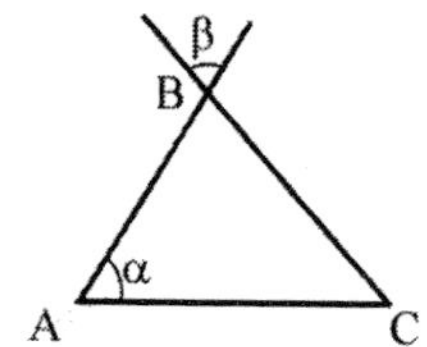

Figure 2. 1

Answer:
Using deduction method:
Vertical angles are ≅:
$\angle ABC \cong \beta$
Given $\angle BAC$, $\alpha = \beta$
Since $\angle ABC = \angle BAC = \beta$
It follows that length of sides AC = BC
Thus, $\angle ABC = \angle BAC$ and length of sides, AC = BC, $\triangle ABC$ is an isosceles triangle.//

2. In Figure 2.2, $\overleftrightarrow{AC}$ and $\overleftrightarrow{DF}$ are parallel lines. Find the value of α.

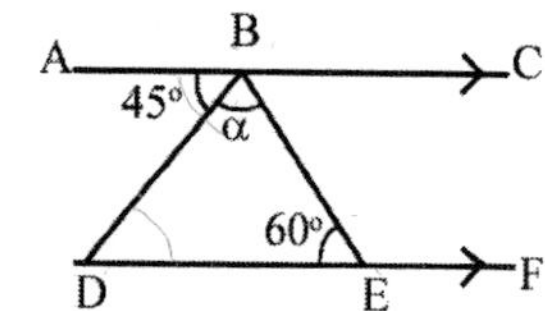

Figure 2. 2

Answer:
Given $\overleftrightarrow{AC} \parallel \overleftrightarrow{DF}$
Alternate interior angles are ≅:
$\angle BDF \cong \angle ABD$
$\angle BDF = 45°$
Sum of interior angles in $\triangle BDE = 180°$
$\angle DBE + \angle BDF + \angle BED = 180°$
$\alpha + 45° + 60° = 180°$
$\therefore \alpha = 75°$ //

3. In Figure 2.3, ABC is a straight line. If BD = CD, find the value of x.

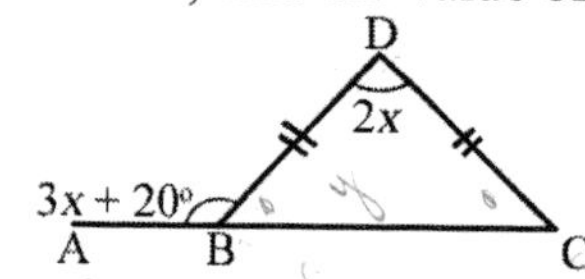

Figure 2. 3

Answer:
Given BD = CD
=> $\triangle BCD$ = isosceles △
=> $\angle CBD \cong \angle BCD$ ...*
Sum of interior angles in $\triangle BCD = 180°$
$\angle CBD + \angle BCD + \angle BDC = 180°$
$2\angle CBD + 2x = 180°$ ⇦ From *
$2\angle CBD = 180° - 2x$
$\angle CBD = 90° - x$ ...(1)
Given ABC = straight line
Sum of straight line angles = 180°
$\angle CBD + \angle ABD = 180°$
$\angle CBD + 3x + 20° = 180°$
$\angle CBD = 160° - 3x$ ...(2)
Substitute (1) into (2):
$90° - x = 160° - 3x$
$2x = 70°$
$\therefore x = 35°$ //

4. $\triangle ABC$ is a right triangle whose hypotenuse is 25″ and its two legs are 24″ and 2x + 1″. Find the possible values of x.

Answer:
Given $\triangle ABC$ = right triangle

Hypotenuse length = 25
Using Pythagorean Theorem:
$c^2 = a^2 + b^2$
$25^2 = 24^2 + (2x + 1)^2$
$625 = 576 + 4x^2 + 4x + 1$
$4x^2 + 4x - 48 = 0$ ⇦ divide by 4
$x^2 + x - 12 = 0$
$(x + 4)(x - 3) = 0$
Critical values of x = 3, – 4
Reject x = –4 as length has to be positive value.
$\therefore x = 3$ //

5. Find the unknown parts of the right triangle in Figure 2.4.

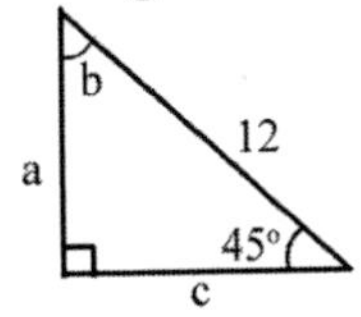

Figure 2. 4

Answers:
Given triangle is a right triangle
Sum of interior angles in a triangle = 180°
$b + 90° + 45° = 180°$
$\therefore b = 45°$ //
Thus Figure 2.4 is a 45°–45°–90° triangle
=> a = c
Applying Pythagorean Theorem:
$12^2 = a^2 + a^2$
$144 = 2a^2$
$a^2 = 72$
$a = \sqrt{72} = \sqrt{36 \times 2} = 6\sqrt{2}$ //
Since a = c, $\therefore c = 6\sqrt{2}$ //

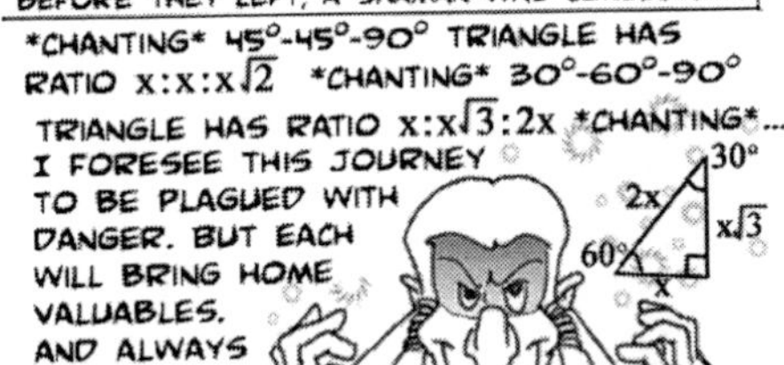

Alternatively:
Let, $x\sqrt{2} = 12 =$ hypotenuse
$$x = \frac{12}{\sqrt{2}}$$
$$x = \sqrt{\frac{144}{2}}$$
$$x = \sqrt{72}$$
$$x = 6\sqrt{2} \ //$$
Based on 45°–45°–90° triangle ratio of sides:
$x = a = c = 6\sqrt{2}$ units //

6. Figure 2.5 is an equilateral triangle. If the altitude is $5\sqrt{3}$, find a, b, and c.

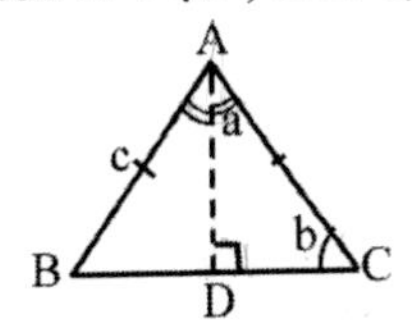

Figure 2. 5

Answers:
Given △ABC = equilateral triangle
All angles are equal in measure:
=> ∠ABC ≅ ∠BAC ≅ ∠ACB = b
=> 3b = 180°
$\therefore b = 60°$ //

Sum of interior angles in △ACD = 180°
$a + b + 90° = 180°$
$a + 60° + 90° = 180°$
$\therefore a = 30°$ //

Given altitude, AD = $5\sqrt{3}$
Using 30°–60°–90° triangle side ratio:
$x : x\sqrt{3} : 2x$

Hence,
AD = $x\sqrt{3} = 5\sqrt{3}$
=> $x = 5$
$\therefore c = 2x = 2(5) = 10$ units //

7. In Figure 2.6, $\triangle ABC$ is an equilateral triangle. $\overline{DE}$ splits $\triangle ABC$ into an equilateral triangle $\triangle ADE$ and a trapezoid BCED. Find θ and the area of the trapezoid.

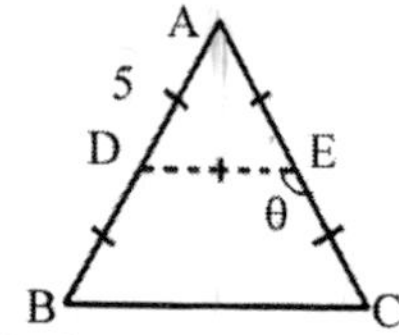

Figure 2. 6

Answers:
Given $\triangle AED$ = equilateral $\triangle$
=> all angles in $\triangle AED = 60°$
=> $\angle AED = 60°$
$\overline{AC}$ = straight line
Sum of straight line angles = 180°
$\angle CED + \angle AED = 180°$
$\theta + 60° = 180°$
$\therefore \theta = 120°$ //
Let m = altitude of $\triangle ADE$
Using 30°–60°–90° triangle:

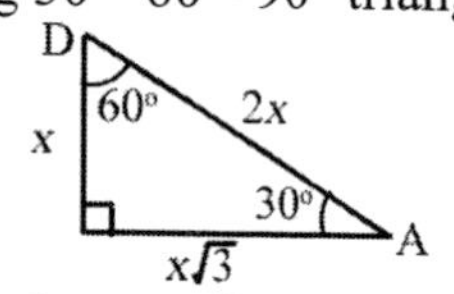

$AD = 2x = 5$
$x = 2.5$
$m = x\sqrt{3} = 2.5\sqrt{3}$
Area of $\triangle ADE$:
$= \frac{1}{2} \times DE \times m$
$= \frac{1}{2} \times 5 \times 2.5\sqrt{3}$
$= 10.8253$ units^2
Altitude of $\triangle ABC$ = 2m
Area of $\triangle ABC$:
$= \frac{1}{2} \times BC \times 2m$
$= \frac{1}{2} \times 10 \times 2(2.5\sqrt{3})$
$= 43.3013$ units^2
Area of trapezoid BCED:
= Area of $\triangle ABC$ – Area of $\triangle ADE$
$= 43.3013 - 10.8253$
$= 32.4760$ units^2 //

Alternatively:
Area of trapezoid BCED:
$= \frac{1}{2} \times 2.5\sqrt{3} \times (5+10)$
$= 32.4760$ units^2 //

8. In Figure 2.7, $\triangle PQR$ is divided by $\overline{ST}$ into two parts; $\triangle PST$ an isosceles triangle and QRST a trapezoid. Find the values of α and θ.

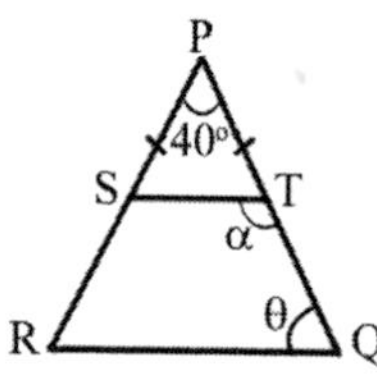

Figure 2. 7

Answers:
Given $\triangle PST$ = isosceles $\triangle$
=> $\angle PST \cong \angle PTS$ ...*
Sum of interior angles in $\triangle PST = 180°$
$\angle PST + \angle PTS + \angle SPT = 180°$
$2\angle PTS + 40° = 180°$ ⇦ From *
$\angle PTS = 70°$
Sum of straight line angles = 180°
$\angle QTS + \angle PTS = 180°$
$\alpha + 70° = 180°$
$\therefore \alpha = 110°$ //
Given QRST = trapezoid
=> $\overline{QR} \parallel \overline{TS}$
Corresponding angles are $\cong$:
$\angle PQR \cong \angle PTS$
$\therefore \theta = 70°$ //

9. In Figure 2.8, find the values of α, β and θ.

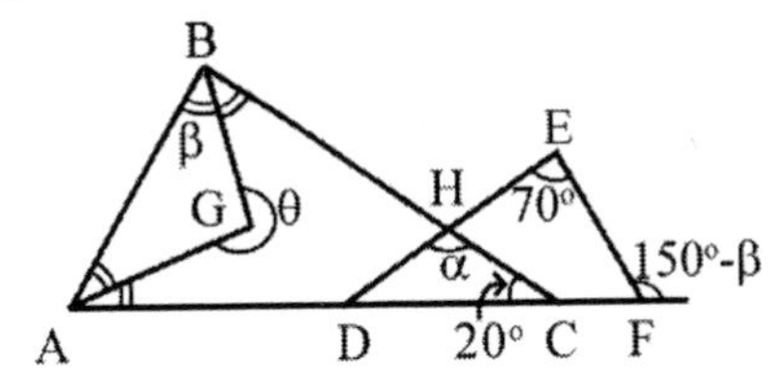

Figure 2. 8

Answers:

Given $\angle ACB = 20°$

Sum of interior angles in $\triangle ABC = 180°$

$\angle ABC + \angle BAC + \angle ACB = 180°$

$2\beta + 2\beta + 20° = 180°$

$4\beta = 180° - 20°$

$4\beta = 160°$

$\beta = \dfrac{160°}{4}$

$\therefore \beta = 40°$ //

Sum of straight line angles = 180°

$\angle EFD = 180° - (150° - \beta)$

$\angle EFD = 180° - 150° + 40° = 70°$

Sum of interior angles in $\triangle DEF = 180°$

$\angle EDF + \angle EFD + \angle DEF = 180°$

$\angle EDF + 70° + 70° = 180°$

$\angle EDF = 40°$

Sum of interior angles in $\triangle CDH = 180°$

$\angle CHD + \angle ACB + \angle EDF = 180°$

$\alpha + 20° + 40° = 180°$

$\therefore \alpha = 120°$ //

Sum of interior angles $\triangle ABG = 180°$

$\angle AGB + \angle ABG + \angle BAG = 180°$

$\angle AGB + \beta + \beta = 180°$

$\angle AGB + 40° + 40° = 180°$

$\angle AGB = 100°$

Reflex angle of $\angle AGB$:

$\theta = 360° - \angle AGB$

$\theta = 360° - 100°$

$\therefore \theta = 260°$ //

10. In Figure 2.9, ABCD is a rectangle while $\triangle ACD$ is a right triangle. If 2AD = DC, prove that the median drawn from vertex A, to the base of $\triangle ACD$ bisects $\angle BAD$.

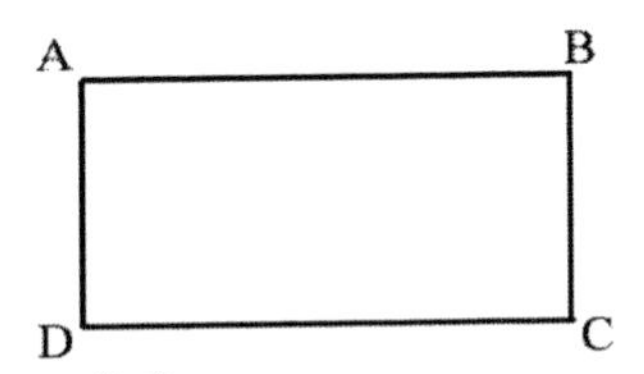

Figure 2. 9

Answer:

Let M = midpoint of DC

=> $\overline{AM}$ = median of $\triangle ACD$

Given 2AD = DC

=> 2AD = DM + MC

=> AD = DM = MC

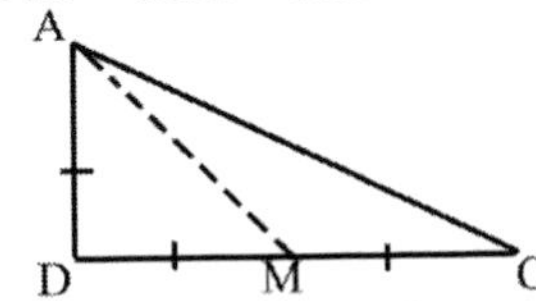

Hence $\triangle ADM$ is an isosceles right $\triangle$

=> $\angle DAM \cong \angle AMD$ ...*

Since ABCD = rectangle

=> $\angle ADC = 90°$

Sum of interior angles in $\triangle ADM$:

$\angle DAM + \angle AMD + \angle ADC = 180°$

$2\angle DAM + 90° = 180°$ ⇦ From *

$\angle DAM = 45°$

From ABCD = rectangle

=> $\angle BAD = 90°$

We have shown $\angle DAM = 45°$

=> $\angle BAM = \angle BAD - \angle DAM$

$= 90° - 45°$

$= 45°$

$\therefore$ The median, $\overline{AM}$, drawn from the vertex A, to the base of $\triangle ADC$ is an angle bisector of $\angle BAD$. Therefore median, $\overline{AM}$, is a bisector. //

11. What is the difference between the altitude and median of a triangle?

Answer:
Altitude is the *perpendicular* distance from the vertex to the base of the triangle. The median is the line drawn from any vertex of a triangle to the *midpoint* of its opposite side. In diagram below, BE = EC, where E is the midpoint. //

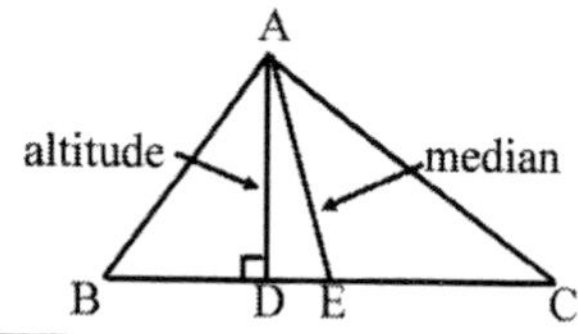

$\overline{AD}$ = altitude of $\triangle ABC$

$\overline{AE}$ = median of $\triangle ABC$

12. In Figure 2.10, $\overrightarrow{BA}$ and $\overrightarrow{DC}$ are parallel lines. $\overline{BE}$ and $\overline{DB}$ are transversals. It is further given that $\overline{BE}$ bisects $\angle ABD$. Prove that $\triangle EBD$ is an isosceles triangle.

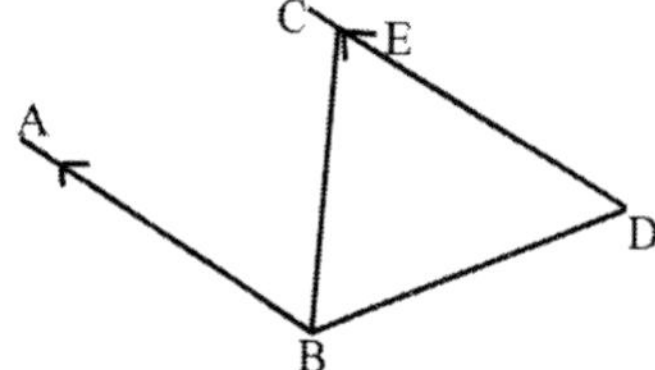

Figure 2. 10

Answer:
Let $\alpha = \angle ABE$
Given $\overline{BE}$ bisects $\angle ABD$
$\Rightarrow \angle DBE \cong \angle ABE$
$\Rightarrow \angle DBE = \alpha$
Given $\overrightarrow{BA} \parallel \overrightarrow{DC}$
Alternate interior angles are $\cong$:
$\angle BED \cong \angle ABE$
$\angle BED = \alpha$
Since $\angle DBE = \angle BED = \alpha$
Reversing Base Angle Theorem:
$\angle DBE \cong \angle BED$
$\Rightarrow \overline{BD} \cong \overline{DE}$
$\therefore$ We have proven that $\triangle EBD$ is an isosceles triangle. //

13. Find the value of x for the following right angle triangles.

i)

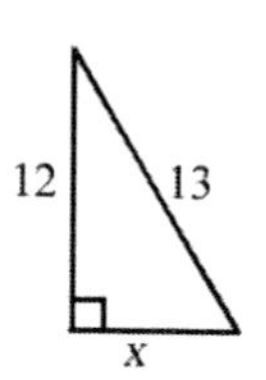

ii)

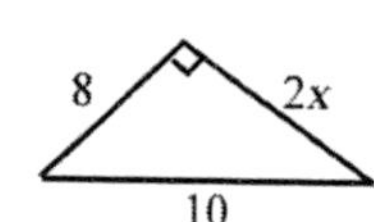

iii)

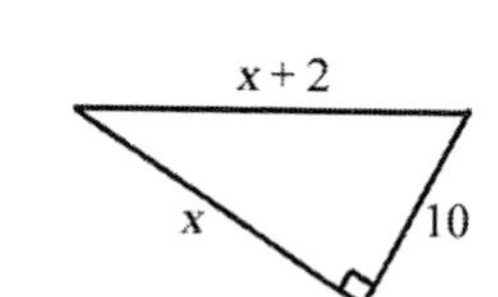

Figure 2. 11

Answers:

i)
Using Pythagorean Theorem:
$13^2 = 12^2 + x^2$
$169 = 144 + x^2$
$x^2 = 25$
$\therefore$ x = 5 units //

ii)
Using Pythagorean Theorem:
$10^2 = 8^2 + (2x)^2$
$100 = 64 + 4x^2$
$4x^2 = 36$
$x^2 = 9$
$\therefore$ x = 3 units //

iii)
Using Pythagorean Theorem:
$(x + 2)^2 = x^2 + 10^2$
$x^2 + 4x + 4 = x^2 + 100$
$4x = 96$
$\therefore x = 24$ units //

14. In $\triangle PQR$, S is the midpoint of $\overline{PR}$ and T is the midpoint of $\overline{PQ}$. Find the values of x and y.

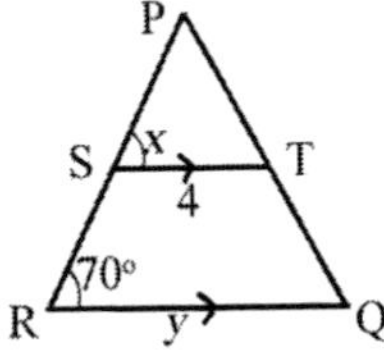

Figure 2. 12

Answers:
Given $\overline{ST} \parallel \overline{RQ}$
Corresponding angles are $\cong$:
$\angle PST \cong \angle PRQ$
$\therefore x = 70°$ //

Given $\overline{ST}$ is drawn from the midpoints S and T.
$\Rightarrow ST = \frac{1}{2} RQ$
$4 = \frac{1}{2} y$
$\therefore y = 8$ units //

15. In Figure 2.13, $\triangle ABE$ and $\triangle ACD$ are overlapping triangles. Further, it is also given that AE = x, BE = 7 cm, CD = 10 cm, AC = y and $\angle CAD = 45°$. Find the values of x and y.

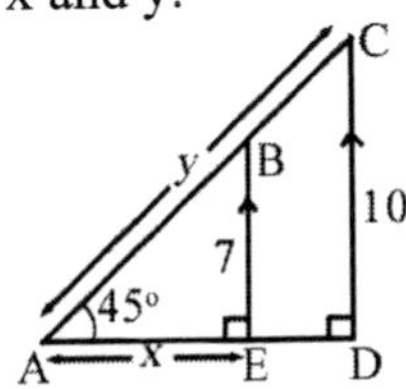

Figure 2. 13

Answers:
Given $\triangle ABE$ and $\triangle ACD$ are overlapping triangles.
$\Rightarrow \triangle ABE$ and $\triangle ACD$ share $\angle CAD = 45°$
$\triangle ABE$ and $\triangle ACD$ are right angle $\triangle$s
$\Rightarrow \angle ABE = 180° - 90° - 45° = 45°$
$\Rightarrow \angle ACD = 180° - 90° - 45° = 45°$

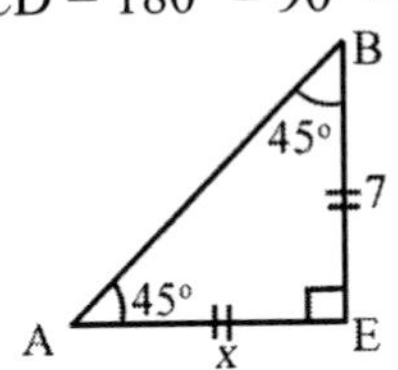

Since $\triangle ABE$ is a 45°–45°–90° triangle:
AE = BE
$\therefore x = 7$ cm //

$\triangle ACD$ is also a 45°–45°–90° triangle:
$AC = CD \times \sqrt{2}$
$y = 10\sqrt{2}$
$\therefore y = 14.1421$ cm //

16. Find the value of x, if the measure of each angle of an equilateral triangle is 3x.

Answer:
Given equilateral triangle
Each $\angle$ measure = 3x
Sum of total angles in equilateral $\triangle = 180°$
$3x \times 3 = 180°$
$9x = 180°$
$\therefore x = 20°$ //

17. Cars exiting Clayton Garage Parking has to go up a ramp (see Figure 2.14) to reach the main street. How long is the ramp?

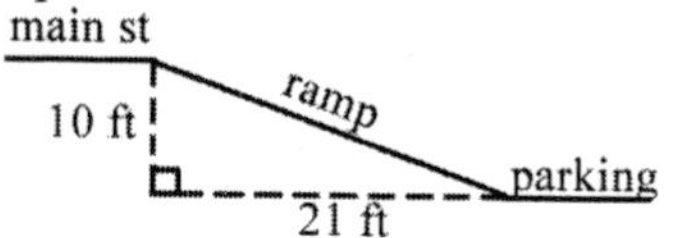

Figure 2. 14

Answer:
Let r = length of ramp
Using Pythagorean Theorem:
$r^2 = 10^2 + 21^2$
$r^2 = 100 + 441$
$r = \sqrt{541}$
r = 23.2594 ft
$\therefore$ The ramp is 23.2594 ft long. //

18. In Figure 2.15, two triangles are connected at $\overline{CQ}$. It is also given that $\overline{AC}$ is perpendicular to $\overline{AB}$ and $\overline{QR}$ is perpendicular to $\overline{PR}$. If $\overline{AB}$ is parallel to $\overline{PR}$ and $\angle ABC$ is congruent to $\angle QPR$. Prove that $\overline{BC}$ is parallel to $\overline{QP}$.

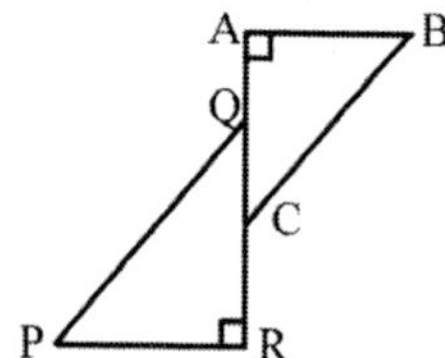

Figure 2. 15

Answer:
Given $\overline{AC} \perp \overline{AB}$
$\overline{QR} \perp \overline{PR}$
=> $\angle BAC = \angle PRQ = 90°$
Given $\angle ABC \cong \angle QPR$
Since sum of interior angles in $\triangle = 180°$:
$\therefore \angle PQR \cong \angle ACB$

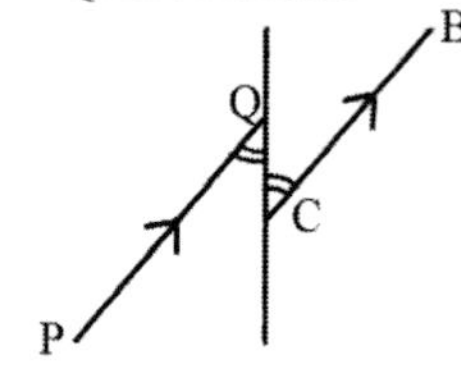

Alternate interior angles: $\angle PQR \cong \angle ACB$
$\therefore \overline{BC}$ and $\overline{QP}$ are parallel lines. //

19. In Figure 2.16, C is the midpoint of $\overline{AD}$ and $\overline{BE}$. Show that $\triangle ABC$ and $\triangle CDE$ are congruent.

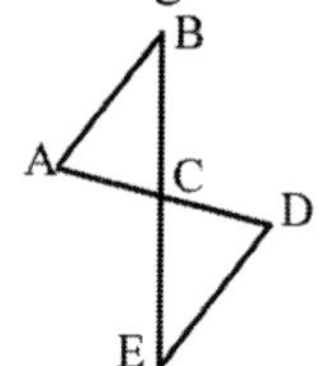

Figure 2. 16

Answer:
Given C = midpoint of $\overline{AD}$ and $\overline{BE}$
=> AC = CD [side]
=> BC = CE [side]

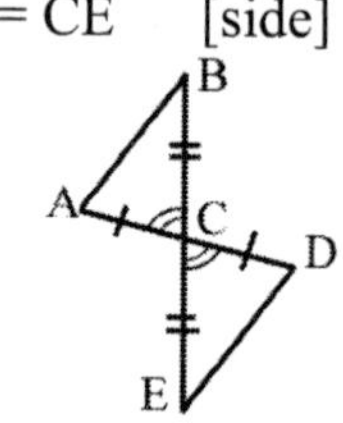

Reproduce Figure 2.16 for labeling
Vertical angles are $\cong$:
$\angle ACB \cong \angle DCE$ [angle]
$\therefore$ Based on S-A-S theorem, $\triangle ABC$ and $\triangle CDE$ are congruent. //

20. Given AB = ED and $\angle ABC \cong \angle DEC$ in Figure 2.17. Hence prove that $\triangle ABC$ and $\triangle EDC$ are congruent.

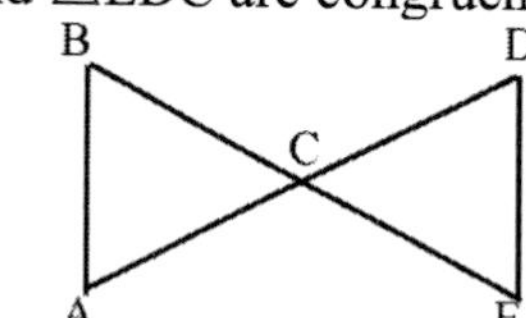

Figure 2. 17

Answer:
Given $\overline{AD}$ and $\overline{BE}$ intersect at C
Vertical angles are congruent:
=> $\angle ACB \cong \angle DCE$ [angle]
Given $\angle ABC \cong \angle DEC$ [angle]
Given AB = ED [side]

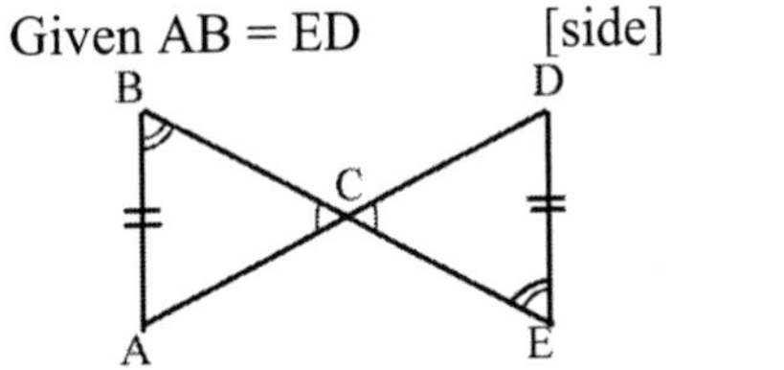

$\therefore$ Based on A–A–S theorem, $\triangle ABC$ and $\triangle EDC$ are congruent. //

21. In Figure 2.18, given $\overline{CD}$ and $\overline{BF}$ are congruent, while AB = EF. If $\overline{AB}$ and $\overline{EF}$ are parallel lines, show that $\triangle ABC$ is congruent to $\triangle EFD$.

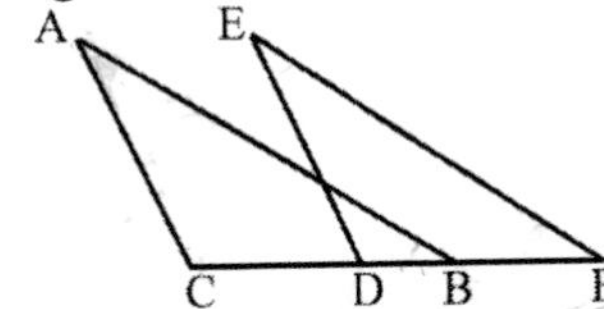

Figure 2. 18

Answer:
Given $\overline{CD} \cong \overline{BF}$
Since $\overline{DB} \cong \overline{BD}$ ⇦ Reflexive property
Thus,
$\overline{CD} + \overline{DB} = \overline{BD} + \overline{BF}$ ⇦ Addition property
$\overline{CB} \cong \overline{DF}$ [side]
Also given $\overline{AB} \parallel \overline{EF}$
Corresponding angles are $\cong$:
$\angle ABC \cong \angle EFD$ [angle]
Also given AB = EF [side]

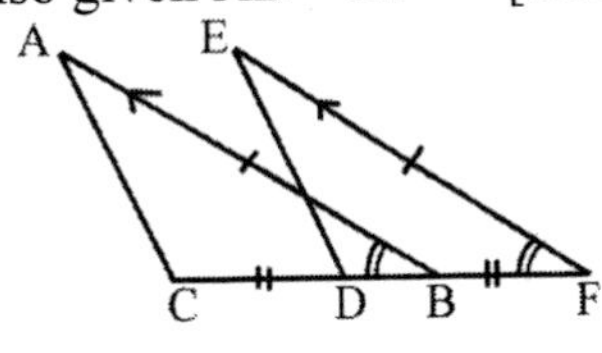

Based on S–A–S theorem, we have therefore shown that $\triangle ABC$ is congruent to $\triangle EFD$. //

22. Figure 2.19 shows two overlapping triangles, $\triangle PQR$ and $\triangle QPS$. If it is further known that QT = PT and QS = PR, prove that the 2 triangles are congruent.

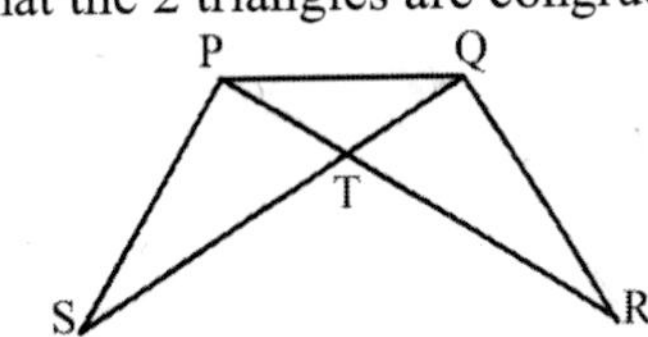

Figure 2. 19

Answer:
Given QS = PR [side]
Also given, QT = PT
=> $\triangle PQT$ = isosceles triangle
=> $\angle QPT \cong \angle PQT$ [angle]
Since $\triangle QPS$ and $\triangle PQR$ overlap
=> $PQ \cong QP$ [side]

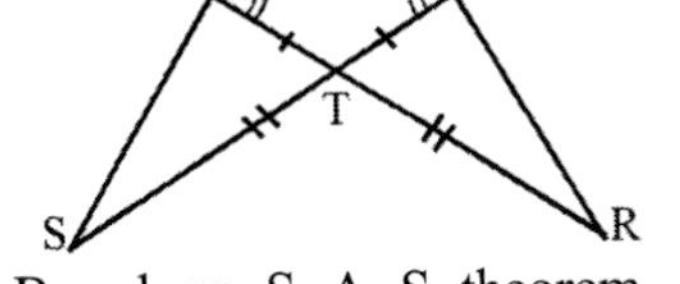

$\therefore$ Based on S–A–S theorem, $\triangle QPS$ and $\triangle PQR$ are congruent triangles. //

23. Figure 2.20, shows 2 adjoining right triangles; $\triangle ABD$ and $\triangle BCD$. If their hypotenuses, $\overline{AD}$ and $\overline{BC}$ are congruent, show that $\triangle ABD$ and $\triangle BCD$ are congruent.

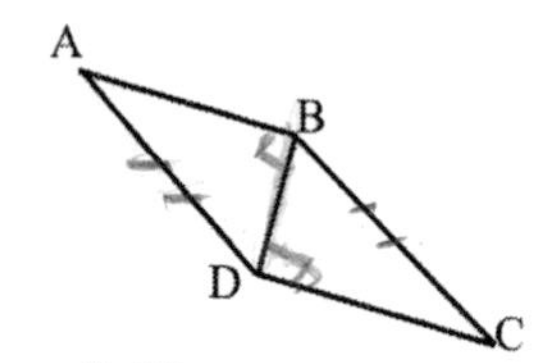

Figure 2. 20

Answer:
Given △ABD and △BCD are right triangles
Angle opposite hypotenuse is 90°
=> ∠ABD ≅ ∠CDB [angle]
Given $\overline{AD} \cong \overline{BC}$ [side]
From reflexive property of equality:
$\overline{BD} \cong \overline{DB}$ [side]

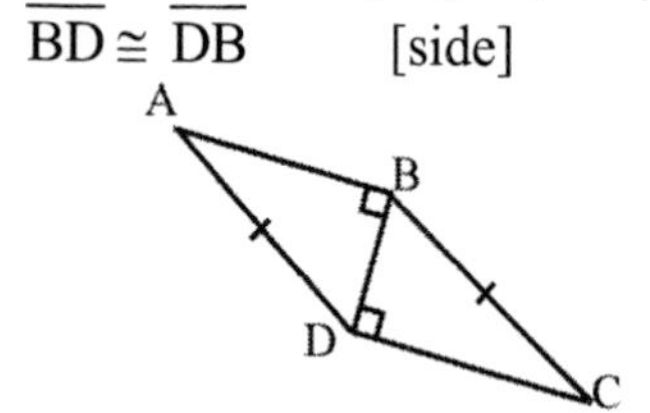

Based on hypotenuse-leg theorem, △ABD and △BCD are congruent. //

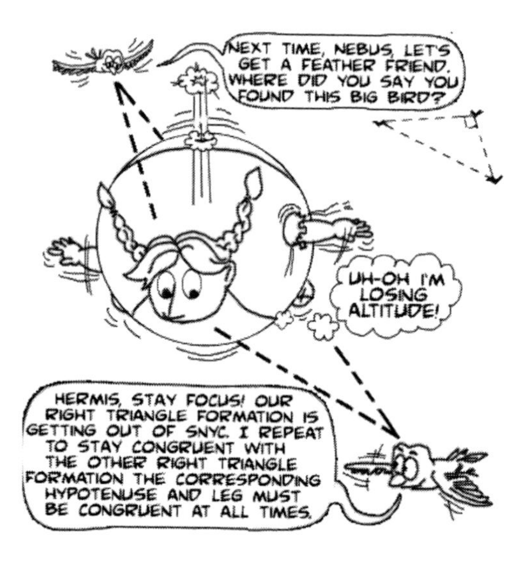

24. In Figure 2.21, △AED is a right triangle. Given that AC = DB and EB = EC, prove that △BAE and △CDE are congruent.

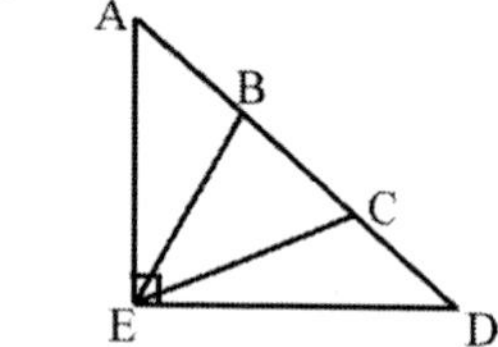

Figure 2. 21

Answer:
Given AC = DB ...(1)
BC = CB ...(2)
From (1) and (2), using subtraction property of equality:
AB + BC = DC + CB
Since BC = CB ⇦ Reflexive property
=> AB = DC [side]
Given EB = EC [side]
=> △BCE = isosceles triangle
=> ∠DBE ≅ ∠ACE ...(3)
Straight line angles = 180°
∠ABE = 180° – ∠DBE ...(4)
∠DCE = 180° – ∠ACE ...(5)
(4) – (5):
∠ABE – ∠DCE = – ∠DBE + ∠ACE
Substitute with (3):
∠ABE – ∠DCE = – ∠DBE + ∠DBE
∠ABE = ∠DCE [angle]

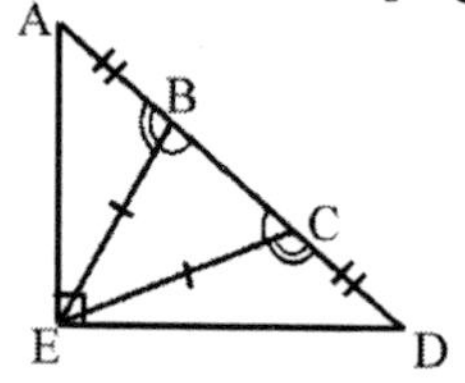

∴ Based on S–A–S theorem, △BAE and △CDE are congruent. //

25. Figure 2.22 shows a trapezoid ABCDE. It is given that ABDE and ABCD are rhombuses. Hence show that △ADE and △BCD are congruent.

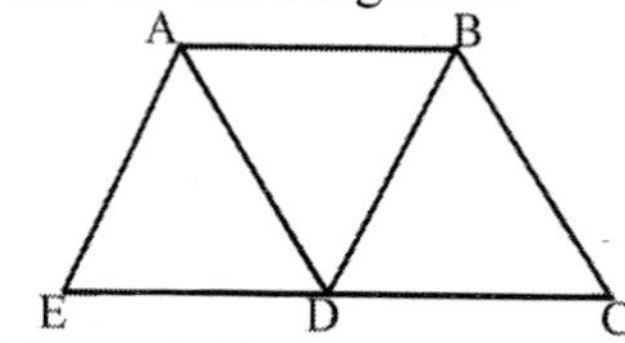

Figure 2. 22

Answer:
Given ABDE = rhombus
$\overline{AE} \cong \overline{BD}$ [side]
$\overline{AB} \cong \overline{ED}$ ...(1)
Also given ABCD = rhombus
$\overline{AD} \cong \overline{BC}$ [side]
$\overline{AB} \cong \overline{DC}$ ...(2)
From (1) & (2) using transitive property:
=> $\overline{ED} \cong \overline{DC}$ [side]

∴ Based on S–S–S theorem, $\triangle ADE$ and $\triangle BCD$ are congruent. //

26. In Figure 2.23, $\triangle ABC$ is an isosceles triangle where AB = AC. $\overline{AD}$ is the angle bisector of $\angle BAC$. If $\overline{DF}$ bisects $\angle ADC$ and $\overline{DE}$ bisects $\angle ADB$, prove that $\triangle CDF$ and $\triangle BDE$ are congruent.

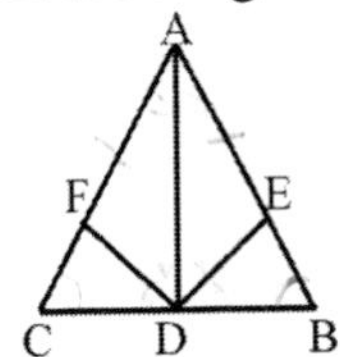

Figure 2. 23

Answer:
Given $\triangle ABC$ = isosceles triangle
$\angle ACD \cong \angle ABD$ [angle]
Given $\overline{AD}$ bisects $\angle BAC$
=> $\overline{CD} = \overline{DB}$ [side]
=> $\angle ADC \cong \angle ADB = 90°$
Since $\overline{DF}$ bisects $\angle ADC$
=> $\angle CDF = 45°$ ...(1)
Since $\overline{DE}$ bisects $\angle ADB = 90°$
=> $\angle BDE = 45°$ ...(2)
From (1) & (2):
=> $\angle CDF = \angle BDE$ [angle]

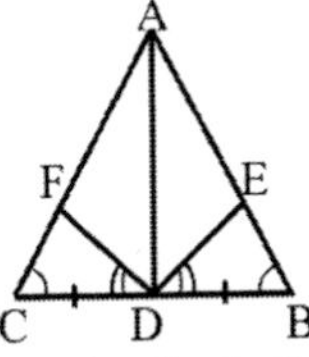

∴ Based on A–S–A theorem, $\triangle CDF$ and $\triangle BDE$ are congruent. //

27. Figure 2.24 depicts a parallelogram ABCD. It is further given that $\overline{FC}$ is parallel to $\overline{AE}$. Prove that $\triangle CBF$ and $\triangle ADE$ are congruent. Subsequently show that $\overline{AE}$ and $\overline{FC}$ are congruent.

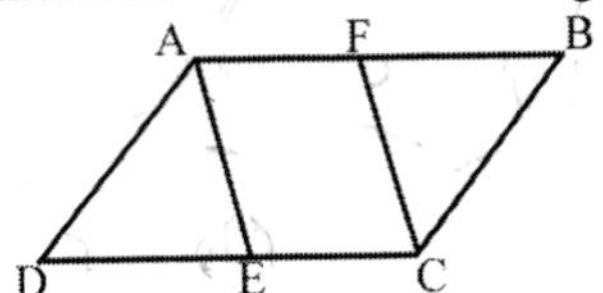

Figure 2. 24

Answers:
Given ABCD = parallelogram
=> $\overline{AD} \cong \overline{BC}$ [side]
Opposite angles in parallelogram are ≅:
=> $\angle ADC \cong \angle ABC$ [angle]
Also given $\overline{AE} \parallel \overline{FC}$
Corresponding angles are ≅:
$\angle BAE \cong \angle BFC$ ...(1)
Since ABCD is parallelogram:
=> $\overline{AB} \parallel \overline{DC}$
Alternate interior angles are ≅:
$\angle BAE \cong \angle AED$ ...(2)
From (1) and (2) transitive property of equality:
$\angle BFC \cong \angle AED$ [angle]

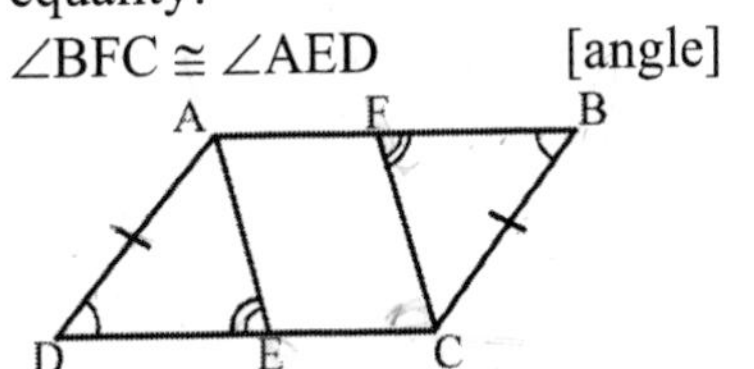

∴Based on A–A–S theorem, $\triangle CBF$ and $\triangle ADE$ are congruent. //

Corresponding parts of congruent triangles are congruent. Since we have proven $\triangle CBF$ and $\triangle ADE$ are congruent triangles therefore we deduce that $\overline{AE}$ and $\overline{FC}$ are congruent. //

28. In Figure 2.25, ABCD is a trapezoid. If AD = BC, show that $\angle DAC$ and $\angle DBC$ are congruent.

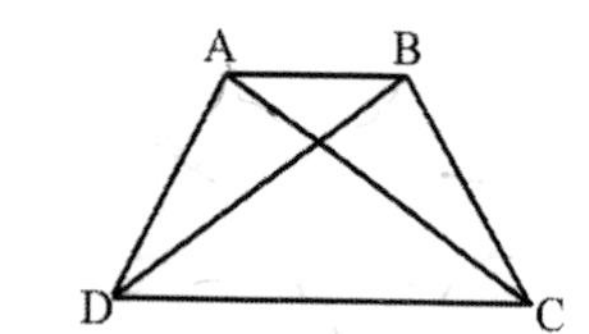

Figure 2. 25

Answer:
Given AD = BC [side]
=> ABCD = isosceles trapezoid
=> $\angle ADC \cong \angle BCD$ [angle]
Based on reflexive property of equality:
DC = CD [side]

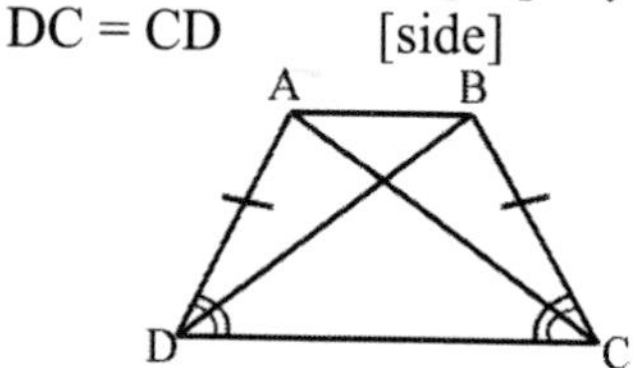

Based on S–A–S theorem, $\triangle ACD$ and $\triangle BCD$ are congruent.

Corresponding parts of congruent triangles are congruent. Since $\triangle ACD$ and $\triangle BCD$ are congruent triangles therefore $\angle DAC$ and $\angle DBC$ must be congruent. //

29. In Figure 2.26, PQRS is a rectangle. It is further given that $\overline{PT}$ and $\overline{QT}$ are angle bisectors of $\angle QPS$ and $\angle PQR$ respectively. Show that T is the midpoint of $\overline{RS}$.

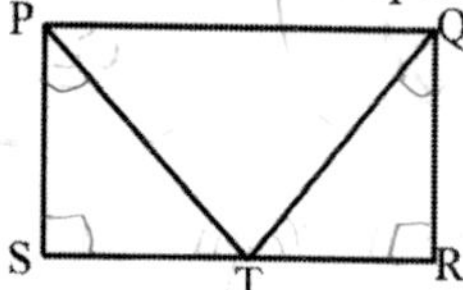

Figure 2. 26

Answer:
Given PQRS = rectangle
=> $\overline{PS} \cong \overline{QR}$ [side]
=> $\angle PSR \cong \angle QRS = 90°$ [angle]
Also given $\overline{PT}$ and $\overline{QT}$ are angle bisectors
=> $\angle SPT \cong \angle RQT = 45°$ [angle]

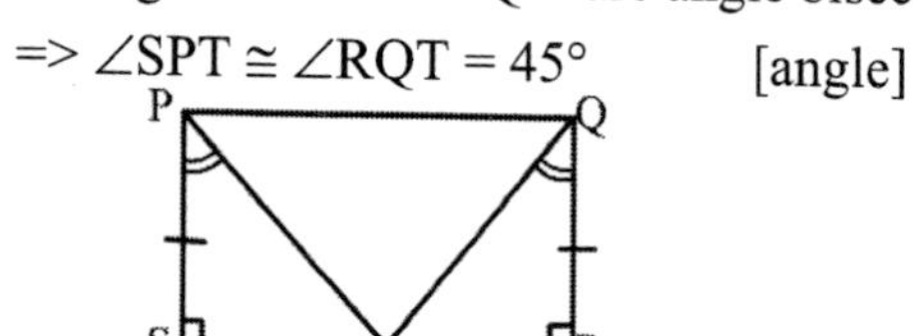

Based on A–S–A theorem, $\triangle PST$ and $\triangle QRT$ are congruent triangles.

Corresponding parts of congruent triangles are congruent. Since $\triangle PST$ and $\triangle QRT$ are congruent, hence corresponding parts:
=> $\overline{ST} \cong \overline{TR}$
And given $\overline{RS}$ = straight line:
$\therefore$ T is the midpoint of $\overline{RS}$. //

30. ABC is a triangle (see Figure 2.27). Given $\overline{AE}$ is a perpendicular bisector of $\angle BAC$. Show that $\triangle BDE$ and $\triangle CDE$ are congruent. Hence show $\triangle ABD$ and $\triangle ACD$ are congruent.

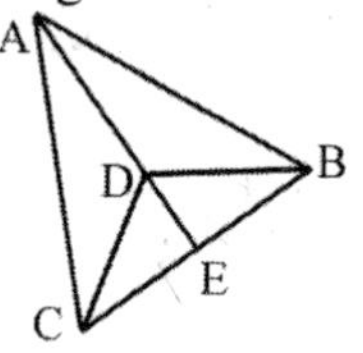

Figure 2. 27

Answers:
Given $\overline{AE}$ bisects $\angle BAC$
=> E is midpoint of CB:
$\overline{CE} \cong \overline{EB}$ [side]
=> $\overline{AE}$ is perpendicular to $\overline{BC}$
$\angle AEB \cong \angle AEC = 90°$ [angle]
From reflexive property of equality:
DE ≅ ED [side]

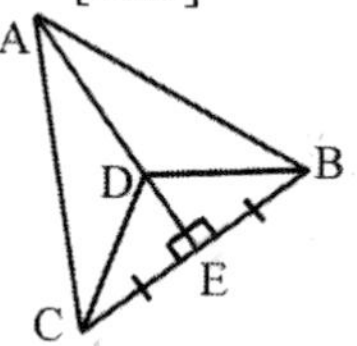

$\therefore$ Based on S–A–S theorem $\triangle BDE$ and $\triangle CDE$ are congruent. //

Since $\triangle BDE \cong \triangle CDE$ and corresponding parts of congruent triangles are congruent:
=> $\overline{CD} \cong \overline{BD}$ [side]
=> $\angle CDE \cong \angle BDE$
Sum of straight line angles = 180°
$\angle ADC + \angle CDE = 180°$ ...(1)
$\angle ADB + \angle BDE = 180°$ ...(2)
(1) – (2):
=> $\angle ADC = \angle ADB$ [angle]
From reflexive property of equality:
AD ≅ DA [side]

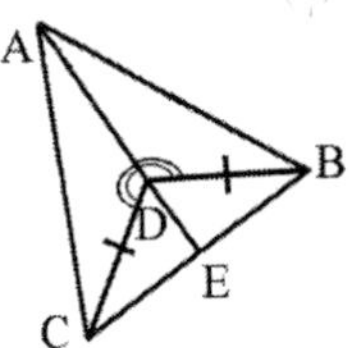

∴ Based on S–A–S theorem, △ABD and △ACD are congruent. //

31. ABCD is a trapezoid and length of AB is greater than length of AD (see Figure 2.28). Prove that ∠ADB is larger than ∠ABD.

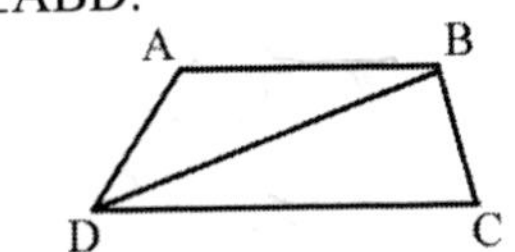

Figure 2. 28

Answer:
Given AB > AD
From properties of triangle, if 2 sides are not congruent, than their *opposite* angles are also not congruent. Hence the larger angle lies opposite the longer side.
Since $\overline{AB}$ is the longer side:
=> Opposite angle of $\overline{AB}$ is ∠ADB
∴ ∠ADB is larger than ∠ABD. //

32. T is the midpoint of the line segments, $\overline{PQ}$ and $\overline{RS}$ (see Figure 2.29). Show that △PTS and △QRT are congruent.

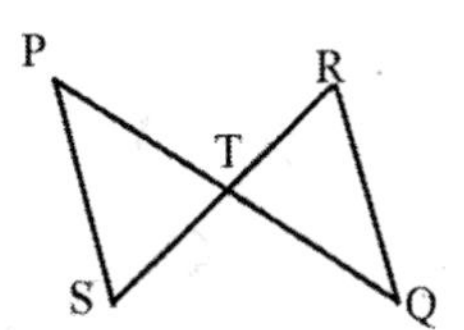

Figure 2. 29

Answer:
Given T = midpoint of $\overline{PQ}$ and $\overline{RS}$
=> $\overline{PT} \cong \overline{TQ}$ [side]
=> $\overline{ST} \cong \overline{TR}$ [side]
Vertical angles are ≅:
∠PTS ≅ ∠RTQ [angle]

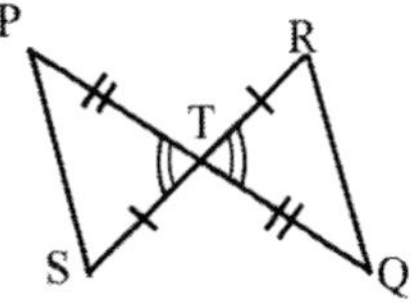

∴ Based on S–A–S theorem, △PTS and △QRT are congruent. //

33. In Figure 2.30, ACD is a triangle. Points F, E and B are the midpoints of $\overline{CD}$, $\overline{AD}$ and $\overline{AC}$ respectively. Prove that △ABE and △BCF are congruent.

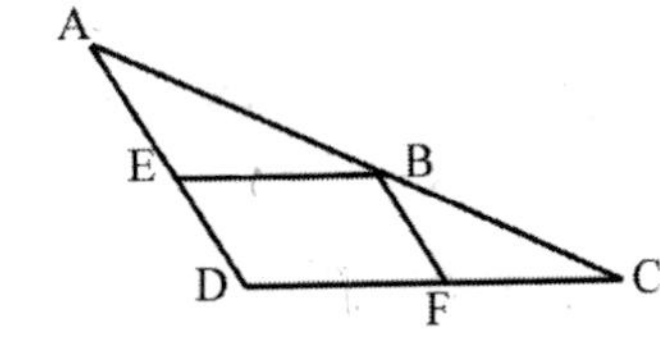

Figure 2. 30

Answer:
Given B is the midpoint of $\overline{AC}$
=> AB = BC [side]
$\overline{EB}$ drawn from the midpoints, E and B is one half the length of base $\overline{CD}$ (triangle midpoint theorem).

=> $\overline{EB} = \frac{1}{2}\overline{CD}$ ...(1)

Given F is midpoint of $\overline{CD}$

=> $\overline{CF} = \frac{1}{2}\overline{CD}$ ...(2)

From (1) and (2):
$\overline{EB} = \overline{CF}$ [side]
From triangle midpoint theorem:
$\overline{EB} \parallel \overline{DC}$

Corresponding angles are ≅:
$\angle ABE \cong \angle ACD$ [angle]

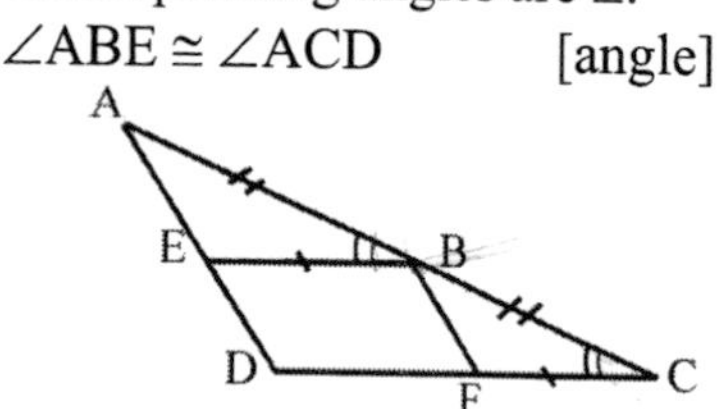

∴ Based on S-A-S theorem, $\triangle ABE$ and $\triangle BCF$ are congruent. //

34. Harry owns a piece of land where he lives with his grandmother (see Figure 2.31). Everyday Harry travels 4 miles to visit his grandmother. If Harry decides to build a new road that reduces his traveling time from his home to his grandmother's, find the length of the shortest road he can build.

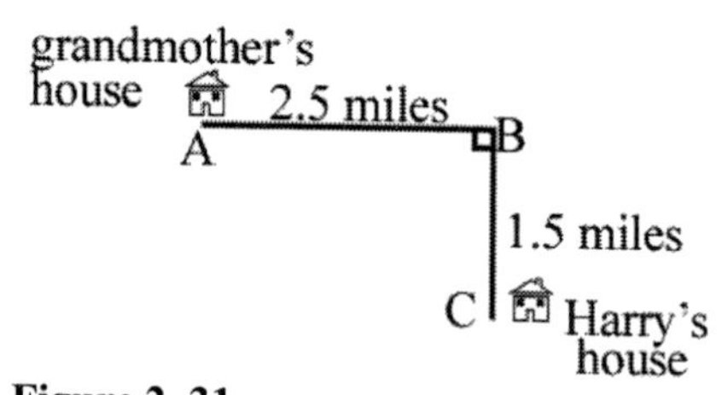

Figure 2. 31

Answer:
Given AB = 2.5
BC = 1.5
Shortest road is the hypotenuse $\overline{AC}$
From Pythagorean Theorem:
$AC^2 = AB^2 + BC^2$
$AC^2 = (2.5)^2 + (1.5)^2$
$= 6.25 + 2.25$
$= 8.5$
$AC = \sqrt{8.5}$
AC = 2.9155 miles //

35. What is the value of x in Figure 2.32?

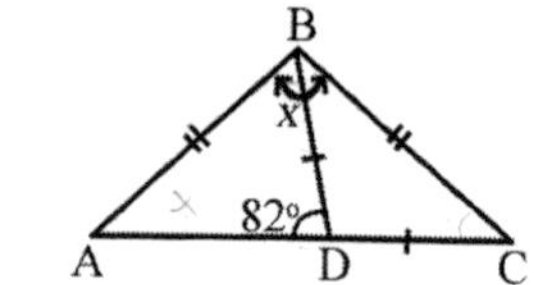

Figure 2. 32

Answer:
Given $\angle ADB = 82°$
Sum of straight line angles = 180°
$\angle BDC + \angle ADB = 180°$
$\angle BDC + 82° = 180°$
$\angle BDC = 98°$
Given BD = DC
=> $\triangle BCD$ = isosceles triangle
=> $\angle DBC \cong \angle BCD$ ...(*)
Sum of interior angles in $\triangle BCD = 180°$
$\angle BDC + \angle DBC + \angle BCD = 180°$
$98° + 2\angle BCD = 180°$ ⇦ from (*)
$2\angle BCD = 82°$
$\angle BCD = 41°$
Also given AB = BC
=> $\triangle ABC$ = isosceles triangle
=> $\angle BAC \cong \angle BCD = 41°$
Sum of interior angles in $\triangle ABC = 180°$
$\angle ABC + \angle BAC + \angle BCD = 180°$
$x + 41° + 41° = 180°$
∴ $x = 98°$ //

36. A boat is at sea and directly 4 miles east (bearing 90°) is a lighthouse. Again from the boat, directly 6 miles south (bearing 180°) is a port. Find the shortest distance from the lighthouse to the port.

Answer:
Let
B = position on the boat
L = position at the lighthouse
P = position at the port
Thus given, BL = 4 miles

BP = 6 miles

Distance from lighthouse to port = $\overline{LP}$

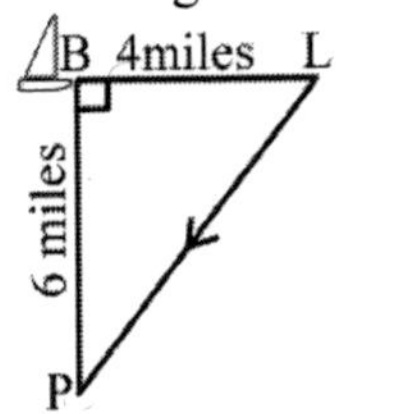

Using Pythagorean Theorem:

$LP^2 = BL^2 + BP^2$

$LP^2 = 4^2 + 6^2$

$= 16 + 36$

$= 52$

LP = 7.2111 miles

∴ Shortest distance from lighthouse to port is 7.2111 miles. //

37. A car heads north 4.8 miles then it turns 90° right and heads east for 3.6 miles. After that it turns 135° and heads southwest back to its starting point. How far has the driver traveled?

Answer:

Let A = starting point

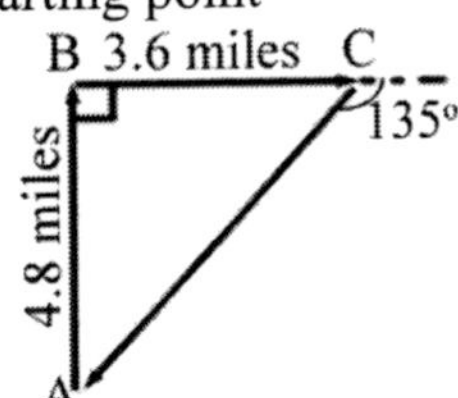

Using Pythagorean Theorem:

$AC^2 = AB^2 + BC^2$

$AC^2 = (4.8)^2 + (3.6)^2$

$AC^2 = 23.04 + 12.96$

$AC = \sqrt{36}$

AC = 6 miles

Total distance traveled:

= AC + AB + BC

= 6 + 4.8 + 3.6

= 14.4 miles //

38. Scott watches a lifeguard on duty from across the swimming pool. The lifeguard sits on an elevated stand 8 feet high. If Scott's angle of elevation is 60° and both Scott and the lifeguard are at the edge of the pool, determine Scott's line of sight and the width of the swimming pool.

Answers:

Given Scott's angle of elevation = 60°

Let w = pool's width

h = Scott's line of sight or hypotenuse length

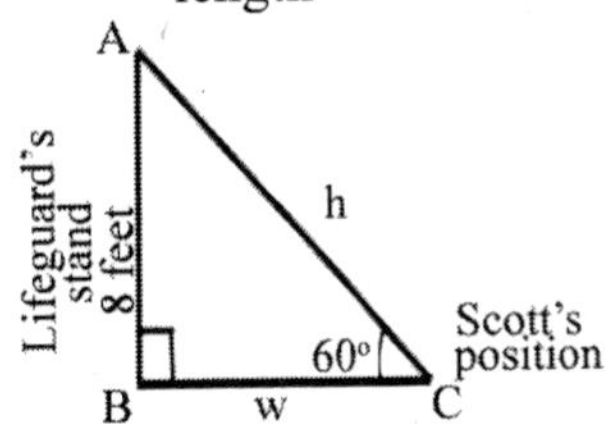

Since △ABC is a 30°–60°–90° triangle:

Using 30°–60°–90° triangle sides ratio:

$\Rightarrow AB = x\sqrt{3} = 8$

$x = \dfrac{8}{\sqrt{3}}$

$AC = h = 2x$

$h = \dfrac{2 \times 8}{\sqrt{3}}$

$h = \dfrac{16}{\sqrt{3}} = 9.2376$ feet

∴ Scott's line of sight is 9.2376 feet. //

$BC = w = x$

$w = \dfrac{8}{\sqrt{3}} = 4.6188$ feet

∴ Width of the pool is 4.6188 feet. //

39. Chad observes his neighbors painting their house from his tree house. If Chad's angle of depression is 30°, and his neighbor's house is 20 feet away from tree house, determine how high Chad is hiding.

Answer:
Let d = distance from Chad's tree house to his neighbor's house.
=> d = 20
Let h = Chad's line of sight and hypotenuse length.
m = Chad's height above ground

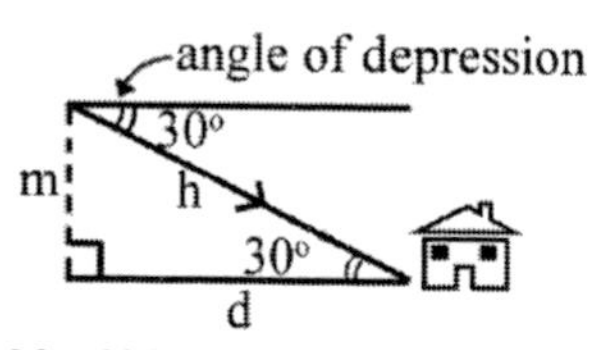

Using 30°–60°–90° triangle sides ratio:

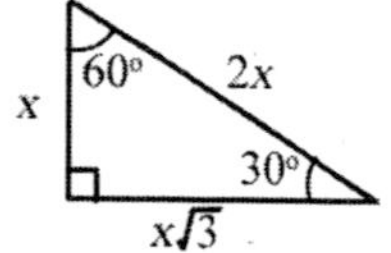

=> $m\sqrt{3} = d$ since x = m

$m\sqrt{3} = 20$

m = 11.5470 feet

∴ Chad is hiding 11.5470 feet high. //

40. On top the crow's nest, a sailor observes his captain waving on the deck. If the sailor's angle of depression is 60° and the crow's nest is 50 feet high, how far is the captain standing from the foot of the crow's nest?

Answer:
Let h = crow's nest height from deck
=> h = 50
Let a = sailor's line of sight
d = distance from foot of crow's nest to captain

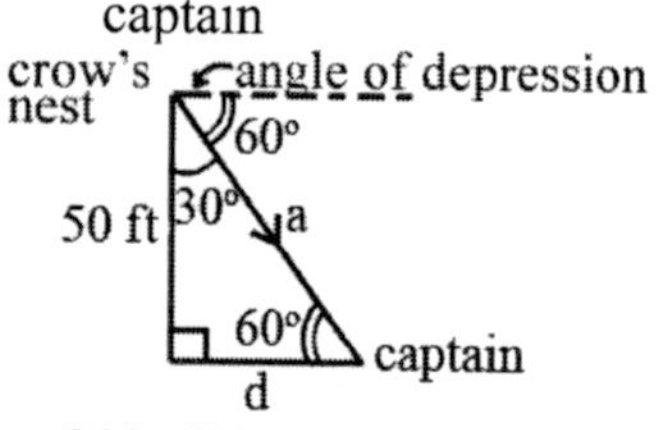

Using 30°–60°–90° triangle sides ratio:

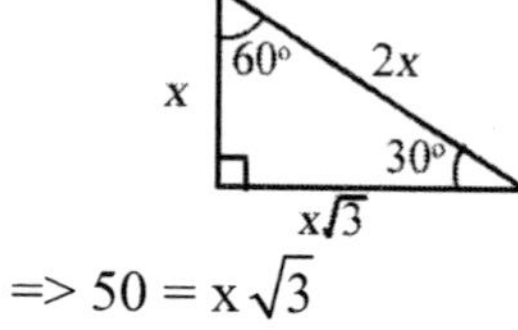

=> $50 = x\sqrt{3}$

$x = \frac{50}{\sqrt{3}} = 28.868$

d = x = 28.868 feet

∴ Distance from the foot of the crow's nest to the captain is 28.868 feet. //

41. △ABC is a scalene triangle. If ∠ABC is 60° and ∠BAC is 70°. Determine which side is the longest.

Answer:
Given ∠BAC = 70°
∠ACB = 180° – 70°– 60° = 50°
∴ ∠BAC is the largest angle in △ABC
Opposite side of largest angle is the longest side, ∴ $\overline{BC}$ is the longest side. //

Chapter 3
Polygons & Quadrilaterals

Polygon – 2-dimension closed figure with at least three sides and angles

Regular polygon – equilateral and equiangular polygon

For regular polygons:

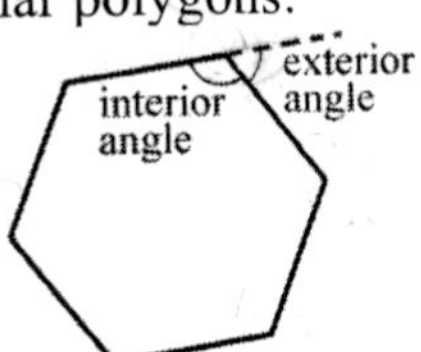

a. Sum of measure of **interior angles** in a polygon can be found through:

i. connecting diagonals to a vertex and multiply the number of triangles by 180°

ii. $(N - 2) \times 180°$ where N = number of sides

b. Sum of **exterior angles** = 360°

c. Measure of interior angle

$$= \frac{(N-2)\times 180°}{N}$$

d. Measure of exterior angle $= \frac{360°}{N}$

Quadrilateral – polygon with 4 sides and sum of measures of angles is 360°

Square – quadrilateral with 4 equilaterals and all angles are 90°

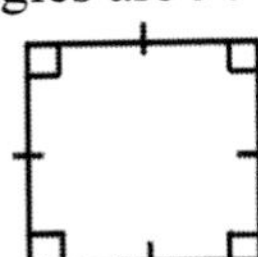

Rhombus – quadrilateral with sides equal in length and opposite sides parallel. The diagonals of a rhombus are perpendicular

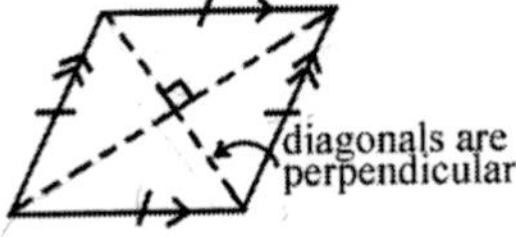

Rectangle – quadrilateral whose opposite sides are equal in length and all angles are 90°

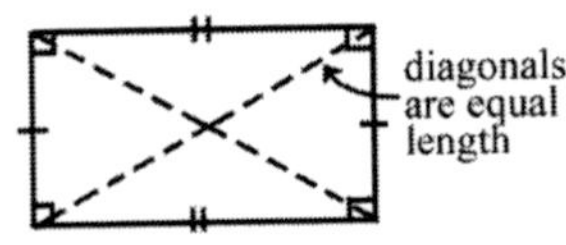

Parallelogram – quadrilateral whose opposite sides are parallel and equal in length. Adjacent angles are supplementary

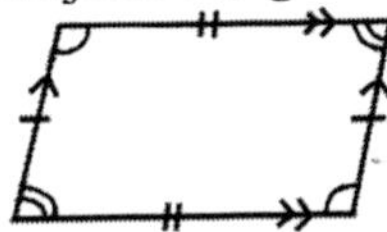

Trapezoid – quadrilateral with only one pair of opposite sides that is parallel

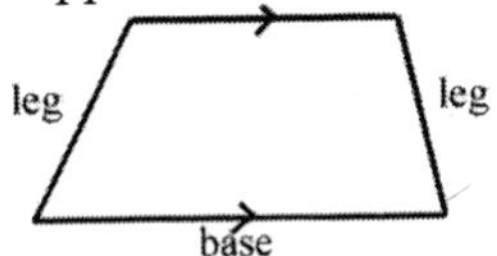

Isosceles trapezoid – trapezoid with equilateral legs

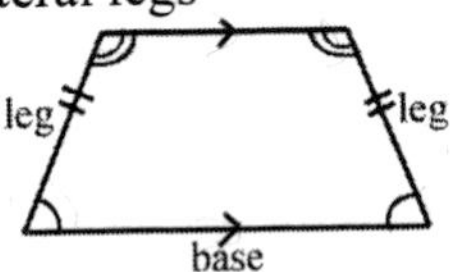

Kite – quadrilateral in the shape of a 'kite' and is symmetrical about one diagonal

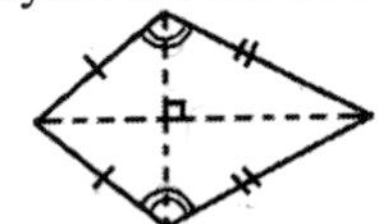

Trapezium – quadrilateral with no two sides parallel

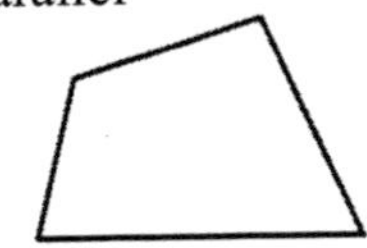

1. In Figure 3.1, PQRS is a trapezium. Find the value of y in terms of a, b and c.

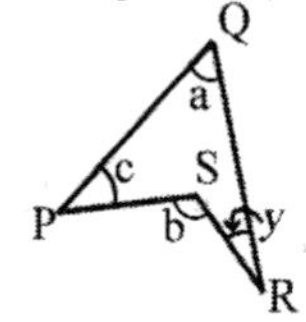

Figure 3. 1

Answer:
Let ∠N = reflex ∠PSR = 360° − b
Sum of interior angles in PQRS = 360°
∠PQR + ∠QRS + ∠N + ∠SPQ = 360°
a + y + 360° − b + c = 360°
∴ y = b − a − c //

2. What is the sum of measure of exterior angles of a regular hexagon?

Answer:
360°. Regular hexagon is a 6 sided regular polygon. Since all regular polygons have sum of measure of exterior angles equal to 360°, total measure of exterior angles for a regular hexagon is also 360°. //

3. Figure 3.2, is an inscribed square. If the radius of the circle is 5″, find the length of the sides of the square. Subsequently find the perimeter of the square.

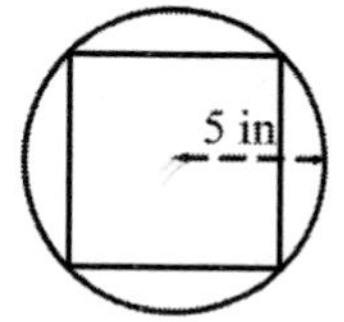

Figure 3. 2

Answers:
Let m = side of inscribed square
Given radius of circle = 5″
=> Diameter of circle = 5 + 5 = 10″
=> Diameter of circle = diagonal of square

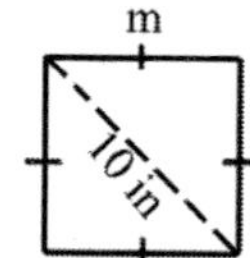

Using Pythagorean Theorem:
$10^2 = m^2 + m^2$
$100 = 2m^2$
$m = \sqrt{50}$
$m \approx 7.0711''$
∴ Length of side of inscribed square is 7.0711″. //
Perimeter of square:
$= 4 \times m$
$= 4 \times 7.0711$
$= 28.2844''$
∴ Perimeter of the inscribed square is 28.2844″. //

Alternatively:
Finding perimeter using inscribed regular polygon formula:

$= 2nr \times \sin\left(\frac{180°}{n}\right)$

$= 2(4)(5) \times \sin\left(\frac{180°}{4}\right)$

$= 40 \times \sin 45°$

$= 40 \times 0.70711$

$= 28.2844''$ //

4. Figure 3.3 shows a circumscribed regular hexagon. If the radius of the circle is 2″, find the perimeter of the polygon.

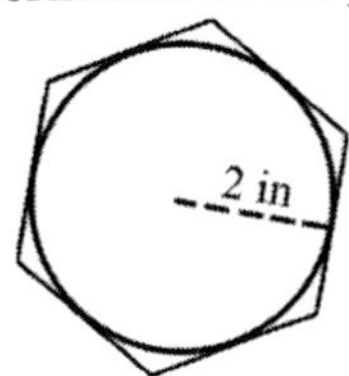

Figure 3. 3

Answer:
Given radius, r = 2
Also given sides of hexagon, n = 6
Perimeter of circumscribed regular hexagon:

$= 2nr \times \tan\left(\frac{180°}{n}\right)$

$= 2(6)(2) \times \tan\left(\frac{180°}{6}\right)$

$= 24 \times \tan 30°$

$= 24 \times 0.5774$

$= 13.8576$ in

$\therefore$ Perimeter of circumscribed regular hexagon is 13.8576 in. //

5. What quadrilateral is equiangular but not equilateral?

Answer:
Rectangle. Equiangular means all angles are congruent. Equilateral means all sides are congruent. A rectangle has congruent angles, 90°, but only opposite sides are equal in length. //

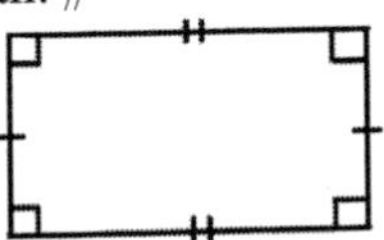

6. Determine the number of diagonals in a decagon.

Answer:
Let N = number of sides
Given decagon
=> N = 10
Number of diagonals:

$= \frac{1}{2} \times N \times (N-3)$

$= \frac{1}{2} \times 10 \times (10-3)$

$= 5 \times 7$

$= 35$ diagonals //

7. Find the sum of measure of the interior angles for the following polygons (see Figure 3.4):

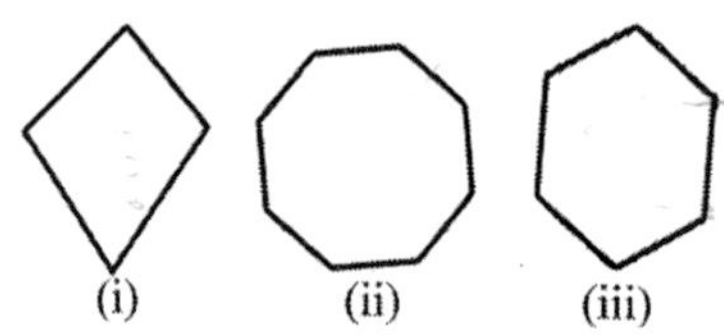

Figure 3. 4

Answers:
Let N = number of sides

i) Figure (i) is a quadrilateral
=> N = 4
Sum of interior angles:
$= 180° \times (N - 2)$
$= 180° \times (4 - 2)$
$= 180° \times 2$
$= 360°$ //

Alternatively:
Solve by dividing polygon into triangles

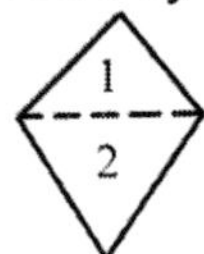

=> Total number of triangles = 2
Sum of interior angles:
$= 180° \times$ number of $\triangle$
$= 180° \times 2$
$= 360°$//

ii) Figure (ii) is an octagon
=> N = 8
Sum of interior angles:
$= 180° \times (N - 2)$
$= 180° \times (8 - 2)$
$= 180° \times 6$
$= 1080°$ //

Alternatively:
Solve by dividing polygon into triangles

=> Total number of triangles = 6
Sum of interior angles:
$= 180° \times$ number of $\triangle$
$= 180° \times 6$
$= 1080°$//

iii) Figure (iii) is a hexagon
=> N = 6
Sum of interior angles:
$= 180° \times (N - 2)$
$= 180° \times (6 - 2)$
$= 180° \times 4$
$= 720°$ //

Alternatively:
Solve by dividing polygon into triangles

=> Total number of triangles = 4
Sum of interior angles:
$= 180° \times$ number of $\triangle$
$= 180° \times 4$
$= 720°$//

8. Figure 3.5 is a regular polygon. Find the value of the exterior angle y.

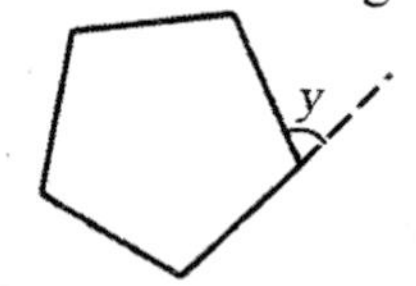

Figure 3. 5

Answer:
Given figure 3.5 is a regular pentagon
Number of sides, N = 5
Measure of exterior angle:

$$= \frac{360°}{N}$$

$= \dfrac{360^\circ}{5}$

$= 72^\circ$

$\therefore y = 72^\circ$ //

Alternatively:

Sum of measure of interior angle:

$= 180^\circ \times (N - 2)$

$= 180^\circ \times (5 - 2)$

$= 180^\circ \times 3$

$= 540^\circ$

Measure of interior angle:

$= \dfrac{540^\circ}{5}$

$= 108^\circ$

Sum of straight line angles = 180°

Interior angle + exterior angle = 180°

$108^\circ + y = 180^\circ$

$\therefore y = 72^\circ$ //

9. Given the exterior angle of a regular polygon is 36°. Hence find the measure of its interior angle. Also find the sum of measure of the interior angles for the given polygon.

Answers:

Given exterior angle = 36°

Since exterior angle and interior angle are supplementary angles:

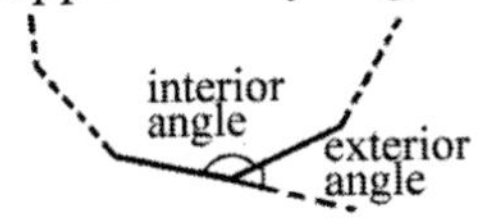

Interior angle + exterior angle = 180°

Interior angle + 36° = 180°

$\therefore$ Interior angle = 144° //

Sum of exterior angles of any regular polygon = 360°

Let N = number of sides

Exterior angle $= \dfrac{360^\circ}{N}$

$36^\circ = \dfrac{360^\circ}{N}$

$N = 10$

Sum of interior angles:

$= 180^\circ \times (N - 2)$

$= 180^\circ \times (10 - 2)$

$= 180^\circ \times 8$

$= 1440^\circ$

$\therefore$ Sum of interior angles = 1440° //

10. Figure 3.6, shows a parallelogram, LMNO. Hence find the values of x and y.

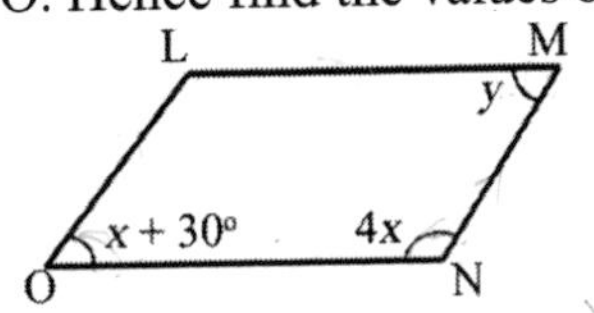

Figure 3. 6

Answers:

Given LMNO = parallelogram

=> $\overline{LO} \parallel \overline{MN}$

Consecutive angles are supplementary:

$\angle LON + \angle MNO = 180^\circ$

$x + 30^\circ + 4x = 180^\circ$

$5x = 150^\circ$

$\therefore x = 30^\circ$ // ...(*)

Since LMNO is a parallelogram,

Opposite angles are ≅:

$\angle LMN = \angle LON$

$y = x + 30^\circ$

Substitute (*):

$y = 30^\circ + 30^\circ$

$\therefore y = 60^\circ$ //

11. In Figure 3.7, given ABCD is a parallelogram. Express a and b in terms of x, y and/or z. Hence prove that △ABC and △ADC are congruent.

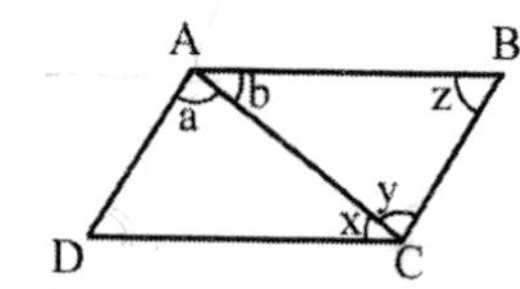

Figure 3. 7

Answers:
Given ABCD = parallelogram
=> $\overline{AB} \parallel \overline{DC}$
Alternate interior angles are $\cong$:
$\angle BAC \cong \angle ACD$
$\therefore$ b $\cong$ x //
=> $\overline{AD} \parallel \overline{BC}$
Alternate interior angles are $\cong$:
$\angle CAD \cong \angle ACB$
$\therefore$ a $\cong$ y //

From above:
$\angle ACD \cong \angle BAC$ [angle]
$\angle ACB \cong \angle CAD$ [angle]
Reflexive property of equality:
$\overline{AC} \cong \overline{CA}$ [side]
$\therefore$ Based on A–S–A postulate, $\triangle ABC$ and $\triangle ADC$ are congruent. //

12. Figure 3.8, ABCD is a parallelogram. And $\angle BAC$ is 30° while $\angle BDC$ is 40°. Show that $\overline{AE} = \overline{EC}$ and $\overline{BE} = \overline{ED}$.

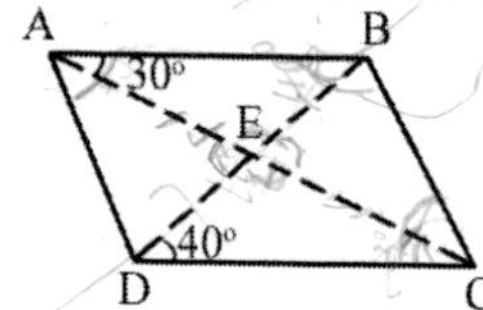

Figure 3. 8

Answer:
Given ABCD = parallelogram
=> $\overline{AB} \parallel \overline{CD}$
Alternate interior angles are $\cong$:
$\angle ACD \cong \angle BAC$ [angle]
$\angle ACD = 30°$
=> $\overline{AD} \parallel \overline{BC}$
Alternate interior angles are $\cong$:
$\angle ABD \cong \angle BDC$ [angle]
$\angle ABD = 40°$
Opposite sides of parallelogram are $\cong$:
$\overline{AB} \cong \overline{DC}$ [side]
Based on A–S–A postulate, $\triangle ABE$ and $\triangle CDE$ are congruent.
Since corresponding parts of congruent triangles are congruent.
Thus,

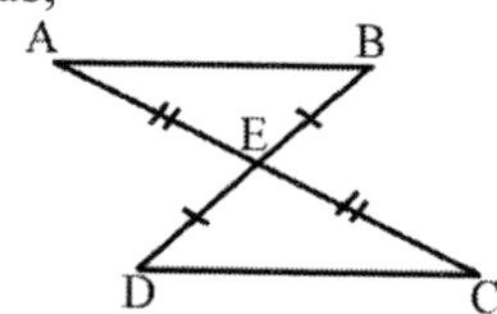

$\triangle ABE$ and $\triangle CDE$ are congruent triangles, corresponding parts $\overline{AE}$ equals $\overline{EC}$, and $\overline{BE}$ equals $\overline{ED}$. //

13. Figure 3.9, shows a rhombus, ABCD. Given $\angle ABD = 35°$, $\angle BAD = y$ and $\angle ADC = x$. Hence find the values of x and y.

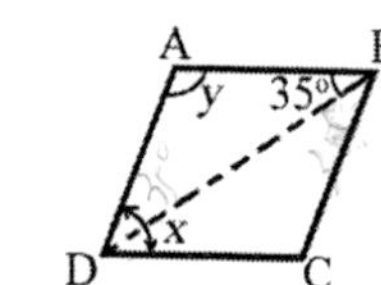

Figure 3. 9

Answers:
Given ABCD = rhombus
$\overline{BD}$ = diagonal
Since diagonals in a rhombus are angle bisectors:
$\angle ADC = 2 \times \angle ABD$
$x = 2 \times 35°$
$\therefore x = 70°$ //
Opposite sides of rhombus are parallel:
$\overline{AB} \parallel \overline{DC}$
Consecutive angles are supplementary:
$\angle BAD + \angle ADC = 180°$

$y + x = 180°$
$y + 70° = 180°$
$\therefore y = 110°$ //

Alternatively:

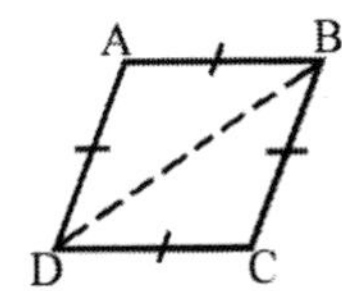

ABCD is a rhombus:
=> All sides of a rhombus are equal in length
=> BD = diagonal
=> $\triangle ABD$ is an isosceles triangle
=> $\angle ABD = \angle ADB = 35°$ ...(1)
Sum of interior angles in $\triangle ABD = 180°$
$\angle BAD + \angle ABD + \angle ADB = 180°$
$y + 35° + 35° = 180°$
$\therefore y = 180° - 70° = 110°$ //
=> $\overline{AB} \parallel \overline{DC}$
Alternate interior angles are ≅:
$\angle ABD \cong \angle BDC = 35°$ ...(2)
Using addition property of equality:
$\angle ADC = \angle ADB + \angle BDC$
$x = 35° + 35°$ ⇦ From (1) and (2)
$\therefore x = 70°$ //

14. In Figure 3.10, PQRS is a rhombus. What is the value of a + b?

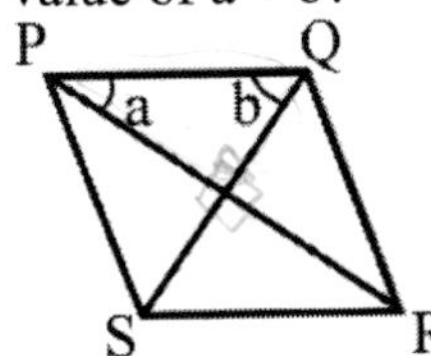

Figure 3. 10

Answer:
Given PQRS = rhombus
=> $\overline{PR}$ and $\overline{QS}$ are diagonals
=> $\overline{PR} \perp \overline{QS}$
Let E = intersection of $\overline{PR}$ and $\overline{QS}$
=> $\angle PEQ = 90°$
Sum of interior angles in $\triangle PEQ = 180°$
$\angle QPR + \angle PQS + \angle PEQ = 180°$
$a + b + 90° = 180°$
$\therefore a + b = 90°$ //

15. In Figure 3.11, WXYZ is a square. If $\overline{UV}$ bisects $\angle YVZ$, find the values of β, θ, and α.

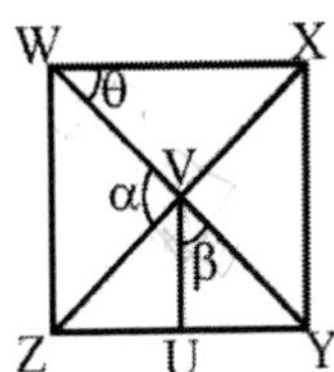

Figure 3. 11

Answers:
Given WXYZ = square
=> All angles are congruent and 90°
=> $\overline{WY}$ and $\overline{XZ}$ are diagonals
=> $\overline{WY}$ and $\overline{XZ}$ are angle bisectors:

$$\angle XWY = \frac{90°}{2}$$

$\therefore \theta = 45°$ //
=> $\overline{WY} \perp \overline{XZ}$
$\angle WVZ = 90°$
$\therefore \alpha = 90°$ //
Since $\overline{WY} \perp \overline{XZ}$:
=> $\angle YVZ = 90°$
Given $\overline{UV}$ = angle bisector of $\angle YVZ$

$$\angle UVY = \frac{\angle YVZ}{2}$$

$$\therefore \beta = \frac{90°}{2} = 45°$$ //

16. In Figure 3.12, ABCD is a rectangle. Given $\overline{EF}$ bisects $\angle CED$ and $\overline{EF}$ is perpendicular to $\overline{DC}$. If $\angle BEC$ is 50°, find the values of x, y and z.

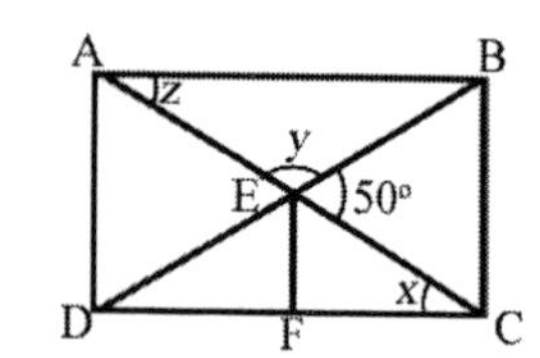
Figure 3. 12

Answers:
Given $\angle BEC = 50°$
Straight line angles are supplementary:
$\angle AEB + \angle BEC = 180°$
$y + 50° = 180°$
$\therefore y = 130°$ //
Vertical angles are ≅:
$\angle CED \cong \angle AEB$
$\angle CED = y = 130°$
Given $\overline{EF}$ bisects $\angle CED$
$\angle CEF = \dfrac{\angle CED}{2}$
$\angle CEF = \dfrac{130°}{2}$
$\angle CEF = 65°$
Given $\overline{EF} \perp \overline{DC}$
$\angle CFE = 90°$
Sum of interior angles of $\triangle CEF = 180°$
$\angle ACD + \angle CEF + \angle CFE = 180°$
$x + 65° + 90° = 180°$
$\therefore x = 25°$ //
Since ABCD is a rectangle
=> $\overline{AB} \parallel \overline{DC}$
Alternate interior angles are ≅:
$\angle BAC \cong \angle ACD$
$z = x$
$\therefore z = 25°$ //

17. In Figure 3.13, PQRS is a trapezoid. Express x and y in terms of θ.

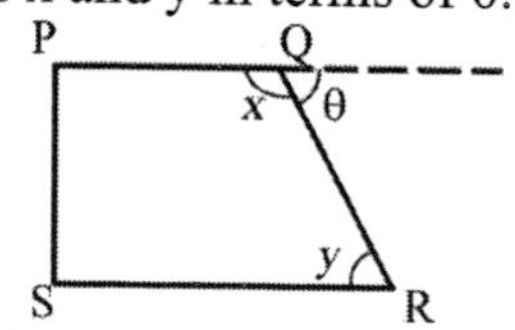
Figure 3. 13

Answers:
Given PQRS = trapezoid
=> $\overline{PQ} \parallel \overline{SR}$

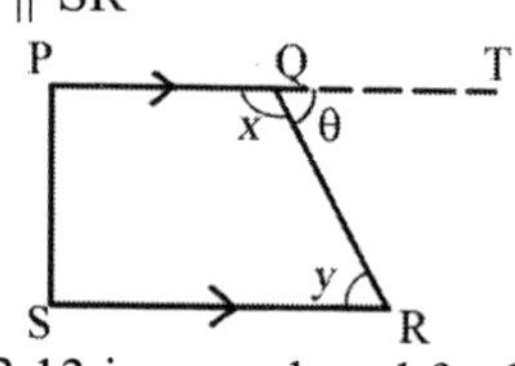
Figure 3.13 is reproduced for labeling
=> $\overline{PT} \parallel \overline{SR}$
Sum of straight line angles = 180°
$\angle PQR + \angle RQT = 180°$
$x + \theta = 180°$
$\therefore x = 180° - \theta$ //
Alternate interior angles are ≅:
$\angle QRS \cong \angle RQT$
$\therefore y = \theta$ //

18. Figure 3.14, represents a quadrilateral, ABCD. Given that AB = AD and BC = DC, find the value of x.

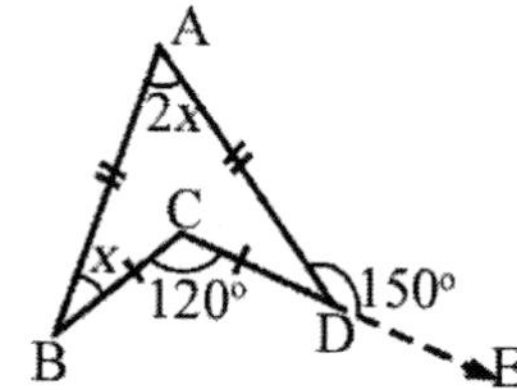
Figure 3. 14

Answer:
Sum of straight line angles = 180°
$\angle ADC + \angle ADE = 180°$
$\angle ADC + 150° = 180°$
$\angle ADC = 30°$
Given $\angle BCD = 120°$
Reflex $\angle BCD = 360° - 120° = 240°$
Sum of interior angles in quadrilateral = 360°
$\angle BAD + \angle ABC + \text{reflex } \angle BCD + \angle ADC = 360°$

$2x + x + 240° + 30° = 360°$
$3x + 270° = 360°$
$3x = 90°$
$\therefore\ x = 30°$ //

19. Figure 3.15 shows a trapezoid ABCD. If $\overline{EF}$ is the median, find the values of x and y.

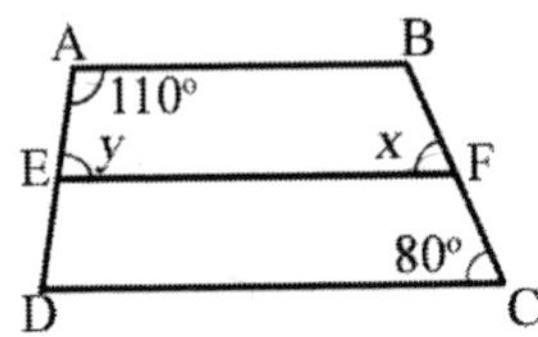

Figure 3. 15

Answers:
Given ABCD = trapezoid
=> $\overline{AB} \parallel \overline{DC}$
Given $\overline{EF}$ = median of trapezoid
=> $\overline{EF} \parallel \overline{AB}$ and $\overline{EF} \parallel \overline{DC}$
Corresponding angles are ≅:
$\angle BFE \cong \angle BCD$
$\therefore\ x = 80°$ //
Consecutive angles are supplementary:
$\angle AEF + \angle BAD = 180°$
$y + 110° = 180°$
$\therefore\ y = 70°$ //

20. In Figure 3.16, ACDF is a trapezoid whose median is $\overline{BE}$. Hence find the values of x and y.

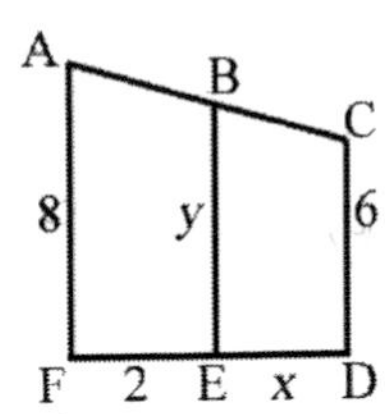

Figure 3. 16

Answers:
Given $\overline{BE}$ = median of trapezoid
Points B and E are midpoints of $\overline{AC}$ and $\overline{FD}$ respectively.
=> $\overline{ED} = \overline{FE}$
$\therefore\ x = 2$ units //
Given BE = y

$$y = \frac{1}{2}(\text{Upper base} + \text{Lower base})$$
$$= \frac{1}{2}(AF + CD)$$
$$= \frac{1}{2}(8 + 6)$$
$$= \frac{14}{2}$$

$\therefore\ y = 7$ units //

21. In Figure 3.17, ABDC is an isosceles trapezoid. If $\angle BAC = 120°$, find the values of x and y.

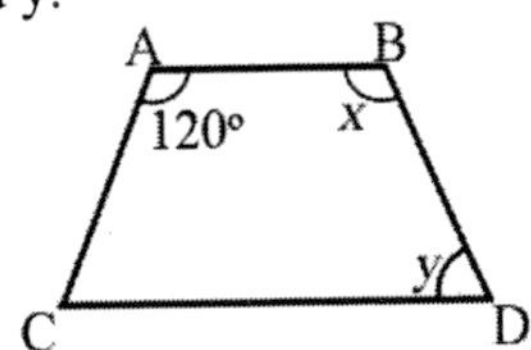

Figure 3. 17

Answers:
Given ABDC = isosceles trapezoid
=> AC = BD
$\angle ABD \cong \angle BAC$
$\therefore\ x = 120°$ //
Since ABDC is a trapezoid
=> $\overline{AB} \parallel \overline{CD}$
Consecutive angles are supplementary:
$\angle BDC + \angle ABD = 180°$
$y + x = 180°$
$y + 120° = 180°$
$\therefore\ y = 60°$ //

22. Figure 3.18 shows ABCD an isosceles trapezoid whose diagonals are $\overline{AC}$ and $\overline{BD}$. Hence find the values of x and y in terms of a.

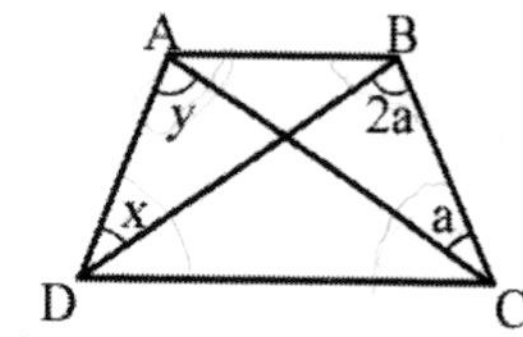

Figure 3. 18

Answers:
Given ABCD = isosceles trapezoid
=> AD = BC [side]
=> Diagonals $\overline{AC} \cong \overline{BD}$
Figure 3.18 is reproduced for labeling:

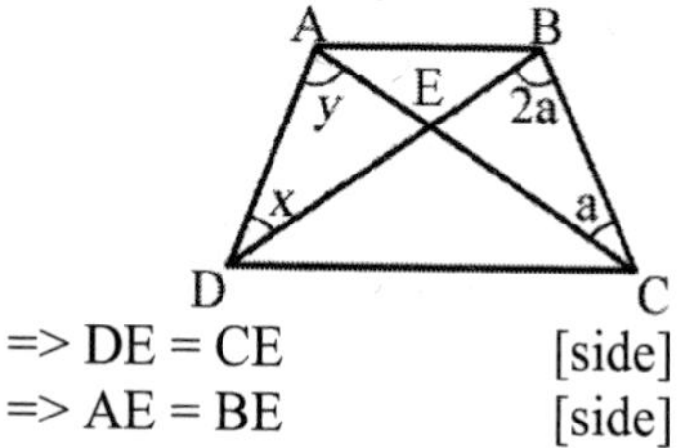

=> DE = CE [side]
=> AE = BE [side]
Thus, from S–S–S postulate, $\triangle$AED and $\triangle$BEC are congruent.
Since corresponding parts of congruent triangles are congruent:
=> $\angle CAD \cong \angle CBD$
$\therefore$ y = 2a //
=> $\angle ADB \cong \angle ACB$
$\therefore$ x = a //

23. In Figure 3.19, EFGHI is part of a regular polygon. Determine the number of sides the polygon has.

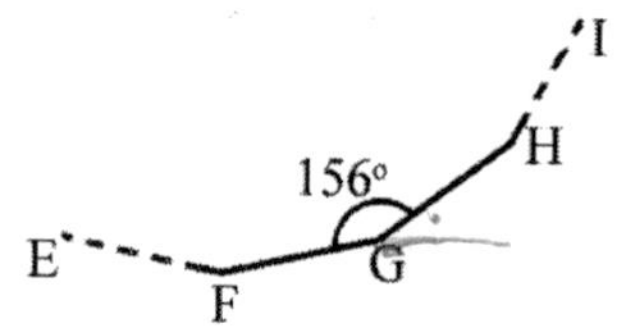

Figure 3. 19

Answer:
Given EFGHI = regular polygon
=> All sides are equal in length
Also given internal angle, $\angle FGH = 156°$
=> External angle = 180° – 156°
= 24°
Let N = number of sides of regular polygon
Measure of external angle:

$$24° = \frac{360°}{N}$$

$$N = \frac{360°}{24°}$$

$\therefore$ N = 15 sides //

24. Find the number of diagonals in a regular heptagon. Hence determine the measure of its external angle.

Answers:
Let N = number of sides of polygon
Given regular heptagon
=> N = 7
Number of diagonals:

$$= \frac{1}{2} \times N \times (N-3)$$

$$= \frac{1}{2} \times 7 \times (7-3)$$

= 14 diagonals //
Sum of external angles for regular polygon = 360°
Measure of external angle:

$$= \frac{360°}{N}$$

$$= \frac{360°}{7}$$

= 51.4286° //

25. In Figure 3.20, ABCDEFG is a regular heptagon while BCQRS is a regular pentagon. Hence find the value of x.

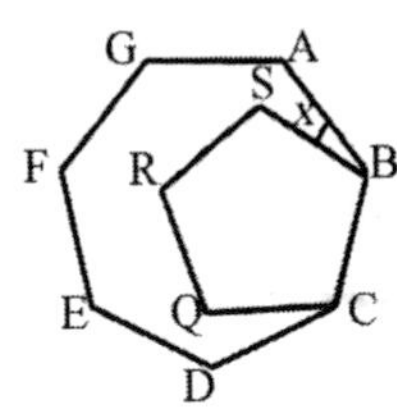

Figure 3. 20

Answer:
Let N = number of sides of polygon
Given ABCDEFG = regular heptagon
=> $N_7 = 7$
Internal angle of ABCDEFG, $\angle ABC$:

$= 180° - \dfrac{360°}{N_7}$

$= 180° - \dfrac{360°}{7}$

$= 180° - 51.43°$
$= 128.57°$

Also given BCQRS = regular pentagon
=> $N_5 = 5$
Internal angle of BCQRS, $\angle CBS$:

$= 180° - \dfrac{360°}{N_5}$

$= 180° - \dfrac{360°}{5}$

$= 180° - 72°$
$= 108°$

Thus,
$\angle ABS$ = internal angle of ABCDEFG – internal angle of BCQRS
$\angle ABS = \angle ABC - \angle CBS$
$x = 128.57° - 108°$
$\therefore x = 20.57°$ //

26. In Figure 3.21, ABCDEF is an irregular hexagon and AGHEF is an irregular pentagon. Find:
a) y
b) x + y

Figure 3. 21

Answers:
a) Given $\angle ABC = 110°$
Sum of straight line angles = 180°
$\angle CBG + \angle ABC = 180°$
$y + 110° = 180°$
$\therefore y = 70°$ //
b) Given $\angle BCD = 130°$
Sum of angles at a point = 360°
Reflex $\angle BCD = 360° -$ obtuse $\angle BCD$
$= 360° - 130°$
$= 230°$
Sum of interior angles in BCDHG pentagon:
$= 180° \times (N - 2)$
$= 180° \times (5 - 2)$ ⇦Pentagon has 5 sides, N=5
$= 180° \times 3$
$= 540°$
Sum of interior angles in BCDHG = 540°
$\angle CBG + \angle BGH + \angle DHG + \angle CDH +$ reflex $\angle BCD = 540°$
$y + y + 90° + x + 230° = 540°$
$70° + 70° + 90° + x + 230° = 540°$
$x = 80°$
Thus,
$x + y = 80° + 70°$
$= 150°$ //

27. Figure 3.22, shows a rhombus, ABCD. If $\overline{AF}$ and $\overline{BF}$ are straight lines, find the value of x.

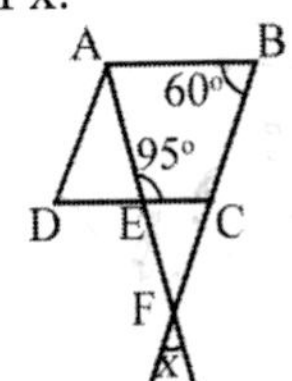

Figure 3. 22

Answer:
Given ABCD = rhombus
=> $\overline{AB} \parallel \overline{DC}$
Corresponding angles are ≅:
$\angle DCF \cong \angle ABC$
$\angle DCF = 60°$
Sum of straight line angles = 180°
$\angle CEF + \angle AEC = 180°$
$\angle CEF + 95° = 180°$
$\angle CEF = 85°$
Sum of interior angles in $\triangle CEF = 180°$
$\angle CFE + \angle CEF + \angle DCF = 180°$
$\angle CFE + 85° + 60° = 180°$
$\angle CFE = 35°$
Vertical angles are ≅:
$x = \angle CFE$
$\therefore x = 35°$ //

28. In Figure 3.23, ABCDEF is a regular polygon. Given $\overline{AG}$ and $\overline{DG}$ are straight lines. Hence find the value of x.

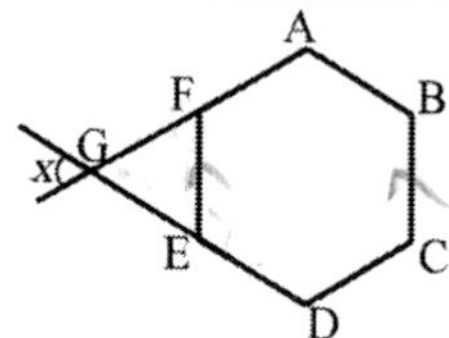

Figure 3. 23

Answer:
Let N = number of sides
Given ABCDEF = regular hexagon
=> N = 6
Measure of external angle:
$= \dfrac{360°}{N}$
$= \dfrac{360°}{6}$
$= 60°$
Given $\angle GFE$ and $\angle FEG$ = external angles
Sum of interior angles of $\triangle EFG = 180°$
$\angle AGD + \angle GFE + \angle FEG = 180°$
$\angle AGD + 60° + 60° = 180°$
$\angle AGD = 60°$
Vertical angles are ≅:
$x = \angle AGD$
$\therefore x = 60°$ //

29. In Figure 3.24, ABCDH is a regular polygon and FD = FE. Find the value of x.

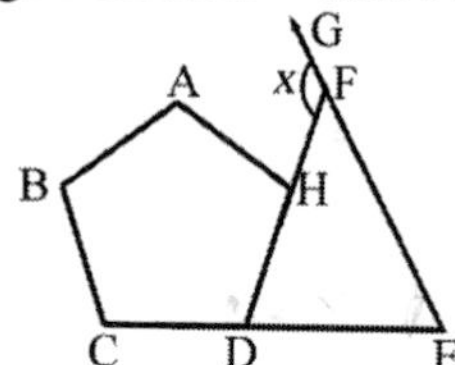

Figure 3. 24

Answer:
Let N = number of sides of polygon
Given ABCDH = regular pentagon
=> N = 5
Measure of external angle:
$= \dfrac{360°}{N}$
$= \dfrac{360°}{5}$
$= 72°$
=> External angle, $\angle EDF = 72°$
Given FD = FE
=> $\triangle EDF$ = isosceles triangle
=> $\angle EDF \cong \angle DEG = 72°$
Thus,
$\angle DFG = \angle EDF + \angle DEG$
$x = 72° + 72°$
$\therefore x = 144°$ //

30. In Figure 3.25, $\triangle ADE$ is an equilateral triangle and CDEF is a rhombus. Find the values of:
a) y
b) z
c) x

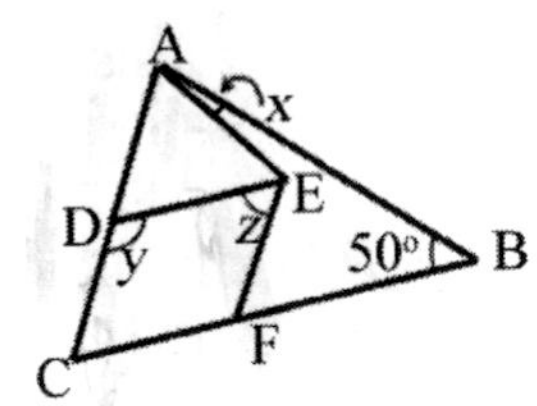

Figure 3. 25

Answers:
a) Given $\triangle ADE$ = equilateral triangle
=> all angles are congruent
=> $\angle ADE = \dfrac{180°}{3} = 60°$
Sum of straight line angles = 180°
$\angle CDE + \angle ADE = 180°$
$y + 60° = 180°$
$\therefore\ y = 120°$ //

b) Given CDEF = rhombus
=> $\overline{DC} \parallel \overline{EF}$
Consecutive angles are supplementary:
$\angle CDE + \angle DEF = 180°$
$y + z = 180°$
$120° + z = 180°$
$\therefore\ z = 60°$ //

c) Since CDEF = rhombus
=> Opposite angles are ≅
$\angle ACB \cong \angle DEF$
$\angle ACB = z = 60°$
Since $\triangle ADE$ = equilateral triangle
=> $\angle CAE = 60°$
Sum of interior angles in $\triangle ABC$ = 180°
$\angle BAE + \angle CAE + \angle ACB + \angle ABC = 180°$
$x + 60° + 60° + 50° = 180°$
$x = 180° - 170°$
$\therefore\ x = 10°$ //

31. In Figure 3.26, ABCDE is a regular pentagon and FAGHCI is part of a regular polygon whose measure of internal angle is 144°. Hence find the values of x and y.

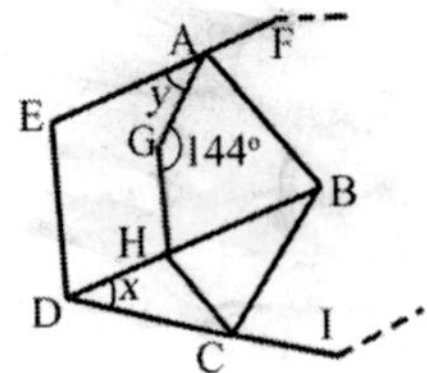

Figure 3. 26

Answers:
Given ABCDE = regular pentagon
Measure of internal angle:
$\angle BCD = 180° - \dfrac{360°}{5}$
$\angle BCD = 180° - 72°$
$\angle BCD = 108°$
Since ABCDE is a regular pentagon:
=> BC = CD
=> $\triangle BCD$ = isosceles triangle
=> $\angle CBD \cong \angle BDC = x$ ...*
Sum of internal angles in $\triangle BCD$ = 180°
$\angle CBD + \angle BDC + \angle BCD = 180°$
$x + x + 108° = 180°$ ⇦From *
$2x = 72°$
$\therefore\ x = 36°$ //
Also given FAGHCI = part of a regular polygon
All interior angles are ≅
=> $\angle AGH = 144°$
=> $\angle FAG = \angle AGH = 144°$
Sum of straight line angles = 180°
$\angle FAG + \angle EAG = 180°$
$144° + y = 180°$
$\therefore\ y = 36°$ //

32. ABCD is an isosceles trapezoid, and $\triangle ABE$ is an isosceles triangle whose sides AB = AE. HAFGB is part of a regular octagon. Find the values of x and y.

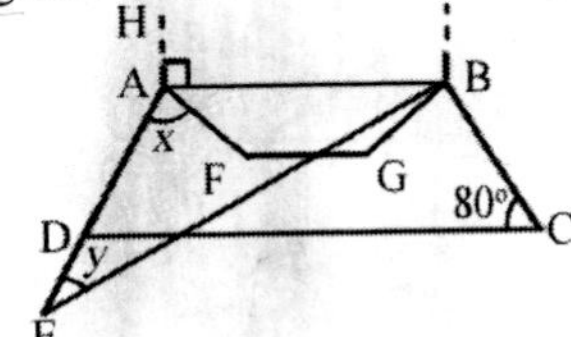

Figure 3. 27

Answers:
Let N = number of sides
Measure of internal angle of regular octagon, HAFGB (N = 8):
$= 180° - \dfrac{360°}{N}$
$= 180° - \dfrac{360°}{8}$
$= 180° - 45°$
$= 135°$
=> $\angle FAH = 135°$

Since given $\angle BAH = 90°$
=> $\angle BAF + \angle BAH = \angle FAH$
$\angle BAF + 90° = 135°$
$\angle BAF = 45°$
Given ABCD = isosceles trapezoid
=> Opposite angles of isosceles trapezoid are supplementary:
$\angle BAE + \angle BCD = 180°$
$\angle EAF + \angle BAF + 80° = 180°$
$x + 45° + 80° = 180°$
$\therefore x = 55°$ //

Given $\triangle ABE$ = isosceles triangle
Also given AB = AE
=> $\angle ABE \cong \angle AEB = y$
Sum of interior angles in $\triangle ABE = 180°$
$\angle BAE + \angle ABE + \angle AEB = 180°$
$\angle EAF + \angle BAF + y + y = 180°$
$x + 45° + 2y = 180°$
$55° + 45° + 2y = 180°$
$2y = 80°$
$\therefore y = 40°$ //

33. In Figure 3.28, $\triangle ABC$ and $\triangle CDE$ are isosceles triangles. It is further given that BCDG is a trapezium and $\triangle GDH$ is a scalene triangle. Hence find the values of x, y and z.

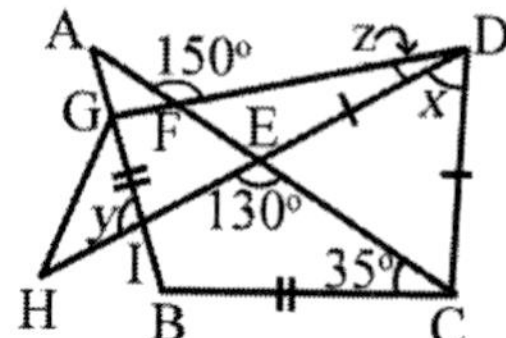

Figure 3. 28

Answers:
Given $\angle CEH = 130°$
Sum of straight line angles = 180°
$\angle CED + \angle CEH = 180°$
$\angle CED + 130° = 180°$
$\angle CED = 50°$
Given $\triangle CDE$ = isosceles triangle
=> CD = ED
=> $\angle DCE \cong \angle CED = 50°$
Sum of interior angles in $\triangle CDE = 180°$
$\angle CDE + \angle DCE + \angle CED = 180°$
$x + 50° + 50° = 180°$
$\therefore x = 80°$ //

Given $\triangle ABC$ = isosceles triangle
=>AB = CB
=> $\angle ACB \cong \angle BAC = 35°$
Sum of interior angles in $\triangle ABC = 180°$
$\angle ABC + \angle BAC + \angle ACB = 180°$
$\angle ABC + 35° + 35° = 180°$
$\angle ABC = 110°$
Sum of interior angles in quadrilateral, BCEI = 360°
$\angle BIE + \angle ABC + \angle ACB + \angle CEI = 360°$
$\angle BIE + 110° + 35° + 130° = 360°$
$\angle BIE = 85°$
Vertical angles are $\cong$:
$\angle AIH \cong \angle BIE$
$\therefore y = 85°$ //

Given $\angle AFD = 150°$
Sum of straight line angles = 180°
$\angle DFC + \angle AFD = 180°$
$\angle DFC + 150° = 180°$
$\angle DFC = 30°$
Vertical angles are $\cong$:
$\angle AED \cong \angle CEH$
$\angle AED = 130°$
Sum of interior angles in $\triangle DEF = 180°$
$\angle GDH + \angle AED + \angle DFC = 180°$
$z + 130° + 30° = 180°$
$\therefore z = 20°$ //

34. Figure 3.29, shows ABCD a kite where AB = AD and CB = CD. If $\triangle BCE$ and $\triangle CEG$ are isosceles triangles, find the values of x, y and z.

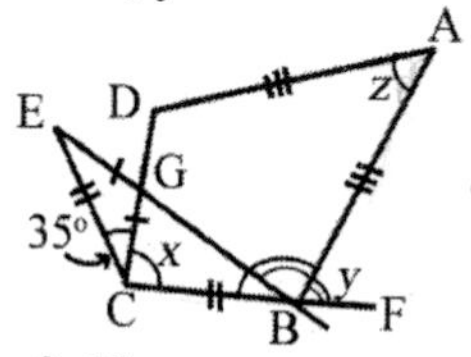

Figure 3. 29

Answers:
Given $\angle DCE = 35°$
Given $\triangle CEG$ = isosceles triangle
=> $\angle BEC \cong \angle DCE = 35°$
Also given $\triangle BCE$ = isosceles triangle
=> $\angle BEC \cong \angle CBE = 35°$
Sum of interior angles in $\triangle BCE = 180°$
$\angle ECF + \angle BEC + \angle CBE = 180°$

$\angle DCE + \angle DCF + 35° + 35° = 180°$
$35° + x + 35° + 35° = 180°$
$\therefore x = 75°$ //

Given $\angle ABE = \angle ABF = y$
Sum of straight line angles = 180°
$\angle CBE + \angle ABE + \angle ABF = 180°$
$35° + y + y = 180°$
$2y = 145°$
$\therefore y = 72.5°$ //

Given ABCD = kite
=> $\angle ADC \cong \angle ABC$
$\angle ADC = \angle CBE + \angle ABE$
$\angle ADC = 35° + y$
$\angle ADC = 35° + 72.5°$
$\angle ADC = 107.5°$
Sum of interior angles in kite, ABCD = 360°
$\angle BAD + \angle DCF + \angle ADC + \angle ABC = 360°$
$z + x + 107.5° + 107.5° = 360°$
$z + 75° + 107.5° + 107.5° = 360°$
$\therefore z = 70°$ //

35. In Figure 3.30, shows ABCEF a regular pentagon. Find the value of x.

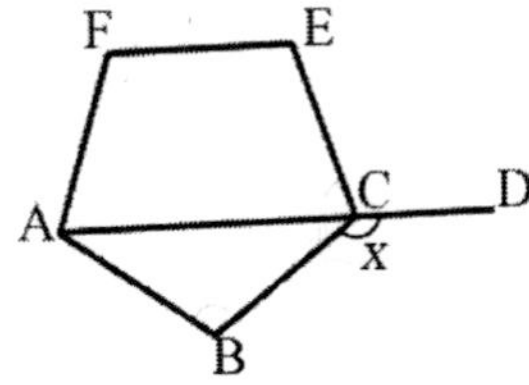

Figure 3. 30

Answer:
Let N = number of sides
Given ABCEF = regular pentagon
=> N = 5
Measure of internal angle, $\angle ABC$:
$= 180° - \dfrac{360°}{N}$
$= 180° - \dfrac{360°}{5}$
$= 180° - 72°$
$= 108°$
Since ABCEF = regular pentagon
=> AB = BC
=> $\triangle ABC$ = isosceles triangle
=> $\angle ACB \cong \angle BAC$ ...*
Sum of interior angles in $\triangle ABC = 180°$
$\angle ACB + \angle BAC + \angle ABC = 180°$
$2\angle ACB + 108° = 180°$ ⇦From *
$2\angle ACB = 72°$
$\angle ACB = 36°$
Sum of straight line angles = 180°
$\angle BCD + \angle ACB = 180°$
$x + 36° = 180°$
$\therefore x = 144°$ //

36. ABCD is part of a regular pentagon, while DCEF is part of a regular octagon. Find the value of x.

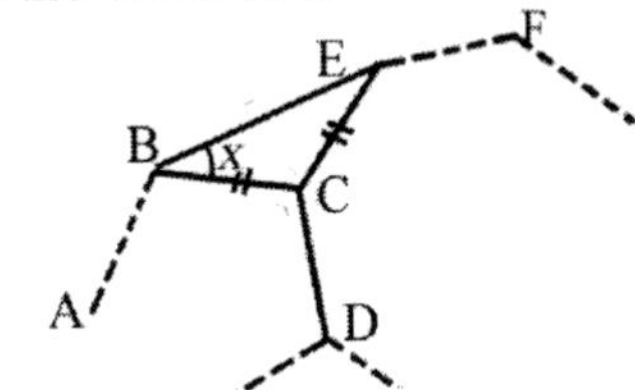

Figure 3. 31

Answer:
Let N = number of sides
Given ABCD = regular pentagon
=> N = 5
Measure of interior angle, $\angle BCD$:
$= 180° - \dfrac{360°}{N}$
$= 180° - \dfrac{360°}{5}$
$= 180° - 72°$
$= 108°$ ...(1)
Given DCEF = regular octagon

$\Rightarrow N = 8$

Measure of interior angle, $\angle DCE$:

$= 180° - \dfrac{360°}{8}$

$= 180° - 45°$

$= 135° \quad ...(2)$

Thus sum of angles at point C = 360°

$\angle BCE + \angle BCD + \angle DCE = 360°$

Substitute (1) and (2):

$\angle BCE + 108° + 135° = 360°$

$\angle BCE = 117°$

Since BC = EC

$\Rightarrow \triangle BCE$ = isosceles triangle

$\Rightarrow \angle EBC \cong \angle BEC = x$

Sum of interior angles in $\triangle BCE = 180°$

$\angle EBC + \angle BEC + \angle BCE = 180°$

$x + x + 117° = 180°$

$2x = 63°$

$\therefore x = 31.5°$ //

37. ABCDEF is a regular hexagon. Find the value of x + y.

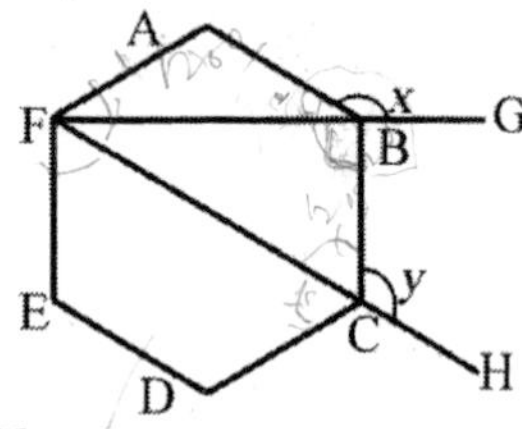

Figure 3. 32

Answer:

Let N = number of sides

Given ABCDEF = regular hexagon

$\Rightarrow N = 6$

Measure of interior angle, $\angle BAF$:

$= 180° - \dfrac{360°}{N}$

$= 180° - \dfrac{360°}{6}$

$= 180° - 60°$

$= 120°$

Since ABCDEF = regular hexagon:

$\Rightarrow AB = AF$

$\Rightarrow \triangle ABF$ = isosceles triangle

$\Rightarrow \angle AFB \cong \angle ABF \quad ...(1)$

Sum of interior angles in $\triangle ABF = 180°$

$\angle ABF + \angle AFB + \angle BAF = 180° \quad ...(2)$

Substitute (1) into (2):

$2\angle ABF + 120° = 180°$

$2\angle ABF = 60°$

$\angle ABF = 30°$

Sum of straight line angles = 180°

$\angle ABG + \angle ABF = 180°$

$x + 30° = 180°$

$\therefore x = 150°$

Since $\overrightarrow{FH}$ is an angle bisector for $\angle BCD$

$\angle BCF = \dfrac{1}{2} \times \angle BCD$

$\angle BCF = \dfrac{1}{2} \times 120° \quad \Leftarrow \angle BCD = \angle BAF$

$\angle BCF = 60°$

Sum of straight line angles = 180°

$\angle BCH + \angle BCF = 180°$

$y + 60° = 180°$

$\therefore y = 120°$

Thus,

$x + y = 150° + 120°$

$= 270°$ //

Chapter 4
Circles

Circle – a round close figure with circumference equidistant from its center

r, **Radius** – straight line from the center of circle to its circumference (plural radii)

d, **Diameter** – straight line that passes through the center point of the circle and with endpoints on the circumference

Diameter = 2 × radius

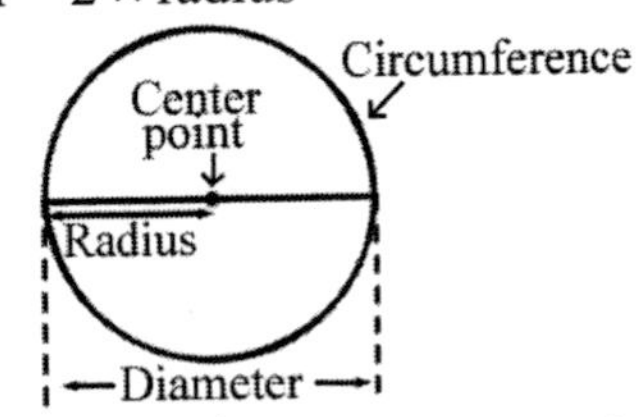

Angles in a circle are measured in rad (called radian) or ° (degrees)

π, pi – ratio of the circumference of circle to its diameter ($\pi = \frac{22}{7} \approx 3.142$)

Circle can be divided into 4 quadrants:

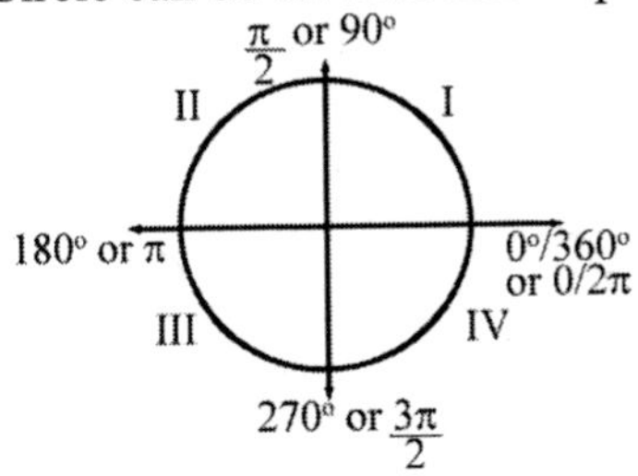

$$1 \text{ rad} = \frac{360°}{2\pi} = \frac{180°}{\pi}$$

$$1° = \frac{2\pi}{360°} = \frac{\pi}{180°}$$

Central angle = ∠AOB:

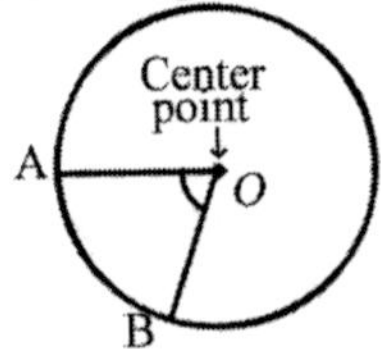

Inscribed angles = ∠ACB and ∠ADB

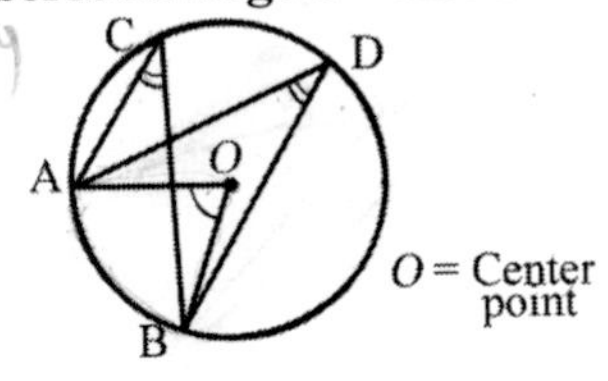

Inscribed angles formed from the same arc, $\overset{\frown}{AB}$ are congruent. ∠ACB ≅ ∠ADB

Inscribed angle is $\frac{1}{2}$ central angle if formed from the same arc. Thus, 2∠ACB = ∠AOB

Chord – straight line whose endpoints are on the circumference of a circle

Tangent – line that touches circumference of a circle at exactly one point and is perpendicular to its radius at that point

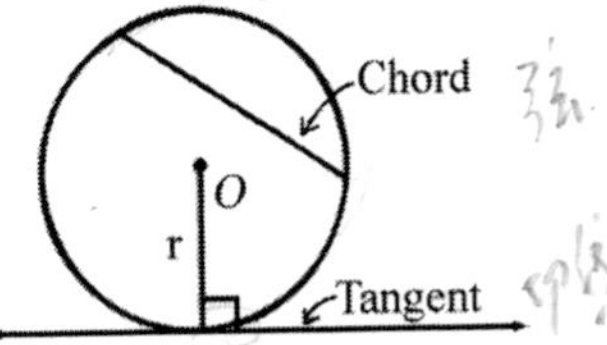

Secant – line that intersects circle at two points. $\overrightarrow{CA}$ and $\overrightarrow{CE}$ are secants.

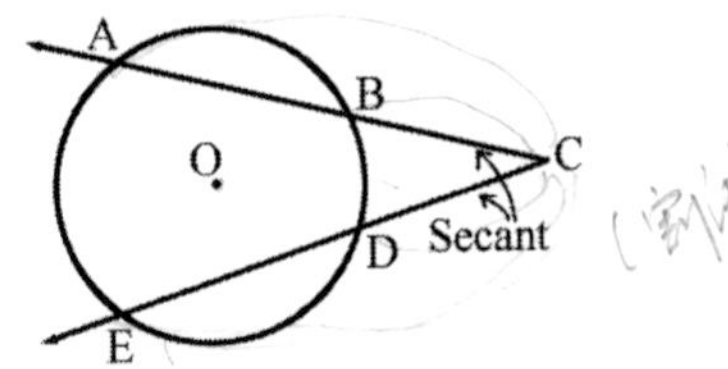

Intersecting secants: CA × CB = CE × CD

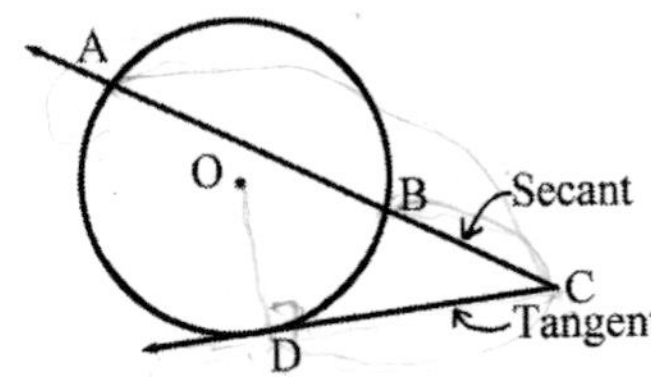

Intersecting secant–tangent:

$DC^2 = CA \times CB$

Circles

1. Given θ = 150°. Determine the quadrant θ is located. Subsequently state the positive coterminal angle.

Answers:
Given θ = 150°

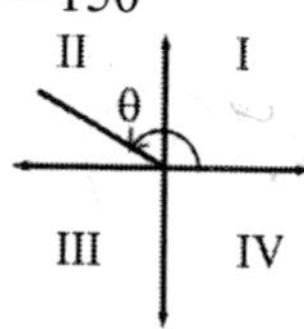

∴ θ is located in the second quadrant. //

Positive coterminal for θ:
$= \theta + 360°$
$= 150° + 360°$
$= 510°$ //

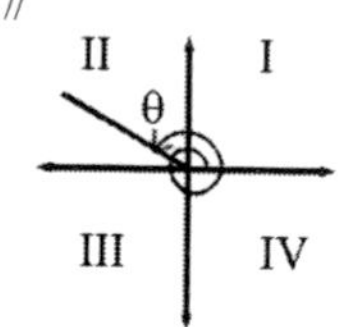

2. Given a circle, whose diameters are AB = 3n – 3 and CD = 2n + 2. What is the length of the circle's radius?

Answer:
Given AB = 3n – 3 ...(1)
CD = 2n + 2 ...(2)
Since all diameters in a circle are ≅:
=> AB = CD
Substitute (1) and (2):
$3n - 3 = 2n + 2$
$n = 5$
Substitute n = 5 into (1):
$AB = 3(5) - 3$
$= 15 - 3$
$AB = 12$
Length of diameter = 2 × length of radius
=> Length of radius:
$= \frac{1}{2} \times$ length of diameter
$= \frac{1}{2} \times AB$
$= \frac{1}{2} \times 12$
= 6 units
∴ Circle's radius is 6 units. //

3. In Figure 4.1, A, B, and C are points on the circumference of circle O. It is further given that $\overline{OB}$ bisects the major arc, $\widehat{ABC}$. Hence show that chord $\overline{AB}$ is congruent to chord $\overline{BC}$.

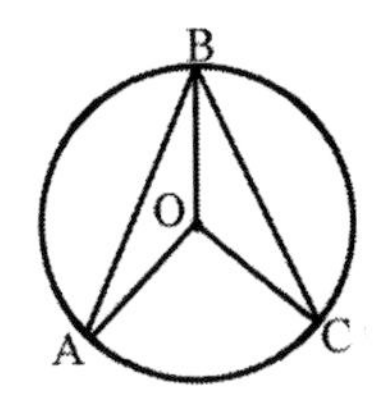

Figure 4. 1

Answer:

Given $\overline{OB}$ = bisects major arc $\overset{\frown}{ABC}$

=> $\angle AOB \cong \angle BOC$ [angle]

All radii in a circle are $\cong$:

=> OA = OC [side]

From reflexive property of equality:

OB = BO [side]

$\therefore$ Based on S–A–S postulate, $\triangle AOB$ and $\triangle BOC$ are congruent.

Since corresponding parts of congruent triangles are congruent:

$\therefore$ We have shown that chord $\overline{AB}$ is congruent to chord $\overline{BC}$. //

4. In circle O, (see Figure 4.2) chords $\overline{AB}$ and $\overline{DC}$ are congruent. Show that $\overset{\frown}{AB}$ is congruent to $\overset{\frown}{CD}$.

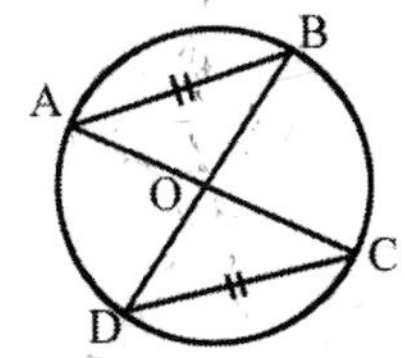

Figure 4. 2

Answer:

Given $\overline{AB}$ and $\overline{DC}$ are $\cong$:

=> AB = DC [side]

All radii in a circle are $\cong$:

OA $\cong$ OC [side]

OB $\cong$ OD [side]

$\therefore$ Based on S–S–S postulate $\triangle AOB$ and $\triangle COD$ are congruent.

Since corresponding parts of congruent triangles are congruent:

$\angle AOB \cong \angle COD$

Based on theorem that states, in the same circle, congruent central angles have congruent arc length.

$\therefore$ We have shown $\overset{\frown}{AB}$ and $\overset{\frown}{CD}$ are congruent. //

5. In Figure 4.3, shows circle O whose chord $\overline{BC}$ is parallel to its radius $\overline{OA}$. Hence express y and z in terms of x.

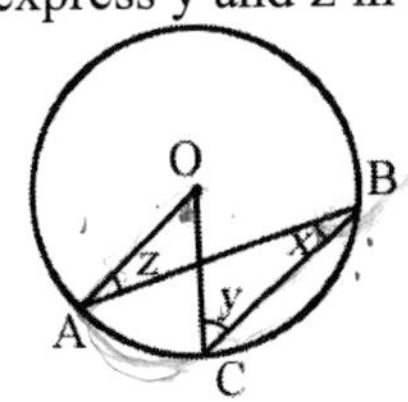

Figure 4. 3

Answers:

Given $\angle AOC$ and $\angle ABC$ share the same arc, $\overset{\frown}{AC}$

=> $\angle ABC$ = inscribed angle

=> $\angle AOC$ = central angle

Thus,

$\angle AOC = 2 \times \angle ABC$

$\angle AOC = 2 \times x$

$\angle AOC = 2x$

Given $\overline{OA} \parallel \overline{BC}$

Alternate interior angles are $\cong$:

$\angle BCO \cong \angle AOC$

$\therefore$ y = 2x //

Alternate interior angles are $\cong$:

$\angle BAO \cong \angle ABC$

$\therefore$ z = x //

6. Figure 4.4, shows circle O and parallel secants $\overline{AB}$ and $\overline{CD}$. If it is known that the measure of $\overset{\frown}{AB} = 110°$, $\overset{\frown}{AC} = x$ and $\overset{\frown}{CD} = 140°$, find the value of x.

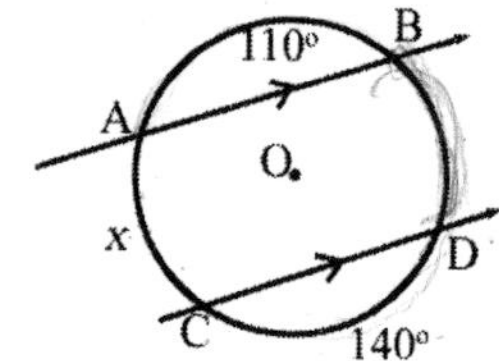

Figure 4. 4

Answer:

Given $\overline{AB} \parallel \overline{CD}$

In a circle, parallel chords cut off equal arcs

$\Rightarrow m\overset{\frown}{AC} = m\overset{\frown}{BD} = x$

Sum of measure of arcs in a circle = 360°

$m\overset{\frown}{AB} + m\overset{\frown}{AC} + m\overset{\frown}{CD} + m\overset{\frown}{BD} = 360°$

$110° + x + 140° + x = 360°$

$2x = 110°$

$\therefore x = 55°$ //

7. Figure 4.5, shows circle O whose diameter is $\overline{AOB}$. Also given, $\angle ABC$ is the inscribed angle. What is the measure of arc x?

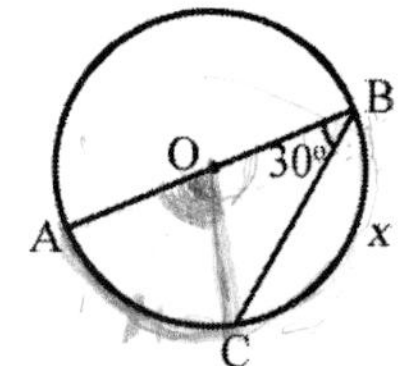

Figure 4. 5

Answer:

Given inscribed angle, $\angle ABC = 30°$

Central angle:

$\angle AOC = 2 \times \angle ABC$

$= 2 \times 30°$

$= 60°$

$\Rightarrow m\overset{\frown}{AC} = 60°$

Given $\overline{AB}$ = diameter of circle O

$\Rightarrow \overset{\frown}{ACB}$ = semicircle

$m\overset{\frown}{ACB} = 180°$

$m\overset{\frown}{AC} + m\overset{\frown}{CB} = 180°$

$60° + x = 180°$

$\therefore x = 120°$ //

8. Figure 4.6, depicts circle O whose inscribed angle, $\angle ABC$ is 35°. Hence find the values of x and y.

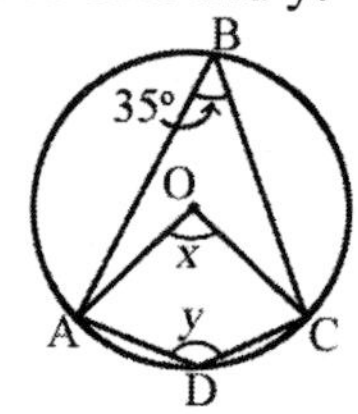

Figure 4. 6

Answers:

Given inscribed angle, $\angle ABC = 35°$

Central angle, $\angle AOC$:

$\angle AOC = 2 \times \angle ABC$

$x = 2 \times 35°$

$\therefore x = 70°$ //

Sum of angles at a point, O = 360°

$m\overset{\frown}{ABC} = 360° - \angle AOC$

$m\overset{\frown}{ABC} = 360° - x$

$= 360° - 70°$

$= 290°$

Inscribed angle of $\overset{\frown}{ADC}$ is $\angle ADC$

$$\angle ADC = \frac{m\overset{\frown}{ABC}}{2}$$

$$y = \frac{290°}{2}$$

$\therefore y = 145°$ //

9. In Figure 4.7, shows circle O whose central angle is 70°. Hence find the sum of x, y and z.

Figure 4. 7

Answer:
Given $\angle AOB$ = central angle
$\angle AOB = 70°$
Since measure of central angle = 2 × measure of inscribed angle
Angles x, y and z are drawn from $\overset{\frown}{AB}$
=> x = y = z

$$\Rightarrow x = \frac{1}{2} \times \angle AOB$$
$$= \frac{1}{2} \times 70°$$
$$= 35°$$

$\therefore x + y + z = 35° + 35° + 35°$
$= 105°$ //

10. Figure 4.8 shows circle with center point O. Given $\overline{AC}$ and $\overline{BD}$ are straight lines. Hence find the value of x.

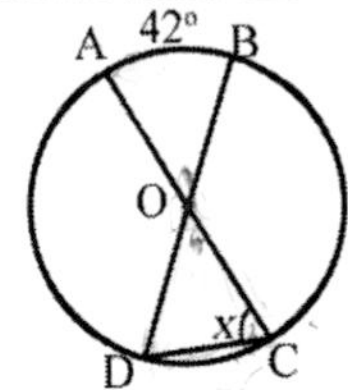

Figure 4. 8

Answer:
Given $\overline{AC}$ and $\overline{BD}$ = diameters of circle O
Also given $m\overset{\frown}{AB} = 42°$
=> $\angle AOB = 42°$
Vertical angles are ≅:
=> $\angle COD \cong \angle AOB = 42°$

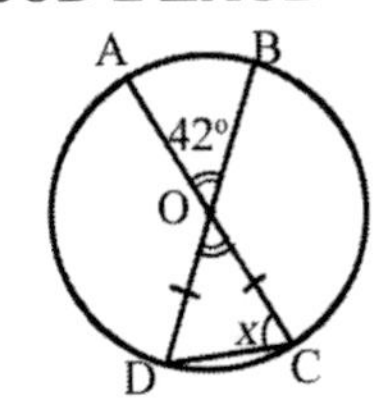

All radii in a circle are ≅:
=> OC = OD
=> $\triangle COD$ = isosceles triangle
=> $\angle ACD \cong \angle BDC = x$ ...*
Sum of interior angles in $\triangle COD = 180°$
$\angle ACD + \angle BDC + \angle COD = 180°$
$x + x + 42° = 180°$ ⇦ From *
$2x = 138°$
$\therefore x = 69°$ //

11. Figure 4.9, a circle whose center point is O. $\overrightarrow{BA}$ is a tangent and B is the point of contact. If $\overline{CB}$ is a chord, find the value of x.

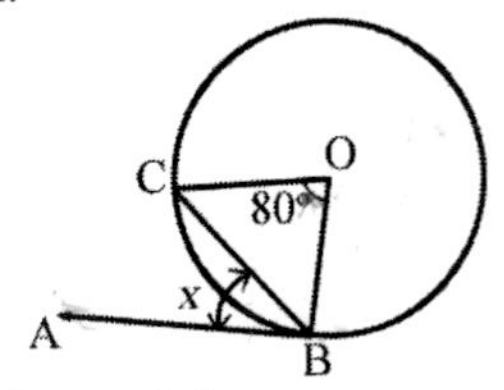

Figure 4. 9

Answer:
Given $\overrightarrow{BA}$ = tangent
$\overline{CB}$ = chord
Central angle, $\angle BOC = 80°$
Since angle between chord and tangent is half the measure of central angle:

$$\angle ABC = \frac{1}{2} \times \angle BOC$$
$$x = \frac{1}{2} \times 80°$$

$\therefore x = 40°$ //

12. Figure 4.10, shows a circle with center, O.

a) Convert central angle, ∠AOC from radian to degree
b) Find the value of inscribed angle, ∠ABC.

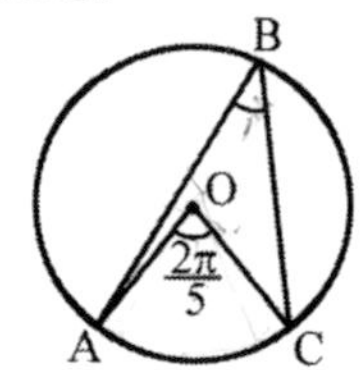

Figure 4. 10

Answers:

Given central angle, $\angle AOC = \frac{2\pi}{5}$ rad

a) Since $\pi = 180°$
Convert from radian to degrees:

$\angle AOC = \frac{2\pi}{5} \times \frac{180°}{\pi}$
$= 72°$ //

b) ∠AOC and ∠ABC are from the same arc, $\overset{\frown}{AC}$
Thus, inscribed angle, ∠ABC:
$= \frac{1}{2} \times \angle AOC$
$= \frac{1}{2} \times 72°$
$= 36°$ //

13. In Figure 4.11, $\overline{AB}$ and $\overrightarrow{BE}$ are common tangents to circle O. Find the values of x, y and z.

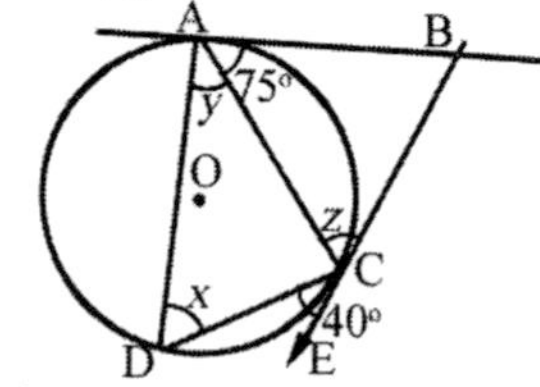

Figure 4. 11

Answers:
Given ∠BAC = 75°
Since ∠BAC is between tangent, $\overline{AB}$ and chord, $\overline{AC}$, ∠BAC equals to angle in alternate segments, $\overline{AD}$ and $\overline{DC}$:
∠ADC = ∠BAC
∴ x = 75° //
Also given ∠DCE = 40°
Since ∠DCE is between tangent, $\overrightarrow{BE}$ and chord $\overline{CD}$, ∠DCE equals to angle in alternate segments, $\overline{AD}$ and $\overline{AC}$:
∠CAD = ∠DCE
∴ y = 40° //
Also given ∠ACB = z
∠ACB is between tangent $\overrightarrow{BE}$ and chord $\overline{AC}$, thus ∠ACB is equal to angle in alternate segments, $\overline{AD}$ and $\overline{DC}$:
∠ACB = ∠ADC
z = x
∴ z = 75° //

14. In Figure 4.12, $\overline{BA}$ and $\overline{BC}$ are tangents to circle O. If AOD is a straight line, find the value of θ.

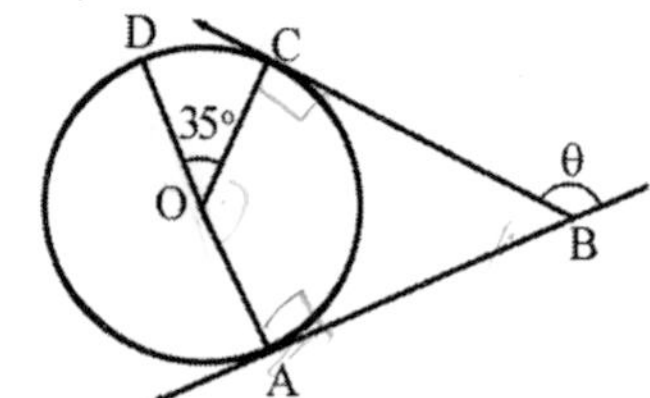

Figure 4. 12

Answer:
Given $\overline{AOD}$ = straight line

=> $\overline{AOD}$ = diameter (since it passes through O)
Sum of straight line angles = 180°
$\angle AOC + \angle COD = 180°$
$\angle AOC + 35° = 180°$
$\angle AOC = 145°$
Also given $\overline{BA}$ and $\overline{BC}$ are tangents
=> A and C are points of contact
Since $\overline{OA}$ and $\overline{OC}$ = radii
=> $\angle BAO = \angle BCO = 90°$
Sum of interior angles in quadrilateral ABCO = 360°
$\angle ABC + \angle BAO + \angle AOC + \angle BCO = 360°$
$\angle ABC + 90° + 145° + 90° = 360°$
$\angle ABC = 35°$
Sum of straight line angles = 180°
$\theta + \angle ABC = 180°$
$\theta + 35° = 180°$
$\therefore \theta = 145°$ //

15. Figure 4.13, shows circle O. $\overline{PQ}$ and $\overline{RQ}$ are common tangents to circle O, while $\overline{PS}$ and $\overline{RS}$ are chords drawn from the tangency points P and R. Find the length, RQ and the values of x and y.

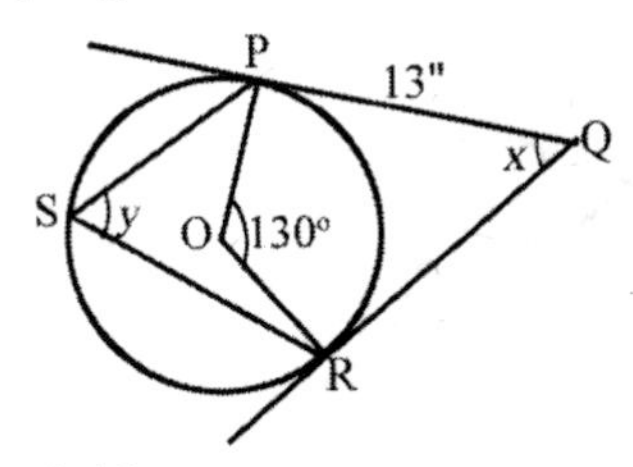

Figure 4. 13

Answers:
Given PQ = 13 ″
Tangent segments $\overline{PQ}$ and $\overline{RQ}$ intersect:
=> RQ = PQ
$\therefore$ RQ = 13 ″ //
Given central angle, $\angle POR = 130°$
=> Inscribed angle, $\angle PSR$

$$\angle PSR = \frac{1}{2} \times \angle POR$$

$$y = \frac{1}{2} \times 130°$$

$\therefore y = 65°$ //

Given P and R = tangency points
Given $\overline{OP}$ and $\overline{OR}$ = radii of circle O
=> $\angle OPQ = \angle ORQ = 90°$
Sum of interior angles in quadrilateral, PQRO = 360°
$\angle PQR + \angle QRO + \angle POR + \angle OPQ = 360°$
$x + 90° + 130° + 90° = 360°$
$\therefore x = 50°$ //

16. In Figure 4.14, $\overline{AB}$ and $\overline{BC}$ are common tangents to circle O. If $\overline{OB}$ bisects $\angle AOC$, find the values of x, y and z.

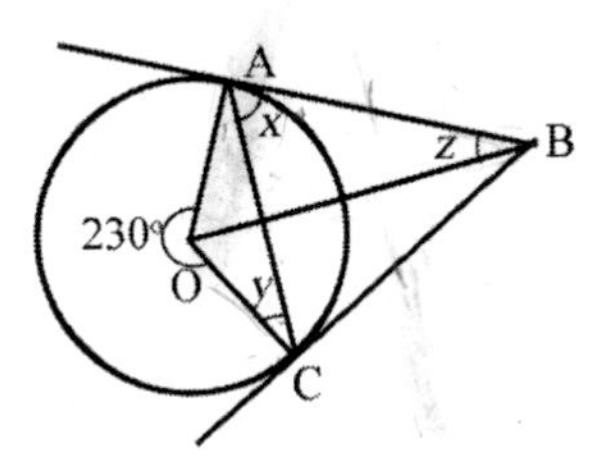

Figure 4. 14

Answers:
Given major arc $m\widehat{AC} = 230°$
=> Minor arc $m\widehat{AC} = \angle AOC$
Sum of angles at a point = 360°
$\angle AOC = 360° - 230°$
$\angle AOC = 130°$
Given $\overline{OA}$ and $\overline{OC}$ = radii of circle O
=> OA = OC
=> $\triangle AOC$ = isosceles triangle
=> $\angle OAC \cong \angle OCA = y$

Sum of interior angles in $\triangle AOC = 180°$
$\angle OAC + \angle OCA + \angle AOC = 180°$
$y + y + 130° = 180°$
$2y = 50°$
$\therefore y = 25°$ //
Since A = point of contact
$\angle OAB = 90°$
$\angle BAC + \angle OAC = \angle OAB$
$x + y = 90°$
$x + 25° = 90°$
$\therefore x = 65°$ //
Given $\overline{OB}$ = bisects $\angle AOC$
Sum of interior angles in $\triangle OAB = 180°$
$\angle OBA + \angle OAB + \angle AOB = 180°$

$$z + 90° + \left(\frac{1}{2} \times \angle AOC\right) = 180°$$

$$z + 90° + \left(\frac{1}{2} \times 130°\right) = 180°$$

$z + 90° + 65° = 180°$
$\therefore z = 25°$ //

17. In Figure 4.15, $\overline{AC}$ and $\overline{BD}$ are secants and intersect at point F. Hence find the value of x.

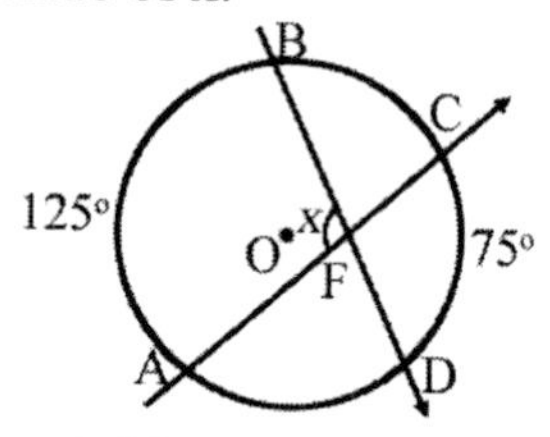

Figure 4. 15

Answer:
Given $\overline{AC}$ and $\overline{BD}$ = secants
Since secants are chords in a circle
=> $\overline{AC}$ and $\overline{BD}$ = chords

Apply chord-chord theorem:

$$\angle AFB = \frac{1}{2} \times \left(m\overset{\frown}{AB} + m\overset{\frown}{CD}\right)$$

$$x = \frac{1}{2} \times (125° + 75°)$$

$$x = \frac{1}{2} \times (200°)$$

$\therefore x = 100°$ //

18. In Figure 4.16, $\overleftrightarrow{AB}$ is a tangent to circle O. If $\overleftrightarrow{CD}$ is a secant, determine the value of x.

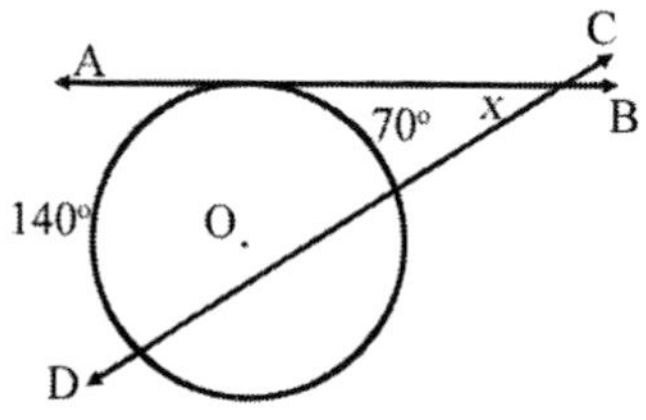

Figure 4. 16

Answer:
Given $\overleftrightarrow{AB}$ = tangent
$\overleftrightarrow{CD}$ = secant

Reproduce Figure 4.16 for labeling:

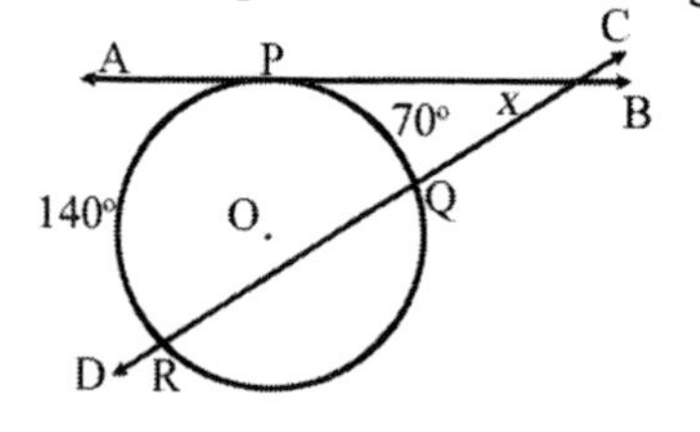

Using secant-tangent angle:

$$x = \frac{1}{2} \times \left(m\overset{\frown}{PR} - m\overset{\frown}{PQ}\right)$$

$$= \frac{1}{2} \times (140° - 70°)$$

$$= \frac{1}{2} \times 70°$$

$\therefore x = 35°$ //

19. In Figure 4.17, $\overline{AB}$ and $\overline{BC}$ are tangents to circle O. The vertex, B is 60°. Hence find the measure of $\widehat{ADC}$, x.

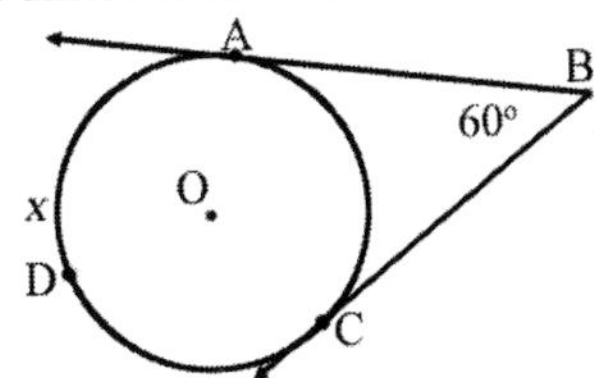

Figure 4. 17

Answer:

Given $\overline{AB}$ and $\overline{BC}$ = tangents

Given $m\widehat{ADC} = x$

Sum of angles at a point, O = 360°

$m\widehat{AC} = 360° - m\widehat{ADC}$

$= 360° - x$

Using tangent-tangent angle:

$\angle ABC = \frac{1}{2} \times \left(m\widehat{ADC} - m\widehat{AC} \right)$

$60° = \frac{1}{2} \times \left(x - (360° - x) \right)$

$120° = x - 360° + x$

$2x = 480°$

$\therefore x = 240°$ //

20. In Figure 4.18, $\overline{AB}$ and $\overline{BC}$ are tangents to circle O. If vertex, B is 62°, find the value of x.

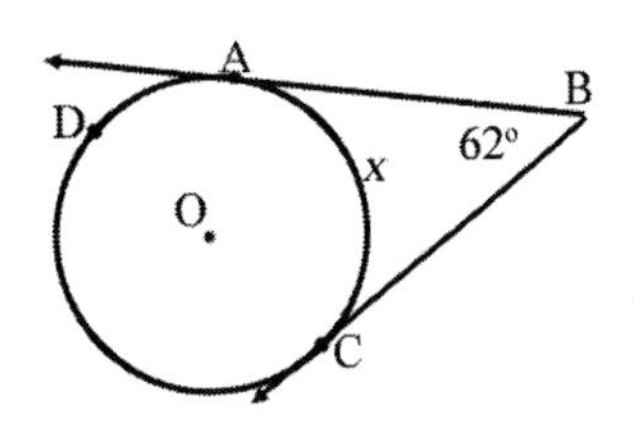

Figure 4. 18

Answer:

Given $\overline{AB}$ and $\overline{BC}$ = tangents

Sum of angles at a point, O = 360°

$m\widehat{ADC} = 360° - m\widehat{AC}$

$= 360° - x$

Using tangent-tangent angle:

$\angle ABC = \frac{1}{2} \times \left(m\widehat{ADC} - m\widehat{AC} \right)$

$62° = \frac{1}{2} \times \left(360° - x - x \right)$

$124° = 360° - 2x$

$2x = 236°$

$\therefore x = 118°$ //

21. In Figure 4.19, $\overline{AB}$ and $\overline{BC}$ are tangents to circle O. If measure of arc, $\widehat{ADC}$ is 260°, find the value of vertex, B, represented by x.

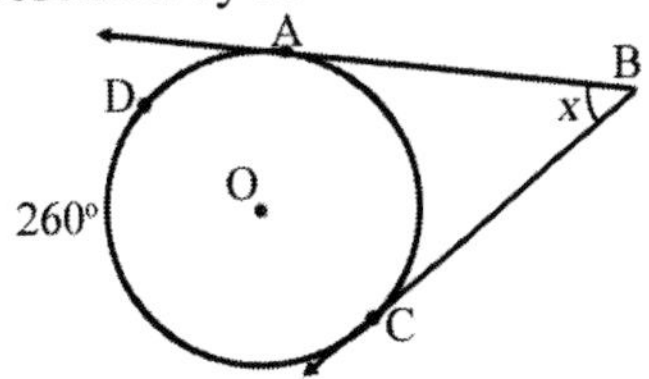

Figure 4. 19

Answer:

Given $\overline{AB}$ and $\overline{BC}$ = tangents

Sum of angles at a point, O = 360°

$m\widehat{AC} = 360° - m\widehat{ADC}$

$= 360° - 260° = 100°$

Using tangent-tangent angle:

$\angle ABC = \frac{1}{2} \times \left(m\widehat{ADC} - m\widehat{AC} \right)$

$x = \frac{1}{2} \times \left(260° - 100° \right)$

$x = \frac{1}{2} \times 160°$

$\therefore x = 80°$ //

22. In Figure 4.20, $\overline{AC}$ and $\overline{BD}$ are secants to circle O. It is further given that, $\overline{CD}$ is a tangent and AB = BD and measure of arc, $\overset{\frown}{AD}$ is 160°. Hence find the values of x and y.

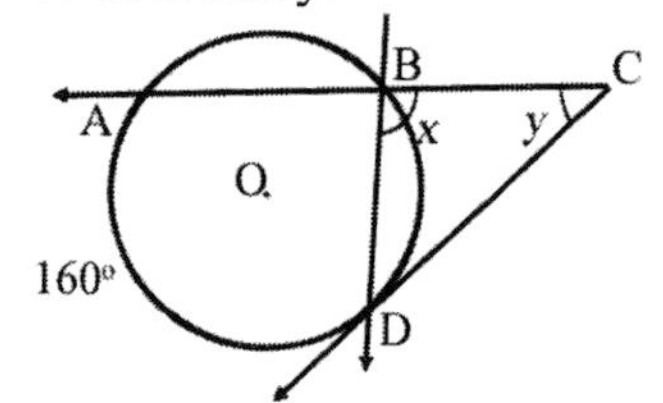

Figure 4. 20

Answers:

Given $m\overset{\frown}{AD} = 160°$

=> Central angle, $\angle AOD = 160°$

=> Let inscribed angle, $\angle ABD = \theta$

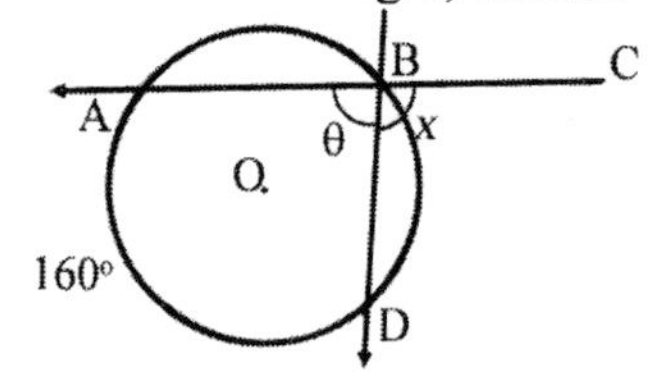

Reproduced Figure 4.20 for labeling

$\angle ABD = \frac{1}{2} \times \angle AOD$

$\theta = \frac{1}{2} \times 160°$

$\theta = 80°$

Sum of straight line angles = 180°

$\angle CBD + \angle ABD = 180°$

$x + \theta = 180°$

$x + 80° = 180°$

$\therefore x = 100°$ //

Given AB = BD

In the same circle, congruent chords have congruent arc length.

=> $m\overset{\frown}{AB} \cong m\overset{\frown}{BD}$ ...*

In a circle, sum of arcs = 360°

$m\overset{\frown}{AD} + m\overset{\frown}{BD} + m\overset{\frown}{AB} = 360°$

$160° + 2m\overset{\frown}{BD} = 360°$ ⇦ From (*)

$2m\overset{\frown}{BD} = 200°$

$m\overset{\frown}{BD} = 100°$

Using secant-tangent angle:

$\angle ACD = \frac{1}{2} \times \left(m\overset{\frown}{AD} - m\overset{\frown}{BD} \right)$

$y = \frac{1}{2} \times (160° - 100°)$

$= \frac{1}{2} \times 60°$

$\therefore y = 30°$ //

23. In Figure 4.21, $\overline{AB}$ is the diameter of circle O. It is further given that, $\overline{CD}$ is a perpendicular chord to $\overline{AB}$ and measure of arc length, $\overset{\frown}{BC}$ is 35°. If $m\overset{\frown}{AD}$ is x, find the value of x.

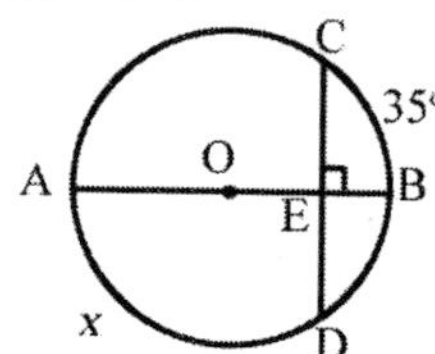

Figure 4. 21

Answer:

Given $\overline{AB} \perp \overline{CD}$

=> $\overline{CE} \cong \overline{DE}$ [side]

All radii in a circle are ≅:

=> OC = OD [side]

Using reflexive property of equality:

OE ≅ EO [side]

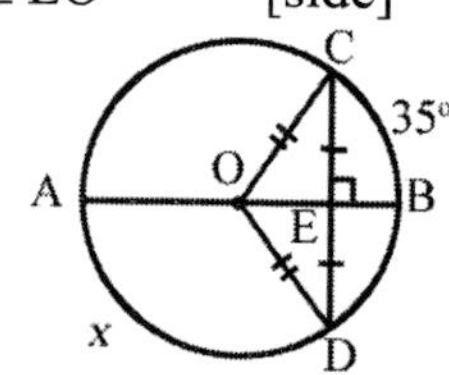

Based on S–S–S postulate, △OCE and △ODE are congruent triangles

Given $m\overset{\frown}{BC} = 35°$

=> $\angle BOC = 35°$
Since corresponding parts of congruent triangles are congruent:
$\angle BOD \cong \angle BOC$
$\angle BOD = 35°$

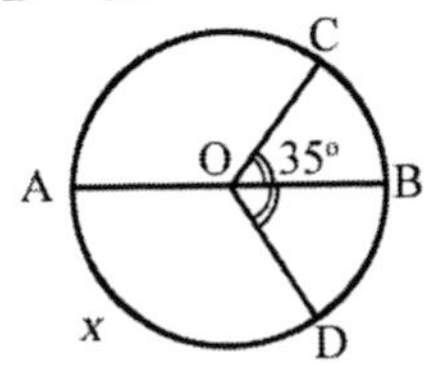

Given $m\overset{\frown}{AD} = x$
=> $\angle AOD = x$
Sum of straight line angles = 180°
$\angle AOD + \angle BOD = 180°$
$x + 35° = 180°$
$\therefore x = 145°$ //

24. In Figure 4.22, $\overline{AC}$ and $\overline{CE}$ are secants to circle O. Also given, $\overline{AD}$ and $\overline{BE}$ are chords, and $\angle AFB$ is 85° in circle O. If the measure of the intercepted arcs, $\overset{\frown}{DE}$ is 80° and $\overset{\frown}{AE}$ is 140°, find the values of x and y.

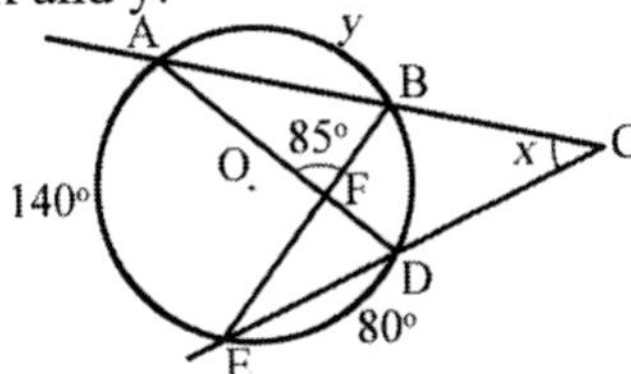

Figure 4. 22

Answers:
Given $\angle AFB = 85°$
Also given $m\overset{\frown}{DE} = 80°$
Using chord-chord angle:

$$\angle AFB = \frac{1}{2} \times \left(m\overset{\frown}{AB} + m\overset{\frown}{DE} \right)$$

$$85° = \frac{1}{2} \times \left(m\overset{\frown}{AB} + 80° \right)$$

$170° = y + 80°$
$\therefore y = 90°$ //
Given $m\overset{\frown}{AE} = 140°$
Sum of measure of arcs in a circle = 360°
$m\overset{\frown}{BD} + m\overset{\frown}{AE} + m\overset{\frown}{AB} + m\overset{\frown}{DE} = 360°$
$m\overset{\frown}{BD} + 140° + 90° + 80° = 360°$
$m\overset{\frown}{BD} = 50°$

Using secant-secant angle:

$$\angle ACE = \frac{1}{2} \times \left(m\overset{\frown}{AE} - m\overset{\frown}{BD} \right)$$

$$x = \frac{1}{2} \times (140° - 50°)$$

$$x = \frac{1}{2} \times 90°$$

$\therefore x = 45°$ //

25. In Figure 4.23, $\overline{PQ}$ and $\overline{QR}$ are tangent segments to circle O. Find the values of α and θ.

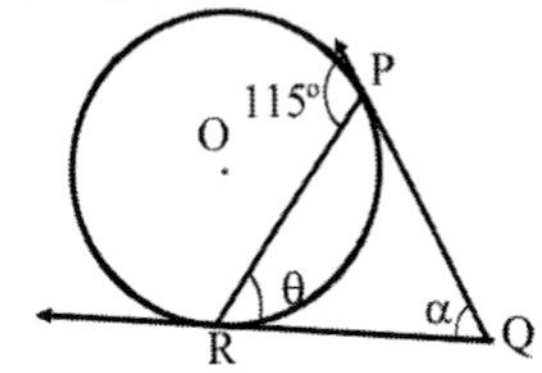

Figure 4. 23

Answers:
Reproduce Figure 4.23 for labeling:

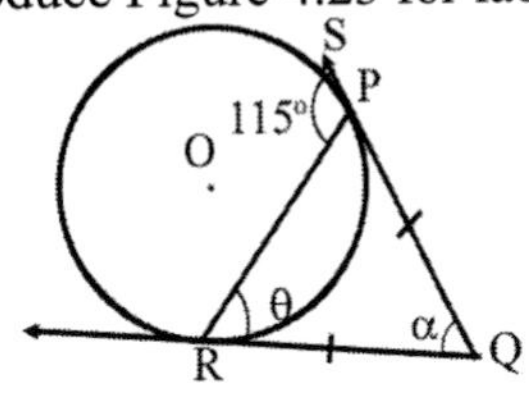

Given $\angle RPS = 115°$
Sum of straight line angles = 180°
$\angle QPR + \angle RPS = 180°$
$\angle QPR + 115° = 180°$
$\angle QPR = 65°$

Given $\overline{PQ}$ and $\overline{QR}$ intersecting tangents of circle O
=> PQ = QR
=> $\triangle$PQR = isosceles triangle
=> $\angle$PRQ $\cong$ $\angle$QPR
$\therefore$ $\theta = 65°$ //
Sum of interior angles in $\triangle$PQR = 180°
$\angle PQR + \angle PRQ + \angle QPR = 180°$
$\alpha + \theta + 65° = 180°$
$\alpha + 65° + 65° = 180°$
$\therefore$ $\alpha = 50°$ //

26. In Figure 4.24, $\overline{AB}$ and $\overline{BC}$ are tangents to circle O. $\overline{AD}$ is a diameter and $\overline{OB}$ is line of symmetry for ABCO. If it is known that BC = 10′ and BE = 5′, determine the length of diameter, AD. Hence find the values of x and y.

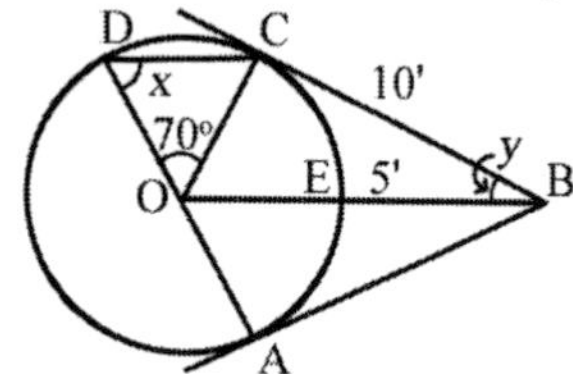

Figure 4. 24

Answers:
Given BC = 10′
BE = 5′
Let r = radius of circle O
Apply Pythagorean Theorem:
$OB^2 = OC^2 + BC^2$
$(r + 5)^2 = r^2 + 10^2$
$r^2 + 10r + 25 = r^2 + 100$
$10r = 75$
$r = 7.5'$
Diameter, $\overline{AD} = 2 \times r$
$= 2 \times 7.5$
$= 15'$ //
Given $\overline{OC}$ and $\overline{OD}$ = radii of circle O
=> OC = OD
=> $\triangle$COD = isosceles triangle
=> $\angle$CDO $\cong$ $\angle$OCD = x
Sum of interior angles in $\triangle$COD = 180°
$\angle CDO + \angle OCD + \angle COD = 180°$
$x + x + 70° = 180°$
$2x = 110°$
$\therefore$ $x = 55°$ //
Sum of straight line angles = 180°
$\angle AOC + \angle COD = 180°$
$\angle AOC + 70° = 180°$
$\angle AOC = 110°$
Given $\overline{OB}$ = line of symmetry of ABCO
=> $\overline{OB}$ = bisects $\angle$AOC
=> $\angle$BOC $\cong$ $\angle$AOB
$\angle BOC = \frac{1}{2} \times \angle AOC$
$= \frac{1}{2} \times 110°$
$= 55°$
Sum of angles in right $\triangle$BCO = 180°
$\angle CBO + \angle BCO + \angle BOC = 180°$
$y + 90° + 55° = 180°$
$\therefore$ $y = 35°$ //

27. In Figure 4.25, $\overrightarrow{CA}$ is a tangent and $\overline{CE}$ is a secant to circle O. If BD = CD and $\overline{ED}$ is a diameter, find the values of x, y and z.

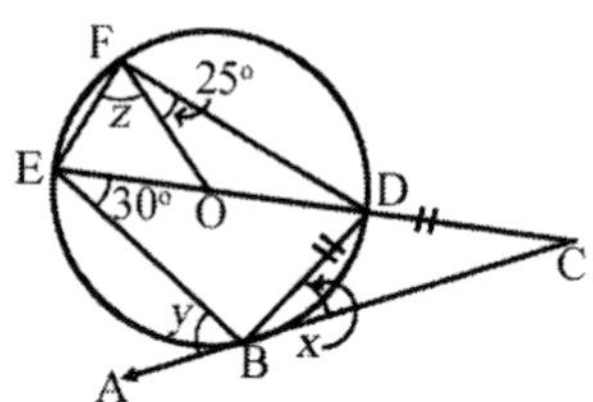

Figure 4. 25

Answers:
Given $\angle$BEC = 30°
Since angle between tangent ($\overrightarrow{CA}$) and chord ($\overline{BD}$) equals to angle in alternate segments $\overline{BE}$ and $\overline{ED}$:
$\angle CBD = \angle BEC$
$\therefore$ $x = 30°$ //

Given BD = CD
=> $\therefore$ $\triangle$BCD = isosceles triangle
=> $\angle$BCD $\cong$ $\angle$CBD
$\angle BCD = 30°$
Sum of interior angles in $\triangle$BCD = 180°
$\angle BDC + \angle BCD + \angle CBD = 180°$
$\angle BDC + 30° + 30° = 180°$
$\angle BDC = 120°$

Sum of straight line angles = 180°
$\angle BDE + \angle BDC = 180°$
$\angle BDE + 120° = 180°$
$\angle BDE = 60°$
Angle between tangent ($\overrightarrow{CA}$) and chord ($\overline{BE}$) equals to angle in alternate segments:
$\angle ABE = \angle BDE$
$\therefore$ y = 60° //

Alternatively:
Angle drawn from the ends of a diameter (points D and E) in a circle = 90°
=> $\angle DBE = 90°$
Sum of straight line angles = 180°
$\angle ABE + \angle DBE + \angle CBD = 180°$
y + 90° + x = 180°
y + 90° + 30° = 180°
$\therefore$ y = 60° //

Since angle drawn from endpoints of a diameter in a circle = 90°
=> $\angle DFE = 90°$
Sum of complementary angles = 90°
$\angle EFO + \angle DFO = 90°$
z + 25° = 90°
$\therefore$ z = 65° //

28. In Figure 4.26, $\overrightarrow{CA}$ is a tangent and $\overline{CE}$ is a secant to circle O. If $\overline{OB}$, $\overline{OE}$ and $\overline{OG}$ are radii, find the values of x, y and z.

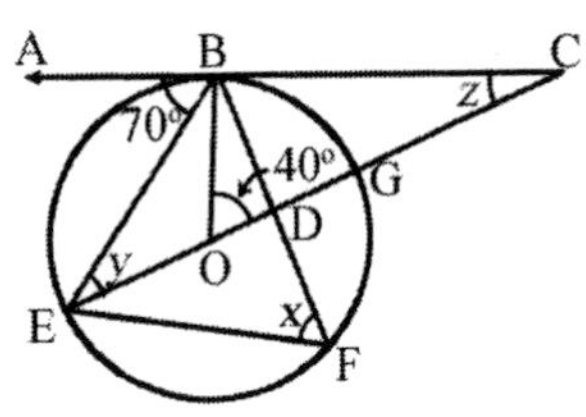

Figure 4. 26

Answers:
Given $\angle ABE = 70°$
Angle between tangent ($\overrightarrow{CA}$) and chord ($\overline{BE}$) = angle in alternate segments ($\overline{BF}$ and $\overline{EF}$)
=> $\angle BFE = \angle ABE$
$\therefore$ x = 70° //

Given central angle, $\angle BOG = 40°$
Since inscribed angle is half the measure of central angle:

$$\angle BEG = \frac{1}{2} \times \angle BOG$$

$$y = \frac{1}{2} \times 40°$$

$\therefore$ y = 20° //

Sum of straight line angles = 180°
$\angle CBE + \angle ABE = 180°$
$\angle CBE + 70° = 180°$
$\angle CBE = 110°$
Sum of internal angles in $\triangle BCE = 180°$
$\angle ACE + \angle CBE + \angle BEG = 180°$
z + 110° + 20° = 180°
$\therefore$ z = 50° //

29. In Figure 4.27, $\overleftrightarrow{AC}$ is a tangent of circle O. If $\overline{BE}$ is a diameter, find the values of x, y and z.

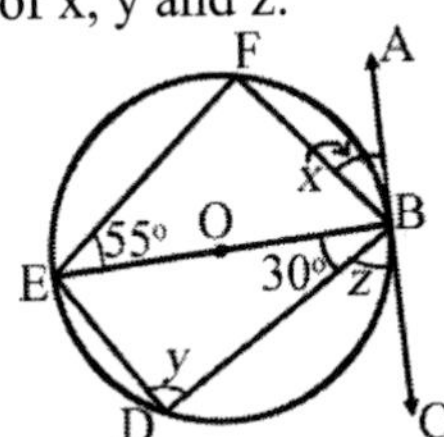

Figure 4. 27

Answers:
Given $\angle BEF = 55°$

Angle between tangent, $\overleftrightarrow{AC}$ and chord, $\overline{BF}$ = angle in alternate segments, $\overline{BE}$ and $\overline{EF}$
=> $\angle ABF = \angle BEF$
$\therefore x = 55°$ //

Angle on the circumference, drawn from the ends of diameter, $\overline{BE}$ in a circle = 90°
=> $\angle BDE = 90°$
$\therefore y = 90°$ //

Sum of interior angles in $\triangle BDE = 180°$
$\angle BED + \angle BDE + \angle DBE = 180°$
$\angle BED + 90° + 30° = 180°$
$\angle BED = 60°$
Angle between tangent, $\overleftrightarrow{AC}$ and chord, $\overline{BD}$ = angle in alternate segments, $\overline{BE}$ and $\overline{DE}$
=> $\angle CBD = \angle BED$
$\therefore z = 60°$ //

30. In Figure 4.28, $\overrightarrow{CA}$ and $\overrightarrow{CE}$ are tangents of circle O. If diameter $\overline{FG}$ is parallel to $\overrightarrow{CA}$, find the values of x and y.

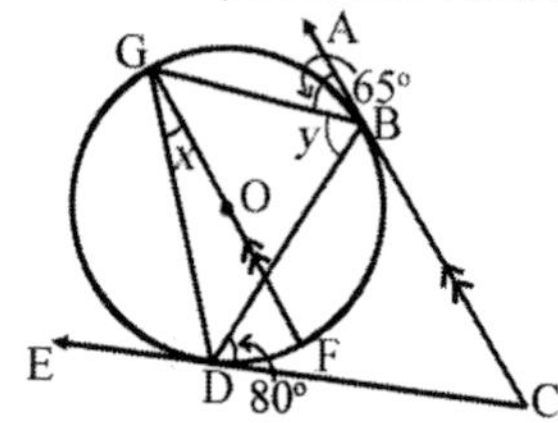

Figure 4. 28

Answers:
Given $\angle BDC = 80°$
Angle between tangent, $\overrightarrow{CE}$ and chord, $\overline{BD}$ = angle in alternate segments, $\overline{DG}$ and $\overline{BG}$
=> $\angle BGD = \angle BDC$
$\angle BGD = 80°$
Given $\overrightarrow{CA} \parallel \overline{FG}$
Alternate interior angles are $\cong$:
$\angle BGF \cong \angle ABG$
$\angle BGF = 65°$
Thus,
$\angle DGF + \angle BGF = \angle BGD$
$x + 65° = 80°$
$\therefore x = 15°$ //

Given $\overrightarrow{CA}$ and $\overrightarrow{CE}$ are intersecting tangents
=> $BC = CD$
=> $\triangle BCD$ = isosceles triangle
=> $\angle CBD \cong \angle BDC$
$\angle CBD = 80°$
Sum of straight line angles =180°
$\angle DBG + \angle CBD + \angle ABG = 180°$
$y + 80° + 65° = 180°$
$\therefore y = 35°$ //

31. In Figure 4.29, circle O and circle P are internally tangent circles, where $\overleftrightarrow{AC}$ is the common tangent. If BFGH is a cyclic quadrilateral and BD = BE, find the values of x and y.

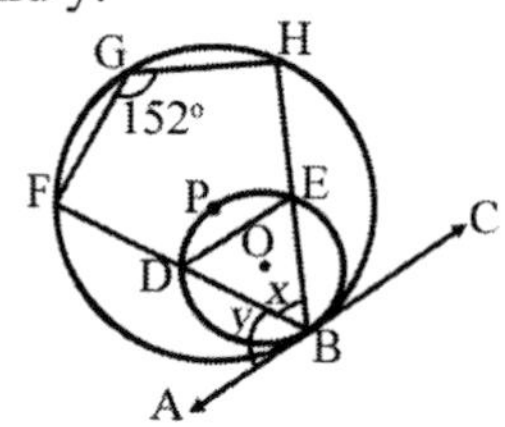

Figure 4. 29

Answers:
Given BFGH = cyclic quadrilateral
Opposite angles are supplementary:
$\angle FBH + \angle FGH = 180°$
$x + 152° = 180°$
$\therefore x = 28°$ //

Given BD = BE
=> $\triangle DBE$ = isosceles triangle
=> $\angle BED = \angle BDE$ ...*
Sum of interior angles in $\triangle BDE = 180°$
$\angle BED + \angle BDE + \angle FBH = 180°$
$2\angle BED + 28° = 180°$ ⇦ From *
$2\angle BED = 152°$
$\angle BED = 76°$
Angle between tangent, $\overleftrightarrow{AC}$ and chord, $\overline{BF}$ = angle in alternate segments $\overline{BE}$ and $\overline{DE}$
=> $\angle ABF = \angle BED$
$\therefore y = 76°$ //

32. In Figure 4.30, ABCD is an inscribed quadrilateral in circle O. Find the values of x and y.

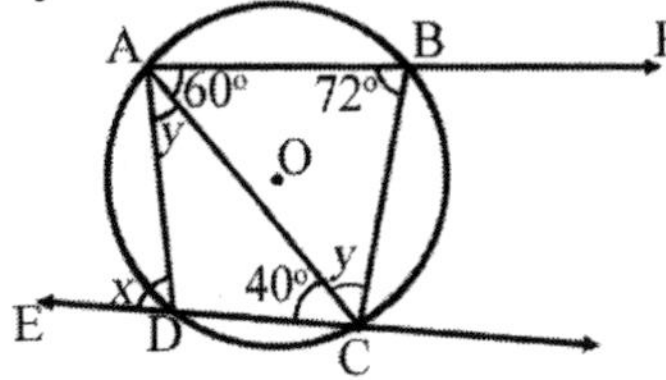

Figure 4. 30

Answers:

Sum of interior angles in $\triangle ABC = 180°$

$\angle BAC + \angle ABC + \angle ACB = 180°$

$60° + 72° + y = 180°$

$\therefore\ y = 48°$ //

Sum of interior angles in $\triangle ACD = 180°$

$\angle ADC + \angle CAD + \angle ACD = 180°$

$\angle ADC + y + 40° = 180°$

$\angle ADC + 48° + 40° = 180°$

$\angle ADC = 92°$

Sum of straight line angles = 180°

$\angle ADE + \angle ADC = 180°$

$x + 92° = 180°$

$\therefore\ x = 88°$ //

33. In Figure 4.31, circle O and circle P are non-intersecting circles. $\overleftrightarrow{AC}$ and $\overleftrightarrow{DE}$ are common internal tangents. Find the values of:

a) x

b) y + z

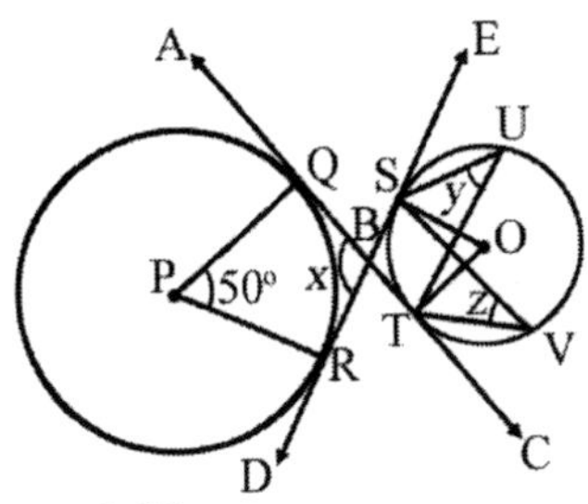

Figure 4. 31

Answers:

a)

Radii of circle P are ≅:

=> PQ = PR

Given $\overleftrightarrow{AC}$ and $\overleftrightarrow{DE}$ = tangents to circle P

=> $\overleftrightarrow{AC} \perp \overline{PQ}$ and $\overleftrightarrow{DE} \perp \overline{PR}$

=> $\angle PQC = \angle PRE = 90°$

Sum of angles in quadrilateral = 360°

$\angle PQC + \angle QPR + \angle ABD + \angle PRE = 360°$

$90° + 50° + x + 90° = 360°$

$\therefore\ x = 130°$ //

b)

Vertical angles are ≅:

$\angle CBE \cong \angle ABD = 130°$

Radii of circle O are ≅:

=> OS = OT

Given $\overleftrightarrow{AC}$ and $\overleftrightarrow{DE}$ = tangents to circle O

=> $\overleftrightarrow{AC} \perp \overline{OT}$ and $\overleftrightarrow{DE} \perp \overline{OS}$

=> $\angle OTA = \angle OSD = 90°$

Sum of angles in quadrilateral = 360°

$\angle SOT + \angle OSD + \angle CBE + \angle OTA = 360°$

$\angle SOT + 90° + 130° + 90° = 360°$

$\angle SOT = 50°$

Since $\angle SOT$ = central angle

From the same arc, inscribed angles are half the measure of central angle

$$\angle SUT = \frac{1}{2} \times \angle SOT$$

$$y = \frac{1}{2} \times 50°$$

$= 25°$

Since $\angle SUT$ and $\angle SVT$ are drawn from the same arc $\overset{\frown}{ST}$, they have the same measure

$\angle SUT = \angle SVT$

$y = z = 25°$

Thus,
$y + z = 25° + 25°$
$= 50°$ //

34. In Figure 4.32, points A, B, D, and E are located on the circumference of circle O. If $\overrightarrow{AC}$ is a straight line, find the value of x.

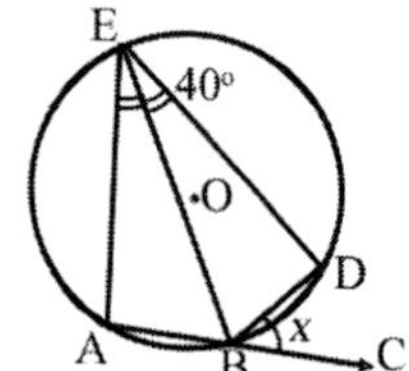

Figure 4. 32

Answer:
Given $\overrightarrow{AC}$ = straight line
Note: $\overline{BE}$ is not a diameter of circle O.
Since A, B, D and E are on circumference of circle O:
ABDE = cyclic quadrilateral
=> Opposite angles are supplementary
$\angle AED + \angle ABD = 180°$
$\angle AEB + \angle BED + \angle ABD = 180°$
$40° + 40° + \angle ABD = 180°$
$\angle ABD + 80° = 180°$
$\angle ABD = 100°$

Sum of straight line angles = 180°
$\angle CBD + \angle ABD = 180°$
$x + 100° = 180°$
$\therefore x = 80°$ //

35. In Figure 4.33, circle O and circle P are intersecting circles with common tangent $\overrightarrow{DA}$. $\overrightarrow{DF}$ is tangent to circle P and secant to circle O. Find the values of x, y and z.

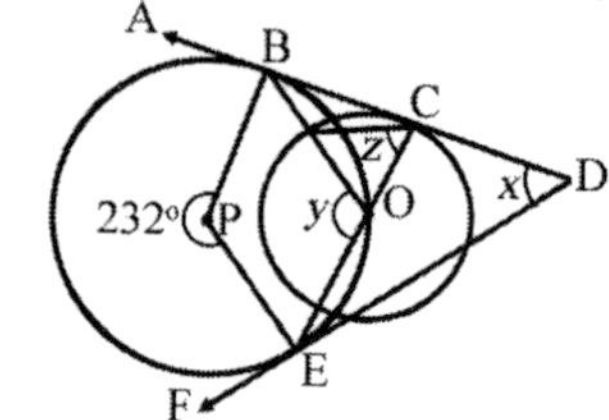

Figure 4. 33

Answers:
Given reflex $\angle BPE = 232°$
=> Obtuse $\angle BPE = 360° - 232°$
$= 128°$
Given $\overrightarrow{DA}$ and $\overrightarrow{DF}$ = tangents of circle P
$\overline{PB}$ and $\overline{PE}$ are radii of circle P
=> $\angle PBD = \angle PED = 90°$
Thus,
BDEP = quadrilateral
Sum of angles in quadrilateral = 360°
Obtuse $\angle BPE + \angle ADF + \angle PBD + \angle PED = 360°$
$128° + x + 90° + 90° = 360°$
$\therefore x = 52°$ //

Given central angle, reflex $\angle BPE = 232°$
Inscribed angle, is half the measure of central angle:

$\angle BOE = \frac{1}{2} \times$ reflex $\angle BPE$

$y = \frac{1}{2} \times 232°$

$\therefore y = 116°$ //

Since $\angle BOE$ = central angle of circle O
Inscribed angle is half the measure of central angle:

$z = \frac{1}{2} \times \angle BOE$

$= \frac{1}{2} \times 116°$

$\therefore z = 58°$ //

36. Figure 4.34, shows two externally tangent circles O and P, whose point of

contact is R. Further given that, $\overleftrightarrow{AB}$ is the common tangent and $\overline{CD}$ is a straight line. If CEFR is a cyclic quadrilateral, find the values of x, y and z.

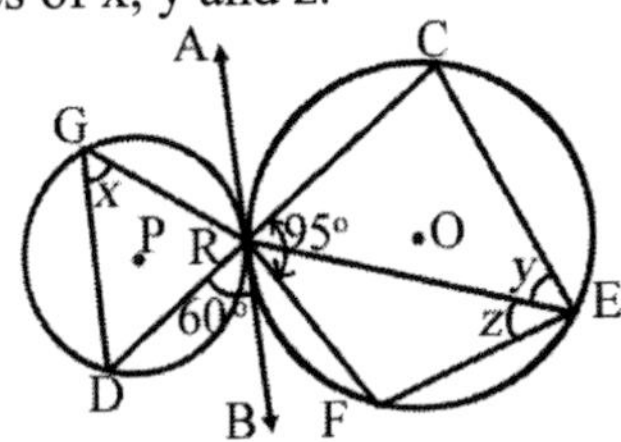

Figure 4. 34

Answers:
Given $\overleftrightarrow{AB}$ = common tangent
Angle between tangent, $\overleftrightarrow{AB}$ and chord, $\overline{DR}$ = angle in alternate segments, $\overline{DG}$ and $\overline{GR}$
=> $\angle DGR = \angle BRD$
$\therefore$ x = 60° //

Given $\overline{CD}$ = straight line
Vertical angles are ≅:
=> $\angle ARC \cong \angle BRD$
$\angle ARC = 60°$
Angle between tangent, $\overleftrightarrow{AB}$ and chord, $\overline{CR}$ = angle between alternate segments, $\overline{CE}$ and $\overline{ER}$
=> $\angle CER = \angle ARC$
$\therefore$ y = 60° //

Given CEFR = cyclic quadrilateral
Opposite angles are supplementary
$\angle CEF + \angle CRF = 180°$
$\angle CER + \angle FER + 95° = 180°$
$60° + z + 95° = 180°$
$\therefore$ z = 25° //

37. In Figure 4.35, $\overline{AC}$ and $\overline{CE}$ are secants of circle O. If CE = x, find the value of x.

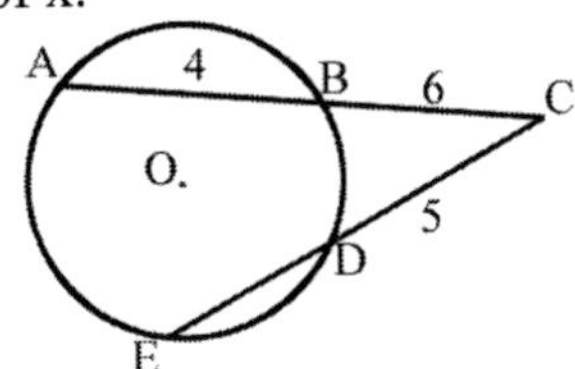

Figure 4. 35

Answer:
Given $\overline{AC}$ and $\overline{CE}$ = secant segments
From secant-secant segment measurements:
$AC \times BC = CE \times CD$
$(4 + 6) \times 6 = x \times 5$
$10 \times 6 = 5x$
$5x = 60$
$\therefore$ x = 12 units //

38. Figure 4.36, shows secants $\overline{AC}$ and $\overline{CE}$ of circle O. Hence find the value of x.

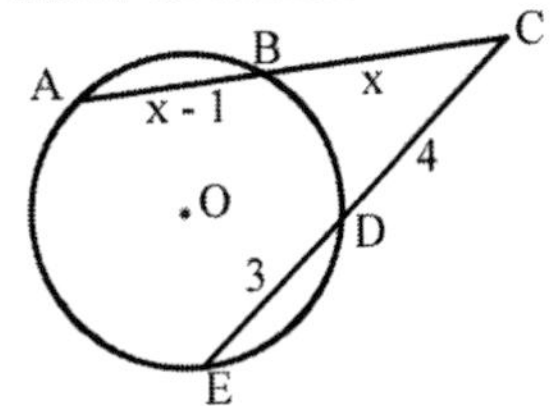

Figure 4. 36

Answer:
Given $\overline{AC}$ and $\overline{CE}$ = secants
From secant-secant segment measurements:
$AC \times BC = CE \times CD$
$\left((x-1)+x\right) \times x = (3+4) \times 4$
$(2x - 1) \times x = 7 \times 4$
$2x^2 - x = 28$
$2x^2 - x - 28 = 0$
$(2x + 7)(x - 4) = 0$
Critical values of $x = \frac{-7}{2}$ and 4
Since x is measure of length, x is always positive.
$\therefore$ x = 4 units //

39. In Figure 4.37, $\overline{AB}$ and $\overline{BC}$ are tangents of a circle whose center is D. Find the length of AB and BD if it is given that the radius of circle D is 5 ″ and BC is 12 ″.

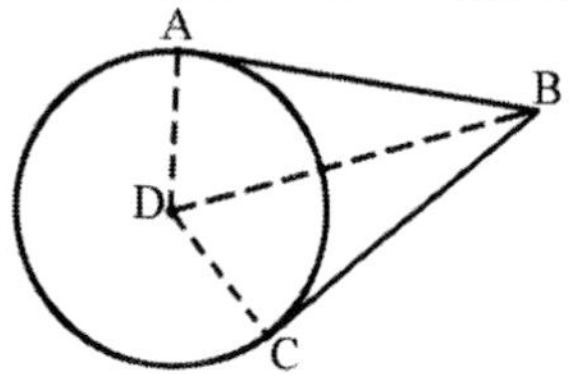

Figure 4. 37

Answers:
Given BC = 12
$\overline{AB}$ and $\overline{BC}$ = intersecting tangents
Theorem: Intersecting tangents from the same circle are equal in length
=> AB = BC
$\therefore$ AB = 12 ″ //
Given radius = 5
=> CD = radius
=> CD = 5

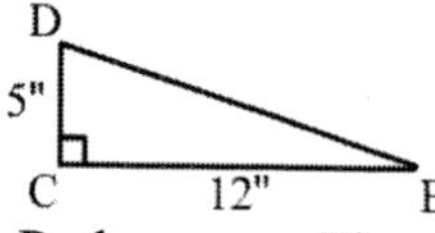

From Pythagorean Theorem:
$BD^2 = CD^2 + BC^2$
$= 5^2 + 12^2$
$= 25 + 144$
$= 169$
$BD = \sqrt{169}$
$= 13$ ″ //

40. Figure 4.38, shows $\overline{AB}$ a tangent and $\overline{BD}$ a secant of circle O. Hence find the value of x.

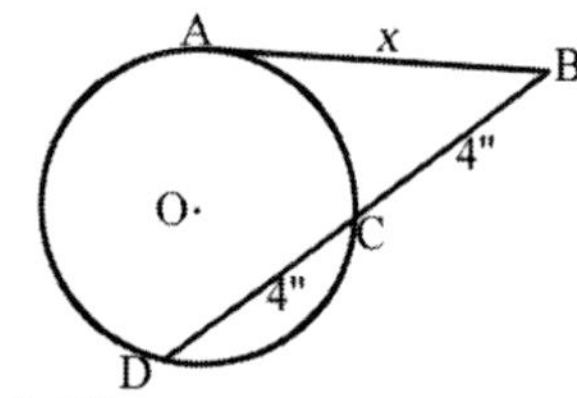

Figure 4. 38

Answer:
Given AB = x
Since $\overline{AB}$ and $\overline{BD}$ are from the same circle and intersect, using tangent-secant segment measurements:

$AB^2 = BD \times BC$
$x^2 = (4 + 4) \times 4$
$= 8 \times 4$
$= 32$
$x = \sqrt{32}$
$\therefore x = 5.6569$ ″ //

41. In Figure 4.39, $\overline{AB}$ is a tangent and $\overline{AD}$ is a secant of circle O. If AB = 4 ″, AC = 2 ″ and CD = x, find the value of x.

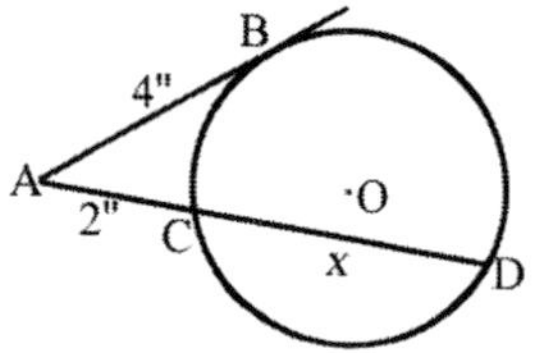

Figure 4. 39

Answer:
Given AB = 4
AC = 2
By tangent-secant segment measurements:
$AB^2 = AD \times AC$
$4^2 = (2 + x) \times 2$
$16 = 4 + 2x$
$2x = 16 - 4$
$2x = 12$
$\therefore x = 6$ ″ //

42. In Figure 4.40, shows circle O and circle P whose common tangent is $\overline{AD}$. If it is further known that $\overline{CF}$ is a secant of

circle O and tangent of circle P, find the values of m and n.

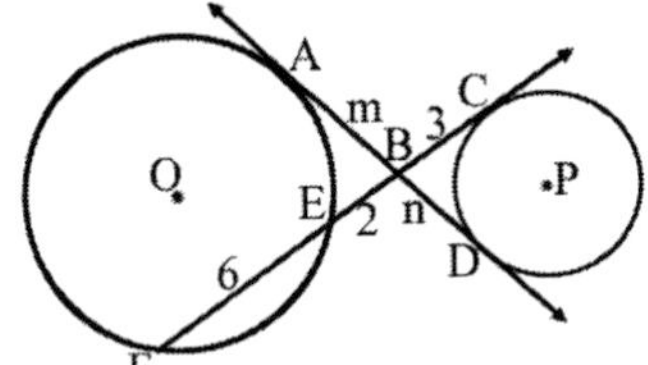

Figure 4. 40

Answers:

Given $\overline{AD}$ = tangent of circle O

$\overline{CF}$ = secant of circle O

By tangent-secant segment measurements:

$AB^2 = BF \times BE$

$m^2 = (2 + 6) \times 2$

$= 8 \times 2$

$= 16$

$m = \sqrt{16}$

$\therefore$ m = 4 units //

Given $\overline{AD}$ and $\overline{CF}$ = tangents of circle P

Theorem: Intersecting tangents from the same circle have equal tangent segment length.

=> BD = BC

$\therefore$ n = 3 units //

43. In Figure 4.41, $\triangle$PQR is circumscribed about circle O. Points A, B and C are points of tangency. Find the perimeter of $\triangle$PQR.

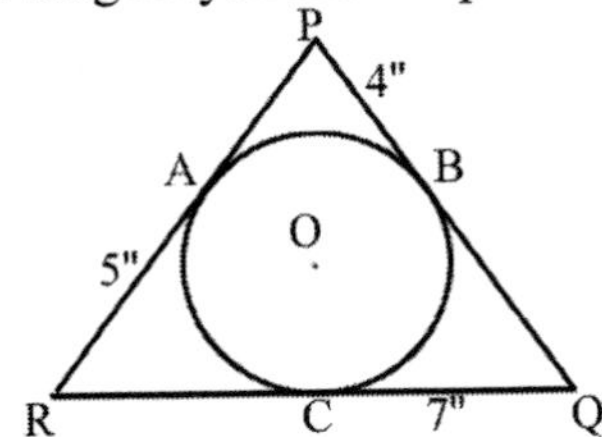

Figure 4. 41

Answer:

Given A, B and C = points of tangency

Theorem: Intersecting tangents from the same circle have equal tangent segment length.

=> AP = BP

$\therefore$ AP = 4 ″

=> BQ = CQ

$\therefore$ BQ = 7 ″

=> CR = AR

$\therefore$ CR = 5 ″

Thus, perimeter $\triangle$PQR:

= AP + BP + BQ + CQ + AR + CR

= 4 + 4 + 7 + 7 + 5 + 5

= 32 ″ //

44. In Figure 4.42, $\overline{AC}$ and $\overline{BD}$ are chords of circle O. Hence find the value of x.

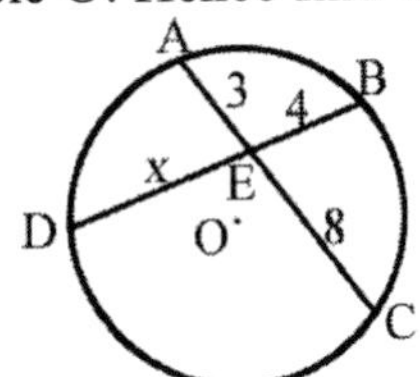

Figure 4. 42

Answer:

Given $\overline{AC}$ and $\overline{BD}$ = chords

Using chord-chord segment measurements:

$AE \times CE = BE \times DE$

$3 \times 8 = 4 \times x$

$24 = 4x$

$\therefore$ x = 6 units //

45. Figure 4.43 depicts circle O whose radius is 5 ″. If arc length, $\overset{\frown}{BC}$ is 8 ″, find the central angle, θ in radians and degrees.

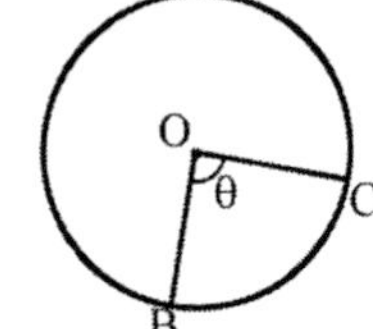

Figure 4. 43

Answers:
Given radius, r = 5
=> OB = OC = 5

Given length of $\widehat{BC}$, = 8

Thus, length of $\widehat{BC}$:

$8 = r \times \theta$

$8 = 5 \times \theta$

$\therefore \theta = 1.6$ rad //

Convert radian to degrees:

$=> \dfrac{180°}{\pi} \times \theta$

$= \dfrac{180°}{\left(\dfrac{22}{7}\right)} \times 1.6$

$= 91.64°$ //

Alternatively:

Length of $\widehat{BC}$:

$8 = 2\pi r \times \dfrac{\angle BOC}{360°}$

$8 = 2\left(\dfrac{22}{7}\right)(5) \times \dfrac{\theta}{360°}$

$\therefore \theta = 91.64°$ //

Convert degrees to radian:

$=> \dfrac{\pi}{180°} \times \theta$

$= \dfrac{\left(\dfrac{22}{7}\right)}{180°} \times 91.64°$

$= 1.6$ rad //

46. In Figure 4.44, OPQ is a sector whose arc length is $\dfrac{3\pi}{2}$ and $\angle POQ = \dfrac{\pi}{3}$. Find the perimeter of sector OPQ.

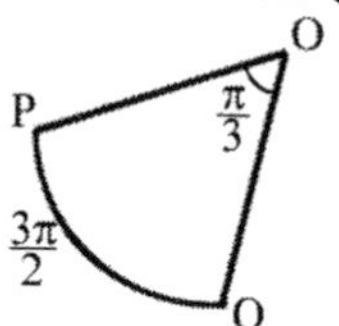

Figure 4. 44

Answer:

Given $\angle POQ = \dfrac{\pi}{3}$ ⇦ Note: Angle given in radian

Length of $\widehat{PQ} = \dfrac{3\pi}{2}$

$= \dfrac{3}{2} \times \dfrac{22}{7} = 4.7143$ units

Since $\overline{OP}$ and $\overline{OQ}$ are radii of circle O
=> radius, r = OP = OQ

Length of $\widehat{PQ}$:

$\widehat{PQ} = r \times \angle POQ$

$\dfrac{3\pi}{2} = r \times \dfrac{\pi}{3}$

$r = \dfrac{9}{2}$

r = 4.5 units
Thus,
=> r = OP = OQ = 4.5 units
Perimeter of sector OPQ:

= OP + OQ + length of $\widehat{PQ}$

= 4.5 + 4.5 + 4.7143
= 13.7143 units //

Alternatively:
Changing radian to degrees:

Given $\angle POQ = \dfrac{\pi}{3} = \dfrac{180°}{3}$

$= 60°$

Length of $\widehat{PQ}$:

$\dfrac{3\pi}{2} = 2 \times \pi \times r \times \dfrac{\angle POQ}{360°}$

$\dfrac{3\pi}{2} = 2 \times \pi \times r \times \dfrac{60°}{360°}$

r = 4.5 units

47. Given a situation where the train driver has a choice of either going around the mountains using semicircle, $\overset{\frown}{ABC}$, or through the mountains using tunnel, $\overline{AC}$. If the semi circular distance, $\overset{\frown}{ABC}$ is 8 miles, determine the length of the tunnel, $\overline{AC}$. Which route is shorter? [Assume O is the center point of the circle]

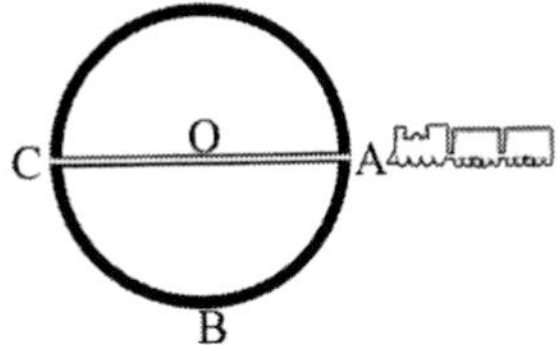

Figure 4. 45

Answers:

Given length of $\overset{\frown}{ABC} = 8$

radius, $r = OA = OC$

Length of semicircle:

Length of $\overset{\frown}{ABC} = r \times \pi$

$8 = r \times \dfrac{22}{7}$

$r = 2.5455$ miles

Diameter $= 2 \times r$

$AC = 2 \times r$

$= 2 \times 2.5455$

$= 5.091$ miles //

$\therefore$ The length of the tunnel, $\overline{AC}$ is 5.091 miles. Therefore, the shorter route is through the tunnel, $\overline{AC}$. //

48. A straight path through the center of a rotunda is 40 feet. How long is the semicircular path around the building? [Let $\pi = \dfrac{22}{7}$]

Answer:

Given diameter, $d = 40$

$\Rightarrow$ radius, $r = \dfrac{1}{2} \times d = \dfrac{1}{2} \times 40$

$= 20$

Length of semicircle arc:

$= \pi \times r$

$= \dfrac{22}{7} \times 20$

$= 62.8571$ feet //

49. Find the area of sector AOB whose radius is 5 ″ and $\angle AOB = 30°$.

Answer:

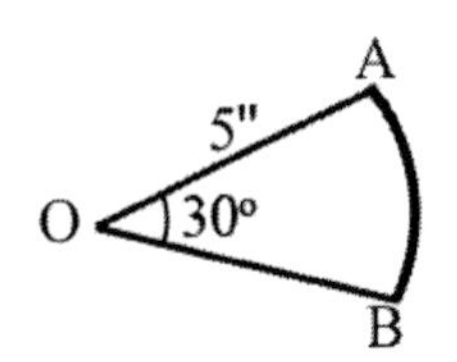

Given $\angle AOB = 30°$

Radius, $r = 5$

$\Rightarrow r = OA = OB = 5$

Area of sector AOB, A:

$A = \dfrac{\angle AOB}{360°} \times \pi \times r^2$

$= \dfrac{30°}{360°} \times \dfrac{22}{7} \times 5^2$

$\approx 6.5476\ in^2$ //

Alternatively:

Changing degrees to radian:

$\angle AOB = 30° \times \dfrac{\pi}{180°}$

$= 0.52381$ rad

Area of sector AOB, A:

$A = \dfrac{1}{2} \times r^2 \times \angle AOB$

$= \dfrac{1}{2} \times 5^2 \times 0.52381$

$\approx 6.5476\ in^2$ //

50. Sector POQ has radius 7 ″ and ∠POQ = 2 rad. What is the area of sector POQ?

Answer:
Given radius, r = 7
Let θ = ∠POQ = 2 rad

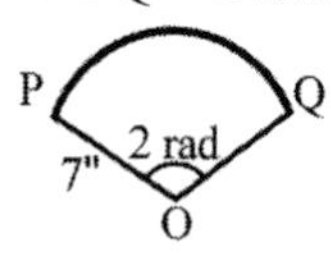

Area of sector POQ, A:

$$A = \frac{1}{2} \times r^2 \times \theta$$
$$= \frac{1}{2} \times 7^2 \times 2$$
$$= 49 \text{ in}^2 {}_{//}$$

Alternatively:

Changing radian to degrees:

$$\angle POQ = \frac{180°}{\pi} \times 2$$
$$= 114.55°$$

Area of sector, POQ, A:

$$A = \pi \times r^2 \times \frac{\angle POQ}{360°}$$
$$= \frac{22}{7} \times 7^2 \times \frac{114.55°}{360°}$$
$$= 49 \text{ in}^2 {}_{//}$$

51. In Figure 4.46, PQRS is a sector whose radius is 5 yd and ∠QPS = 50°. Find the perimeter of the shaded segment area, QRS.

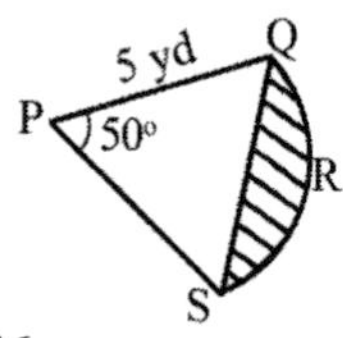

Figure 4. 46

Answer:
Given ∠QPS = 50°
Let T = midpoint of $\overline{QS}$
=> $\overline{PT} \perp \overline{QS}$
=> $\overline{PT}$ = angle bisector of ∠QPS

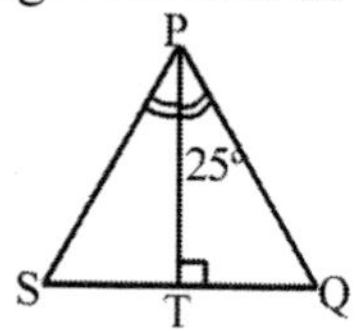

=> ∠QPT ≅ ∠SPT = 25°
=> ST = QT ...(*)

$$\sin \angle QPT = \frac{QT}{PQ}$$
$$\sin 25° = \frac{QT}{5}$$

QT = 5 sin 25° ...(1)
QS = ST + QT
= QT + QT ⇦ From (*)
= 2 QT ...(2)

Substitute (1) into (2):
QS = 2(5 sin 25°)
= 10 × 0.4226 ⇦Note: sin 25° = 0.4226
= 4.226 yd

Length of arc, $\widehat{QRS}$:

$$= \frac{\angle QPS}{360°} \times 2\pi r$$
$$= \frac{50°}{360°} \times 2 \times \frac{22}{7} \times 5$$
$$= 4.3651 \text{ yd}$$

Perimeter of shaded sector, QRS:
= Length of $\widehat{QRS}$ + QS
= 4.3651 + 4.226
= 8.5911 yd $_{//}$

52. What is the area of the shaded segment in Figure 4.47?

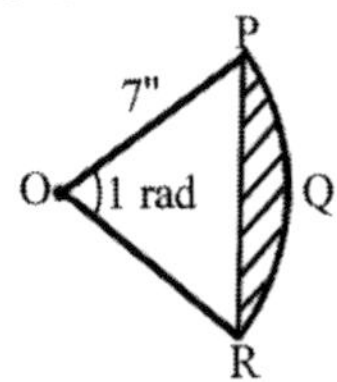

Figure 4. 47

Answer:

Let OS = perpendicular bisector of ∠POR

Given ∠POR = 1 rad

$$=> \angle POR = 1 \times \frac{180°}{\pi} = 1 \times \frac{180°}{3.142}$$

$$= 57.29°$$

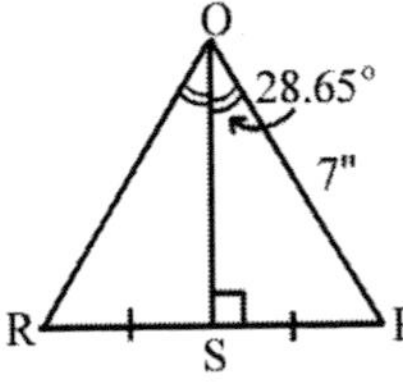

$$=> \angle POS = \frac{1}{2} \times \angle POR$$

$$\angle POS = \frac{1}{2} \times 57.29° = 28.65°$$

Thus,

$PS = OP \times \sin 28.65°$

$= 7 \sin 28.65°$

$= 7 \times 0.4795$ ⇦ Note: $\sin 28.65° = 0.4795$

$= 3.3565''$

$PR = 2 \times PS$

$= 2 \times 3.3565$

$= 6.713''$

$OS = OP \times \cos 28.65°$

$= 7 \times \cos 28.65°$

$= 7 \times 0.8776$ ⇦ Note: $\cos 28.65° = 0.8776$

$= 6.1432''$

$$\text{Area of } \triangle OPR = \frac{1}{2} \times OS \times PR$$

$$= \frac{1}{2} \times 6.1432 \times 6.713$$

$$= 20.6197 \text{ in}^2$$

Area of sector OPR:

$$= \frac{\angle POR}{360°} \times \pi \times OP^2$$

$$= \frac{57.29°}{360°} \times \frac{22}{7} \times 7^2$$

$= 24.5074 \text{ in}^2$

Area of shaded segment, PQR:

= Area of sector OPR – area of △OPR

$= 24.5074 - 20.6197$

$= 3.8877 \text{ in}^2$ //

53. In Figure 4.48, sector POS and sector QOR are overlapping concentric sectors. Also given, ∠POS is 60°, and length of $\overset{\frown}{PS}$ is twice the length of OP. If OQ is 8″, find the perimeter of the shaded area.

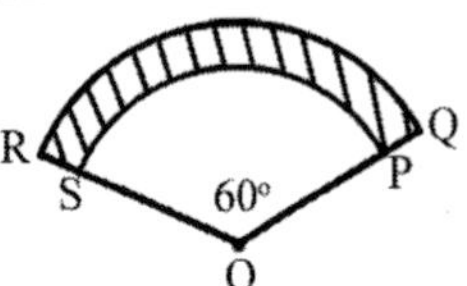

Figure 4. 48

Answer:

Given ∠POS = 60°

Let x = OP

Also given length of $\overset{\frown}{PS}$ = 2OP

=> length of $\overset{\frown}{PS}$ = 2x

Radii of circle are ≅:

PQ = RS

Given OQ = 8

$$\text{Length of } \overset{\frown}{QR} = \frac{60°}{360°} \times 2 \times \pi \times OQ$$

$$= \frac{1}{6} \times 2 \times \frac{22}{7} \times 8$$

$$= 8.3810''$$

Perimeter of shaded area PQRS:
= length of $\widehat{PS}$ + length of $\widehat{QR}$ + 2(OQ – OP)
= 2x + 8.3810 + 2(8 – x)
= 2x + 8.3810 + 16 – 2x
= 24.381 ″ //

54. A child enjoys a swing at an angle, ∠BOA = 95° from point B to point A (see Figure 4.49). If it is known that the arc length, $\widehat{AB}$ is 12 feet, find the length of the chain OB.

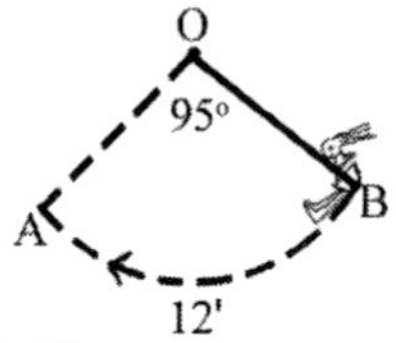

Figure 4. 49

Answer:
Let r = OB
Thus length of arc, $\widehat{AB}$:

$$\text{Length of } \widehat{AB} = \frac{\angle BOA}{360^\circ} \times 2 \times \pi \times r$$

$$12 = \frac{95^\circ}{360^\circ} \times 2 \times \frac{22}{7} \times r$$

r = 7.2344
∴ Length of chain OB is 7.2344 feet. //

55. Circle P whose radius is 8 ″ and circle O whose radius is 5 ″ share a common external tangent (see Figure 4.50). If $\overleftrightarrow{AD}$ is the common tangent and $\overline{EO}$ is parallel to $\overleftrightarrow{AD}$, find the perimeter and area of BCOP.

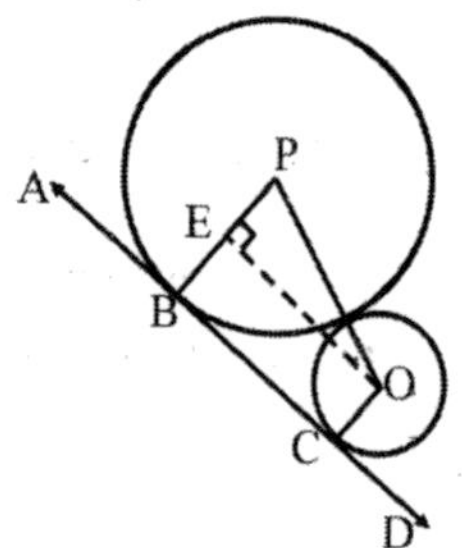

Figure 4. 50

Answers:
Given radius of circle P, PB = 8 ″
Radius of circle O, OC = 5 ″
PO = radius circle P + radius circle O
= 8 + 5
= 13 ″
PE = PB – OC
= 8 – 5 = 3
Using Pythagorean Theorem:
$EO^2 = PO^2 - PE^2$
$= 13^2 - 3^2$
= 169 – 9
= 160
$EO = \sqrt{160}$ ⇦ Note: $\sqrt{160} = \sqrt{16 \times 10}$
$= 4\sqrt{10}$ or 12.6491 ″
Given $\overline{EO} \parallel \overleftrightarrow{AD}$
=> BC = EO
$BC = 4\sqrt{10}$ or 12.6491 ″

Perimeter of BCOP:
= BC + OC + PO + PB
= 12.6491 + 5 + 13 + 8
= 38.6491 ″ //

Area of BCOP:
= area of trapezoid BCOP

$$= \frac{1}{2} \times BC \times (PB + OC)$$

$$= \frac{1}{2} \times 12.6491 \times (8 + 5)$$

$= 82.2192 \text{ in}^2$ //

Chapter 5
Similarities, Ratios, & Proportions

Similarity – objects with the same shape.
Condition for similarity:

1. Corresponding angles are congruent.
2. Lengths of corresponding sides are proportional.

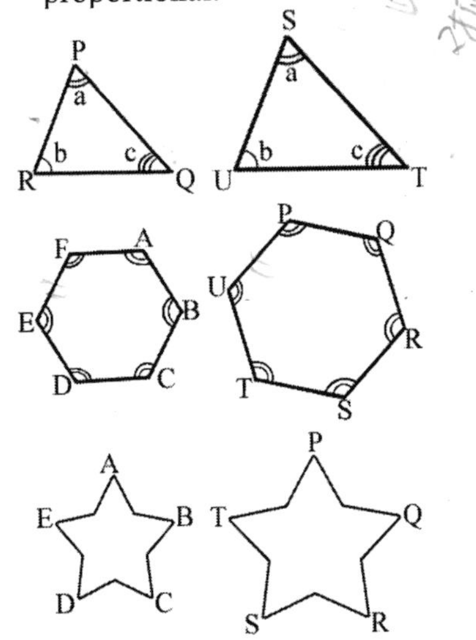

Scale factor – measure of relativity i.e. common ratio of pairs of corresponding sides in similar figures.
Two similar triangles with a scale factor of $a:b$, then the ratio of their areas is $a^2:b^2$.
Prove of similar triangles: **Angle–Angle theorem** (A–A).
If $\triangle A$ and $\triangle B$ are similar triangles:

$$\frac{\text{Area of } \Delta A}{\text{Area of } \Delta B} = \left(\frac{\text{Side of } \Delta A}{\text{Side of } \Delta B}\right)^2$$

If polygon A and polygon B are similar:

$$\frac{\text{Area of polygon A}}{\text{Area of polygon B}} = \left(\frac{\text{Side of polygon A}}{\text{Side of polygon B}}\right)^2$$

Ratio is used to compare 2 or more quantities. Ratio is often expressed in simplified form.

Ratios are expressed as $\frac{a}{b}$, $a \div b$, a to b, or $a:b$.

Proportion – relationship between 2 or more quantities

Direct proportion: $A \propto B$

$$\frac{A_1}{A_2} = \frac{B_1}{B_2}$$

Inverse proportion: $A \propto \frac{1}{B}$

$$\frac{A_1}{A_2} = \frac{B_2}{B_1}$$

1. In Figure 5.1, $\triangle ABC$ and $\triangle PQR$ are two scalene triangles. Determine if $\triangle ABC$ is similar to $\triangle PQR$. Hence find the value of x.

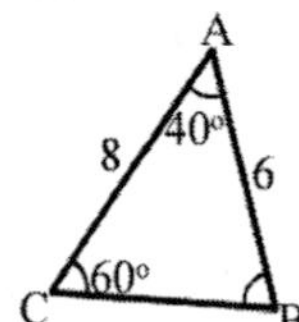

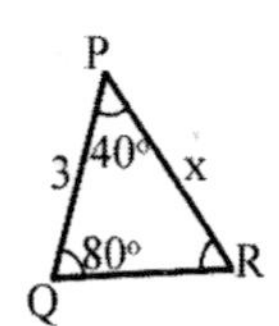

Figure 5. 1

Answers:

Given $\angle BAC = \angle QPR = 40°$ [angle]

Sum of interior angles in $\triangle ABC = 180°$

$\angle ABC + \angle BAC + \angle ACB = 180°$

$\angle ABC + 40° + 60° = 180°$

$\angle ABC = 80°$

=> $\angle ABC = \angle PQR = 80°$ [angle]

Based on A–A theorem, $\triangle ABC$ and $\triangle PQR$ are similar triangles. //

Since corresponding sides of similar triangles are proportional:

Ratio of corresponding sides:

$$\frac{AC}{AB} = \frac{PR}{PQ}$$

$$\frac{8}{6} = \frac{x}{3}$$

$$x = \frac{8}{6} \times 3$$

$\therefore$ x = 4 units //

2. In Figure 5.2, $\overline{AB}$ and $\overline{DE}$ are parallel lines. Prove that $\triangle ABC$ and $\triangle CDE$ are similar triangles.

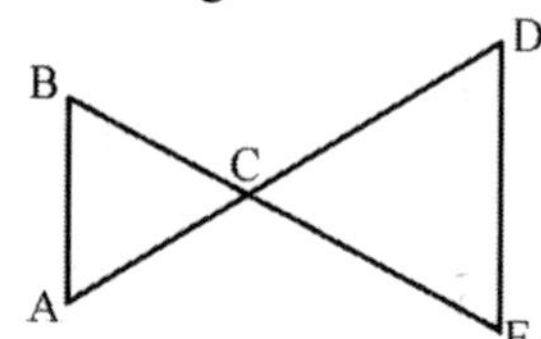

Figure 5. 2

Answer:

Given $\overline{AB} \parallel \overline{DE}$

Alternate interior angles are ≅:

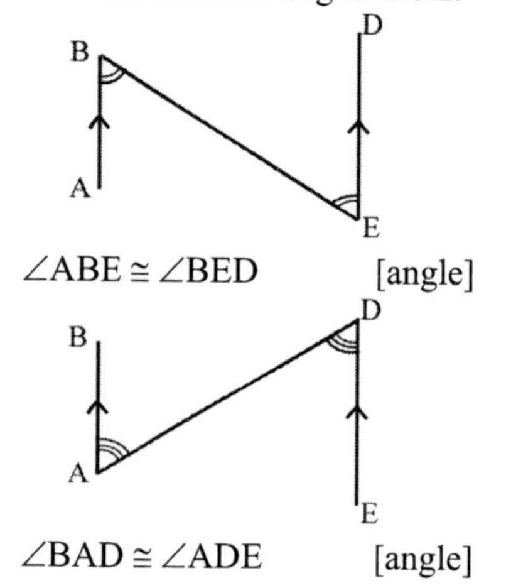

$\angle ABE \cong \angle BED$ [angle]

$\angle BAD \cong \angle ADE$ [angle]

∴Based on A–A theorem, △ABC and △CDE are similar triangles. //

3. Figure 5.3 shows 2 similar polygons ABCDE and PQRST. Hence find the values of x and y.

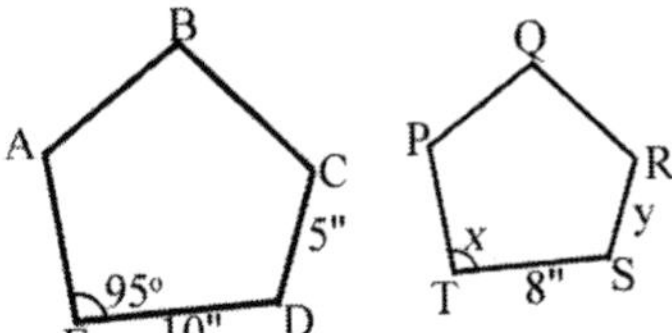

Figure 5. 3

Answers:
Given ABCDE & PQRST = similar pentagons
Since corresponding angles of similar polygons are ≅:
∠PTS ≅ ∠AED
∴ x = 95° //

Corresponding sides of similar polygons are proportional:
Ratio of corresponding sides:

$$\frac{CD}{DE} = \frac{RS}{ST}$$

$$\frac{5}{10} = \frac{y}{8}$$

$$y = \frac{5}{10} \times 8$$

∴ y = 4 " //

4. In Figure 5.4, ABCD is a parallelogram and $\overline{AC}$ is a diagonal. Points E and F are midpoints of $\overline{AB}$ and $\overline{BC}$ respectively. Determine if △ACD and △BEF are similar triangles.

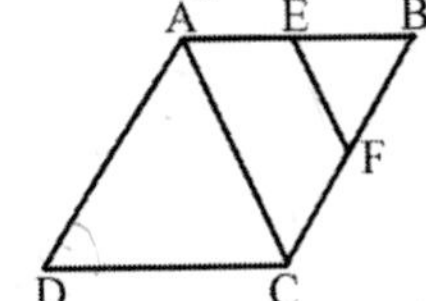

Figure 5. 4

Answer:
Given ABCD = parallelogram
=> Opposite angles are ≅:
∠ADC ≅ ∠ABC [angle]
Given $\overline{AC}$ = diagonal
Diagonal of parallelogram divides the parallelogram into two congruent triangles, △ABC and △ACD.
Also given $\overline{EF}$ is drawn from the midpoints of $\overline{AB}$ and $\overline{BC}$
Hence in △ABC:
From triangle midpoint theorem:
=> $\overline{EF}$ || $\overline{AC}$

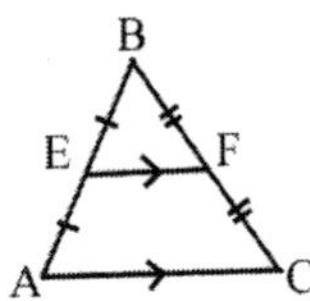

From property of parallelogram:
$\overline{AB}$ || $\overline{DC}$
Alternate interior angles are ≅:
∠BAC ≅ ∠ACD
Corresponding angles are ≅:
∠BAC ≅ ∠BEF

Thus, transitive principle of equality:
=> ∠ACD ≅ ∠BEF [angle]
∴ Based on A–A theorem, △ACD and △BEF are similar triangles. //

5. In Figure 5.5, △ABC and △BDE are similar triangles. $\overline{CQ}$ and $\overline{DR}$ are altitudes

of △ABC and △BDE respectively. If $\overline{CP}$ and $\overline{DS}$ are the medians, find the value of x.

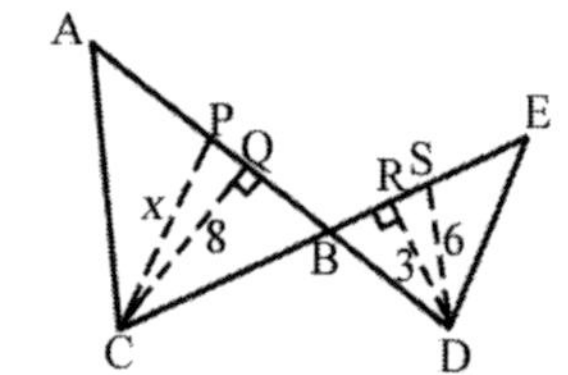

Figure 5. 5

Answer:
Given △ABC and △BDE = similar triangles
Thus, pairs of corresponding medians and altitudes are proportional.
Ratio of corresponding medians and altitudes:

$$\frac{CP}{CQ} = \frac{DS}{DR}$$

$$\frac{x}{8} = \frac{6}{3}$$

$$x = \frac{6}{3} \times 8$$

∴ x = 16 units //

6. Figure 5.6, shows two scalene triangles, △ABC and △BCD that are similar. It is further known that the perimeters of △ABC is 36 ″ and △BCD is 12 ″. If the altitude of △BCD is 4 ″, find the altitude of △ABC.

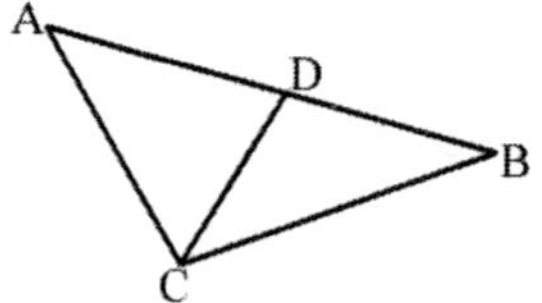

Figure 5. 6

Answer:
Given perimeters:
△ABC = 36
△BCD = 12
Given △BCD's altitude = 4
Let x = △ABC's altitude
Also given △ABC and △BCD = similar triangles
Since similar triangles have proportional pairs of altitudes and perimeters;
Ratio of altitudes and perimeters:

$$\frac{\Delta ABC\text{'s altitude}}{\Delta ABC\text{'s perimeter}} = \frac{\Delta BCD\text{'s altitude}}{\Delta BCD\text{'s perimeter}}$$

$$\frac{x}{36} = \frac{4}{12}$$

$$x = \frac{4}{12} \times 36$$

∴ x = 12 ″ //

7. In Figure 5.7, △ABC is a right triangle. It is further given that △ABC, △BDC and △ADB are similar triangles. If $\overline{AC}$ is 26 ″ and $\overline{BC}$ is 24 ″, find the value of x and the altitude of △ABC.

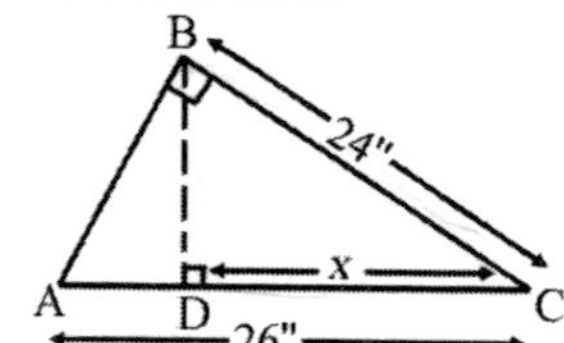

Figure 5. 7

Answers:
Given △ABC = right triangle
AC = 26
BC = 24
Also given △ABC and △BCD = similar triangles

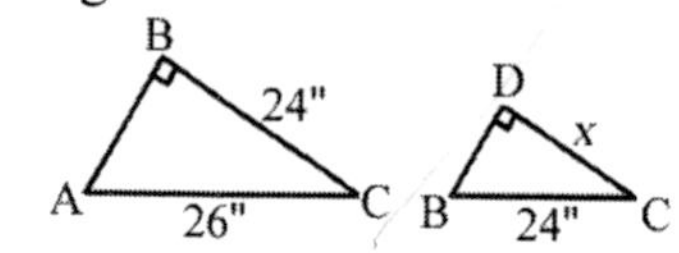

Ratio of corresponding sides:

$$\frac{CD}{BC} = \frac{BC}{AC}$$

$$\frac{x}{24} = \frac{24}{26}$$

$$x = \frac{24}{26} \times 24$$

$\therefore$ x = 22. 1538 ″ //

Let y = altitude of $\triangle$ABC, $\overline{BD}$

Using Pythagorean Theorem:

$BC^2 = BD^2 + CD^2$

$24^2 = y^2 + (22.1538)^2$

$576 = y^2 + 490.7909$

$y^2 = 576 - 490.7909$

$= 85.2091$

$y = \sqrt{85.2091}$

y = 9.2309 ″

$\therefore \triangle$ABC's altitude is 9.2309 ″ //

8. Figure 5.8, shows 2 adjoining triangles, $\triangle$ABC and $\triangle$CDE. Prove that $\triangle$ABC and $\triangle$CDE are similar triangles.

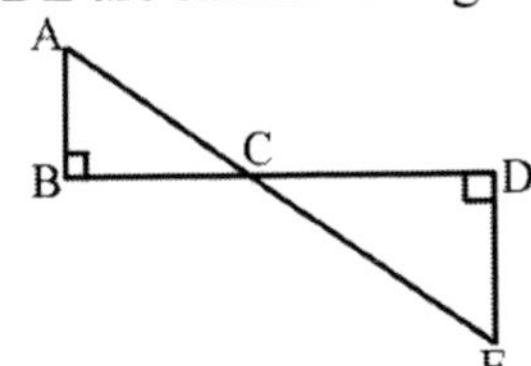

Figure 5. 8

Answer:

Given $\angle ABC = 90°$

$\angle CDE = 90°$

=> $\angle ABC = \angle CDE$ [angle]

Vertical angles are $\cong$:

$\angle ACB \cong \angle DCE$ [angle]

Based on A–A theorem that states if two angles in a triangle are congruent to 2 angles of another triangle, then the 2 triangles are similar triangles.

$\therefore$ We have proven that $\angle$ABC and $\angle$ACB in $\triangle$ABC are congruent to $\angle$CDE and $\angle$DCE in $\triangle$CDE, hence $\triangle$ABC and $\triangle$CDE must be similar triangles. //

9. In Figure 5.9, circle O and circle P are two non-intersecting circles with a common external tangent, $\overline{AC}$. Show that $\triangle$AOC and $\triangle$BPC are similar triangles. Hence find the values of x and y.

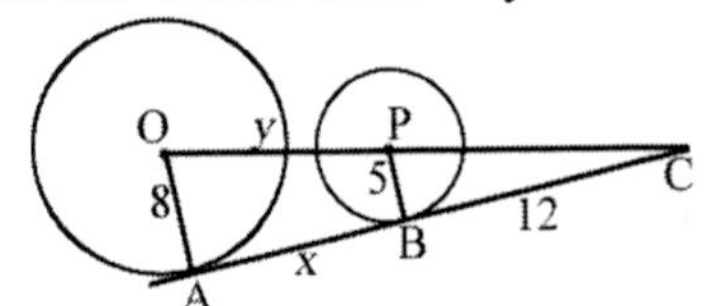

Figure 5. 9

Answers:

Given $\overline{AC}$ = common tangent

=> $\overline{OA} \perp \overline{AC}$

=> $\angle OAC = 90°$

=> $\overline{PB} \perp \overline{BC}$

=> $\angle PBC = 90°$

Thus, $\angle OAC = \angle PBC = 90°$ [angle]

Based on reflexive property of equality:

$\angle ACO = \angle BCP$ [angle]

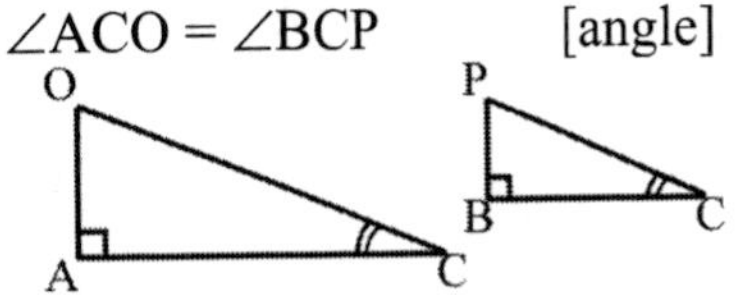

$\therefore$ We have shown based on A–A theorem that $\triangle$AOC and $\triangle$BPC are similar triangles. //

Since $\triangle$AOC and $\triangle$BPC are similar triangles:

Ratio of corresponding sides:

$$\frac{AC}{OA} = \frac{BC}{PB}$$

$$\frac{x+12}{8}=\frac{12}{5}$$

$$x + 12 = \frac{12}{5} \times 8$$

$$= 19.2$$

$\therefore$ x = 7.2 units //

Using Pythagorean Theorem:

$PC^2 = BC^2 + PB^2$
$= 12^2 + 5^2$
$= 144 + 25$
$= 169$

$PC = \sqrt{169}$
= 13 units

Since $\triangle AOC$ and $\triangle BPC$ are similar triangles, hence corresponding sides are proportional.

Ratio of corresponding sides:

$$\frac{OC}{OA}=\frac{PC}{PB}$$

$$\frac{y+13}{8}=\frac{13}{5}$$

$$y + 13 = \frac{13}{5} \times 8$$

$$= 20.8$$

$\therefore$ y = 7.8 units //

10. In Figure 5.10, circle P and circle O are externally tangent circles, where $\overline{AC}$ is the common tangent. Prove that $\triangle BDE$ and $\triangle BFG$ are similar triangles. Hence if BD = x, find the value of x.

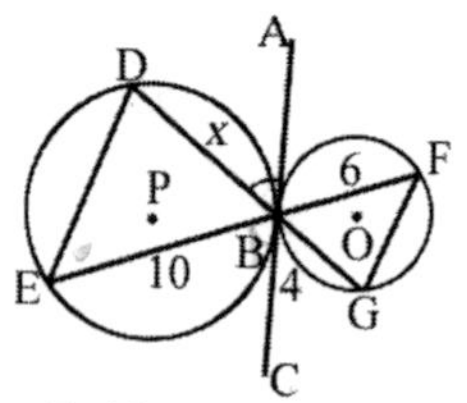

Figure 5. 10

Answers:

Given $\overline{AC}$ = common tangent

Let $\theta = \angle ABD$

Angle between tangent, $\overline{AC}$ and chord, $\overline{BD}$ equals angle in alternate segments $\overline{BE}$ and $\overline{DE}$.

=> $\angle BED = \angle ABD = \theta$

Vertical angles are $\cong$:

$\angle ABD \cong \angle CBG$

$\angle CBG = \theta$

Angle between tangent, $\overline{AC}$ and chord, $\overline{BG}$ equals to angle in alternate segments, $\overline{BF}$ and $\overline{FG}$.

=> $\angle CBG = \angle BFG = \theta$

Thus,

$\theta = \angle BED = \angle BFG$ [angle]

Vertical angles are $\cong$:

$\angle DBE \cong \angle FBG$ [angle]

$\therefore$ Based on A–A theorem, $\triangle BDE$ and $\triangle BFG$ are similar triangles. //

Since corresponding sides of similar triangles are proportional:

Ratio of congruent sides:

$$\frac{BD}{BE}=\frac{BG}{BF}$$

$$\frac{x}{10}=\frac{4}{6}$$

$$x = \frac{4}{6} \times 10$$

$\therefore$ x = 6.6667 units //

11. In Figure 5.11, $\triangle ABC$ and $\triangle PQR$ are similar triangles. It is further given that the area of $\triangle ABC$ is 27 units^2 while area of $\triangle PQR$ is 3 units^2. Find the value of x if the corresponding side of $\overline{PR}$, AC is 9 units.

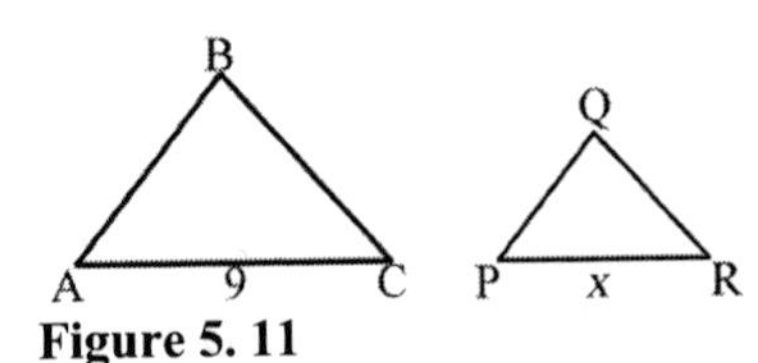

Figure 5. 11

Answer:
Given area of $\triangle ABC = 27$
Area of $\triangle PQR = 3$
Since $\triangle ABC$ and $\triangle PQR$ are similar triangles:

$$\frac{\text{area of } \Delta PQR}{\text{area of } \Delta ABC} = \left(\frac{PR}{AC}\right)^2$$

$$\frac{3}{27} = \left(\frac{x}{9}\right)^2$$

$$\frac{3}{27} = \frac{x^2}{81}$$

$x^2 = 9$
$x = 3$ units //

12. Two triangles, $\triangle ABC$ and $\triangle PQR$ are similar triangles. The area of $\triangle ABC$ is 40 units^2 and area of $\triangle PQR$ is 160 units^2. If the dimension of $\triangle ABC$ is given in Figure 5.12, find the dimension of $\triangle PQR$.

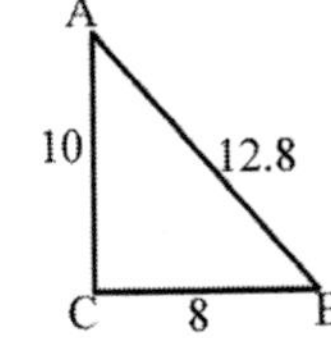

Figure 5. 12

Answers:
Given area of $\triangle ABC = 40$
Area of $\triangle PQR = 160$

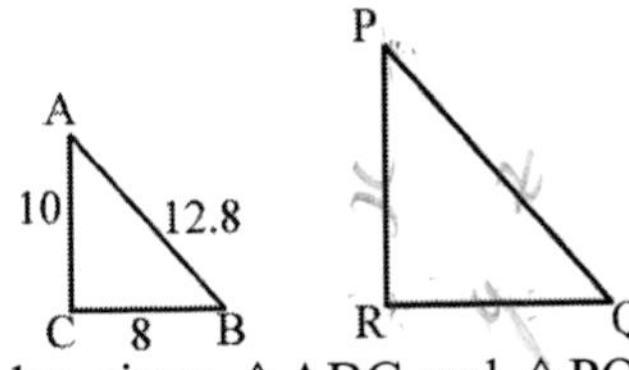

Also given $\triangle ABC$ and $\triangle PQR$ are similar triangles.
Thus area scale factor:

$$\frac{\text{Area of } \Delta ABC}{\text{Area of } \Delta PQR} = \frac{40}{160}$$

$$= \frac{1}{4}$$

Area of $\triangle ABC$: Area of $\triangle PQR = 1:4$
Convert to linear scale factor:
Linear scale factor $= \sqrt{\text{Area scale factor}}$

$$= \sqrt{\frac{1}{4}}$$

$$= \frac{1}{2}$$

Side of $\triangle ABC$: Side of $\triangle PQR = 1:2$
Dimension of $\triangle PQR$:

$PQ = AB \times 2$
$= 12.8 \times 2$
$= 25.6$ units //

$PR = AC \times 2$
$= 10 \times 2$
$= 20$ units //

$QR = BC \times 2$
$= 8 \times 2$
$= 16$ units //

13. In Figure 5.13, $\triangle ABC$ and $\triangle DEF$ are similar triangles with a scale factor of 1:3. Find the sum of their areas.

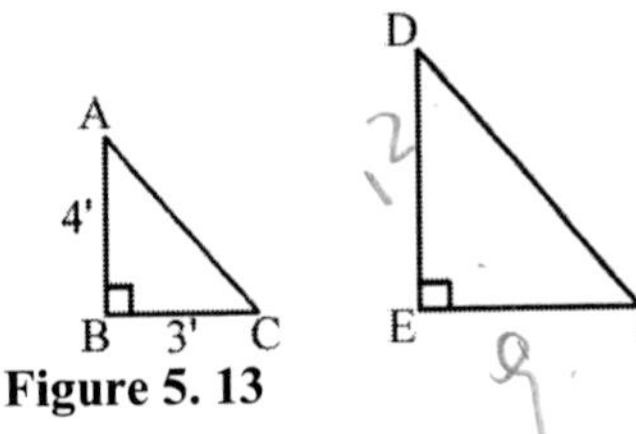

Figure 5. 13

Answer:
Given scale factor $= 1:3$
=> Ratio of areas:

Area of $\triangle ABC$: Area of $\triangle DEF = 1^2 : 3^2$
$= 1:9$

Area of $\triangle ABC$:
$= \frac{1}{2} \times AB \times BC$
$= \frac{1}{2} \times 4 \times 3$
$= 6$ feet^2

Area of $\triangle DEF$:
= (area scale factor) $\times$ (area of $\triangle ABC$)
$= 9 \times 6$
$= 54$ feet^2

Sum of areas:
= area of $\triangle ABC$ + area of $\triangle DEF$
$= 6 + 54$
$= 60$ feet^2 //

14. The perimeters of 2 similar triangles are 30 in. and 40 in. respectively. It is further known that the difference of their areas is 21 in^2. Hence find the area of each triangle.

Answers:
Let $\triangle A$ and $\triangle B$ = the two triangles
Thus, given
Perimeter of $\triangle A = 30$
Perimeter of $\triangle B = 40$
Let $m:n$ = scale factor of $\triangle A$ and $\triangle B$
=> $m:n = 30:40$

$\frac{m}{n} = \frac{30}{40}$
$= \frac{3}{4}$

Ratio of their areas is square of their scale factor:

$\left(\frac{m}{n}\right)^2 = \left(\frac{3}{4}\right)^2$
$= \frac{3^2}{4^2}$
$= \frac{9}{16}$

Thus, $\frac{\text{Area of } \Delta A}{\text{Area of } \Delta B} = \frac{9}{16}$

=> Area of $\triangle A = \frac{9}{16} \times$ Area of $\triangle B$

Let x = Area of $\triangle B$
Given, Area of $\triangle B$ – Area of $\triangle A = 21$

$x - \frac{9}{16}x = 21$

$\frac{16-9}{16}x = 21$

$7x = 336$
$x = 48$
$\therefore$ Area of $\triangle B = x = 48$ in^2 //
$\therefore$ Area of $\triangle A = \frac{9}{16}x$
$= \frac{9}{16} \times 48$
$= 27$ in^2 //

15. Ms Shelby wants to replace the carpets in room A and room B (see Figure 5.14). It is known that the rooms, A and B are similar triangles and the area of room B is 180 square feet. If the carpeting costs \$7 per square feet, how much will it cost Ms Shelby to carpet the 2 rooms?

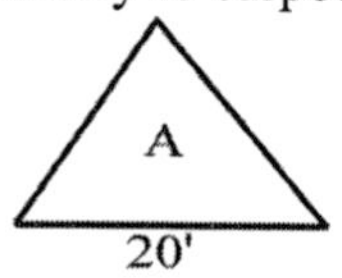

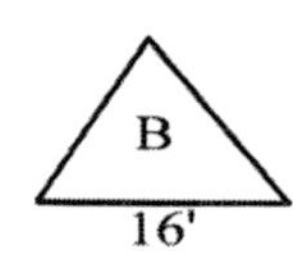

Figure 5. 14

Answer:
Given $\triangle A$ and $\triangle B$ = similar triangles
Also given area of $\triangle B = 180$
Ratio of similar triangles' areas and sides:

$\frac{\text{Area of } \Delta A}{\text{Area of } \Delta B} = \left(\frac{\text{Side of } \Delta A}{\text{Side of } \Delta B}\right)^2$

$$\frac{\text{Area of } \Delta A}{180} = \left(\frac{20}{16}\right)^2$$

$$\text{Area of } \triangle A = \frac{20^2}{16^2} \times 180$$
$$= \frac{400}{256} \times 180$$
$$= 281.25 \text{ feet}^2$$

Sum of areas:
= Area of $\triangle A$ + Area of $\triangle B$
= 281.25 + 180
= 461.25 feet^2
Given carpeting cost = \$7 per square feet
Cost to carpet room A and room B:
= Sum of areas × Carpeting cost
= 461.25 × 7
= \$3228.75

∴ Ms Shelby will have to pay \$3228.75 to carpet her two rooms. //

16. The areas of 2 similar hexagons, A and B, are 64 and 16 units^2. If a side of hexagon B is 5 units, what is the length of the corresponding side in hexagon A?

Answer:
Let x = corresponding side of hexagon A
Given areas:
Hexagon A = 64
Hexagon B = 16
Ratio of similar hexagons' areas and sides:

$$\frac{\text{Area of hexagon A}}{\text{Area of hexagon B}} = \left(\frac{\text{side of hexagon A}}{\text{side of hexagon B}}\right)^2$$

$$\frac{64}{16} = \left(\frac{x}{5}\right)^2$$

$$4 = \frac{x^2}{5^2}$$

$$4 = \frac{x^2}{25}$$

$$x^2 = 4 \times 25$$
$$= 100$$
$$x = \sqrt{100}$$

∴ x = 10 units
∴ Length of corresponding side of hexagon A is 10 units. //

17. Express the ratio of the sum of internal angles in a triangle to the sum of internal angles in a quadrilateral.

Answer:
Sum of internal angles:
Triangle = 180°
Quadrilateral = 360°
Let,
a = sum of internal angles in triangle
b = sum of internal angles in quadrilateral
Thus,
a : b = 180°: 360°
= 1 : 2 //

18. Figure 5.15 shows two similar triangles. Determine the following ratios in the form a : b.

a) $\frac{AB}{PQ}$

b) ∠ACB : ∠PRQ

c) BC ÷ QR

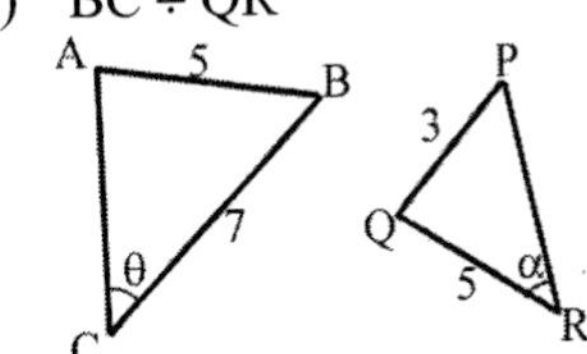

Figure 5. 15

Answers:
a)
Given AB = 5
PQ = 3

$\frac{AB}{PQ} = \frac{5}{3}$

$= 5:3$ //

b)
Given $\angle ACB = \theta$
$\angle PRQ = \alpha$
$\angle ACB : \angle PRQ = \theta : \alpha$ //

c)
Given BC = 7
QR = 5
$BC \div QR = 7 \div 5$
$= 7:5$ //

19. A map has a scale of 1:120. What is the actual length represented by a drawn length of 3 inches? Express answer in feet. (Note: 12 inches = 1 foot)

Answer:
Actual length = drawn length × 120
= 3 × 120
= 360 inches
= 360 ÷ 12
= 30 feet //

20. A map has a scale of 1:3600. The dimension of a rectangular patch of farmland (drawn to scale) on the map is 2″ × 3″. What is the actual dimension of the patch of farmland in yards?

Answer:
Given scale = 1:3600
Dimension of farmland on map = 2″ × 3″
=> width = 2
=> length = 3
Actual width:
= drawn width × 3600
= 2 × 3600
= 7200″
= 7200 ÷ 36
= 200 yards
Actual length:
= drawn length × 3600
= 3 × 3600
= 10800″
= 10800 ÷ 36
= 300 yards
∴ Actual dimension of farmland
= 200 yards × 300 yards //

21. Sum of length of lines, P, Q, and R is 45 inches. Also given, length of line P is 15 inches and length of line R is twice the length of line Q. What is the ratio of lines P, Q, and R?

Answer:
Given P + Q + R = 45
Also given P = 15
Let x = length of line Q
=> R = 2x
Sum of lines Q and R:
x + 2x = 45 − 15
3x = 30
x = 10
Thus,
Length of line Q = x
= 10 inches
Length of line R = 2x
= 2 × 10
= 20 inches
Therefore, ratio $P:Q:R = 15:10:20$
$= 3:2:4$ //

22. A rope whose length is 36 inches was divided into 3 segments; p, q and r in the ratio of 3:4:2. What is the length of segment p?

Answer:
Given total length of rope = 36
$p:q:r = 3:4:2$
Length of segment p:
$= \frac{36}{3+4+2} \times 3$
$= \frac{36}{9} \times 3$
$= 4 \times 3$
= 12 inches //

23. In a circle there are 3 chords, a, b, and c. If $a:b=3:4$ and $b:c=1:2$, find the ratio $a:b:c$.

Answer:
Given $a:b=3:4$
$b:c=1:2$
=> $b:c=4:8$ ⇦ Multiply by 4
Thus,
$a:b:c = 3:4:8$ //

24. Given triangle, ABC, whose sides are represented by $AB:BC:AC=3:5:7$. It is further known that AB = 12 inches. What is the perimeter of triangle ABC?

Answer:
Given $AB:BC:AC=3:5:7$
AB = 12
Perimeter $\triangle ABC$:
$= \frac{3+5+7}{3} \times 12$
= 60 inches //

25. Figure 5.16, shows 2 circles, O and P, whose radii are 4 and 2 units respectively. Find the ratio of their:
a) Radii
b) Areas

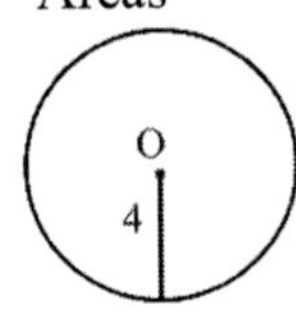

Figure 5. 16

Answers:
Given radius of circle O = 4
Radius of circle P = 2
Let r_O = radius of circle O
r_P = radius of circle P

a)
Ratio of circle O and P radii:
$r_O : r_P = 4:2$
$= 2:1$ //

b)
Area of circle O, A_O:
$A_O = \pi(r_O)^2$
$= \pi(4)^2$
$= 16\pi$
Area of circle P, A_P:
$A_P = \pi(r_P)^2$
$= \pi(2)^2$
$= 4\pi$
Ratio of circle O and P areas:
$A_O : A_P = 16\pi : 4\pi$
$= 4:1$ //

Alternatively:
Area of circle O : Area of circle P
$= (r_O)^2 : (r_P)^2$
$= 2^2 : 1^2$ ⇦ Substitute part (a)
$= 4:1$ //

26. Three containers contain a total of 27 quarts of water. If their ratio is $A:B:C=2:5:k$ and container B contains 15 quarts of water, find the value of k.

Answer:
Given $A:B:C = 2:5:k$
Also given $A + B + C = 27$
$B = 15$
Thus total volume of water in A, B and C:

$$\frac{2+5+k}{5} \times 15 = 27$$

$(7 + k) \times 3 = 27$
$7 + k = 9$
$\therefore k = 2$ //

27. Kyle parked his car between Murray's and Jean's. The 3 cars are parked collinear and the distance between Kyle's car to Jean's car is 25 feet. If the ratio of Kyle's car to Murray's car and Jean's car is $3:5$, find the distance between Kyle's car and Murray's car.

Answer:
Let m = distance between Kyle's and Murray's cars

m — 25 ft
Murray's car — Kyle's car — Jean's car

Given, all 3 cars are parked collinear
Distance between Kyle's car and Jean's car = 25
Ratio of Kyle's to Murray's car : Kyle's to Jean's car = $3:5 = \frac{3}{5}$

Thus,

$$m = \frac{3}{5} \times 25$$
$$= 3 \times 5$$
$$= 15 \text{ feet}$$

$\therefore$ Distance from Kyle's car to Murray's car is 15 feet. //

28. An engineer has carefully drawn an enlarged chipset to scale. If the scale is $1:0.05$ and the actual length of the chipset is 3.1 cm, determine the length of the enlarged chipset drawing.

Answer:
Given scale = $1:0.05$
Also given, actual length = 3.1
Thus,
Actual length = drawn length × scale
3.1 = drawn length × 0.05

$$\text{Drawn length} = \frac{3.1}{0.05}$$

$\therefore$ Drawn length = 62 cm
$\therefore$ Drawn length of the chipset is 62 cm. //

29. A and B are similar triangles. However the lengths of their sides are inversely related. If a side of $\triangle A$ is 4 inches when the corresponding side of $\triangle B$ is 9 inches, find the length of side $\triangle B$ when the corresponding side of $\triangle A$ is 12 inches.

Answer:
Given $A \propto \frac{1}{B}$
Let x = length of B_2
Since $A_1 = 4$ when $B_1 = 9$, for $A_2 = 12$:

$$\frac{\text{Length of } A_1}{\text{Length of } A_2} = \frac{\text{Length of } B_2}{\text{Length of } B_1}$$

$$\frac{4}{12} = \frac{x}{9}$$

$$x = \frac{4}{12} \times 9$$

x = 3 inches
$\therefore$ When side of $\triangle A$ is 12 inches, corresponding side of $\triangle B$ is 3 inches. //

30. A sequoia tree casts a 120 feet shadow. At the same time, a nearby redwood tree that is 350 feet tall casts a 140 feet shadow. If both trees can be presumed to stand 90° from ground level, what is the height of the sequoia tree?

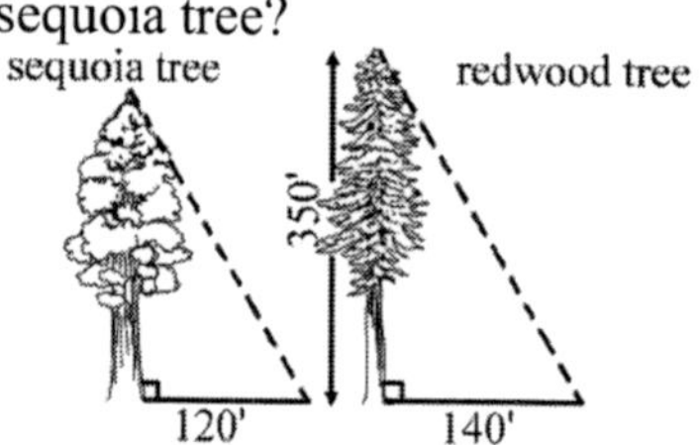

Figure 5. 17

Answer:
Let h = height of sequoia tree
Since the shadows are measured simultaneously and trees are near to each other, the sun ray that touches the peak of the trees is at congruent angles.
Thus,

$$\frac{\text{Height of sequoia}}{\text{Shadow of sequoia}} = \frac{\text{Height of redwood}}{\text{Shadow of redwood}}$$

$$\frac{h}{120} = \frac{350}{140}$$

$$h = \frac{350}{140} \times 120$$

$$= 300 \text{ feet}$$

$\therefore$ Sequoia tree is 300 feet tall. //

31. Find the geometric mean for the following pairs:
a) 4 and 9
b) 5 and 12
c) (a + b) and (a – b)

Answers:
a)Let p = geometric mean

$$4 : p = p : 9$$

$$\frac{4}{p} = \frac{p}{9}$$

$$p^2 = 4 \times 9$$

$$= 36$$

$$p = \sqrt{36}$$

$$\therefore p = 6$$

$\therefore$ Geometric mean of 4 and 9 is 6. //

Alternatively:

$$p = \sqrt{4 \times 9}$$

$$= \sqrt{36}$$

$$= 6 \text{ //}$$

b) Let q = geometric mean

$$5 : q = q : 12$$

$$\frac{5}{q} = \frac{q}{12}$$

$$q^2 = 5 \times 12$$

$$= 60$$

$$q = \sqrt{60} \text{ or } 7.7460 \text{ //}$$

Alternatively:

$$q = \sqrt{5 \times 12}$$

$$= \sqrt{60} \text{ or } 7.7460 \text{ //}$$

c) Let r = geometric mean

$$(a + b) : r = r : (a - b)$$

$$\frac{a+b}{r} = \frac{r}{a-b}$$

$$r^2 = (a + b)(a - b)$$

$$= a^2 - b^2$$

$$r = \sqrt{a^2 - b^2} \text{ //}$$

Alternatively:

$$r = \sqrt{(a+b) \times (a-b)}$$

$$= \sqrt{a^2 - b^2} \text{ //}$$

32. The areas of two rectangular banners A and B are inversely related, as shown in Figure 5.18.

A	6	12
B	9	m

Figure 5. 18

Find the area of banner B when the area of banner A is 12 feet^2.

Answer:

Given $A \propto \frac{1}{B}$ ⇦ Inverse proportion

Since $A_1 = 6$ when $B_1 = 9$, thus $A_2 = 12$:

$$\frac{A_1}{A_2} = \frac{B_2}{B_1}$$

$$\frac{6}{12} = \frac{m}{9}$$

$$m = \frac{6}{12} \times 9$$

$\therefore$ m = 4.5 feet^2

$\therefore$ When area of banner A is 12 feet^2, area of banner B is 4.5 feet^2. //

33. The content of two barrels of wine, V and W is directly related. When barrel V is 12 gal., barrel W is 20 gal. How much wine does barrel V contains when barrel W has 38 gal.?

Answer:

Given $V \propto W$ ⇦Direct proportion

$$=> \frac{V_1}{V_2} = \frac{W_1}{W_2}$$

Since $V_1 = 12$, when $W_1 = 20$. For $W_2 = 38$:

$$\frac{12}{V_2} = \frac{20}{38}$$

$$V_2 = \frac{38}{20} \times 12$$

$$= 22.8 \text{ gal.}$$

$\therefore$ When W contains 38 gal., V will have 22.8 gal. of wine. //

34. Figure 5.19 shows the variable length of 3 line segments that can be represented by the following relationship:

$A \propto BC$

A	16	24
B	4	2
C	p	6

Figure 5. 19

Find the value of p.

Answer:

Given $A \propto BC$ ⇦Direct proportion

$$=> \frac{A_1}{A_2} = \frac{B_1C_1}{B_2C_2}$$

Since $A_2 = 24$ when $B_2 = 2$ and $C_2 = 6$. For $A_1 = 16$ and $B_1 = 4$:

$$\frac{16}{24} = \frac{4 \times p}{2 \times 6}$$

$$\frac{16}{24} = \frac{4p}{12}$$

$$p = \frac{16}{24} \times \frac{12}{4}$$

$\therefore$ p = 2 units //

35. P, Q, and R are variables. If P is directly related to Q and inversely related to R, find the value of m in Figure 5.20.

P	6	10
Q	12	5
R	8	m

Figure 5. 20

Answer:

Given $P \propto \frac{Q}{R}$

$$=> \frac{P_1}{P_2} = \frac{Q_1R_2}{R_1Q_2}$$

Since $P_1 = 6$ when $Q_1 = 12$ and $R_1 = 8$. For $P_2 = 10$ and $Q_2 = 5$:

$$\frac{6}{10} = \frac{12 \times m}{8 \times 5}$$

$$\frac{6}{10} = \frac{12m}{40}$$

$$m = \frac{6}{10} \times \frac{40}{12}$$

$\therefore$ m = 2 //

36. The length of two lines, A and B, can be represented by $A \propto \frac{1}{B^2}$. When A = 4 ″, B = 2 ″. Find length of B when A = 0.01 ″.

Answer:

$$\frac{A_1}{A_2} = \frac{B_2^{\ 2}}{B_1^{\ 2}}$$

$$\frac{4}{0.01} = \frac{B_2^{\ 2}}{2^2}$$

$$B_2^{\ 2} = \frac{4}{0.01} \times 4 = 1600$$

$$B_2 = \sqrt{1600} = 40$$

$\therefore$ When A is 0.01 ″ then B is 40 ″. //

Chapter 6
2D – Perimeters & Areas

Perimeter – circumference or outer boundary of a closed figure

Area – measure of a surface. Units of measurements are expressed in squares.

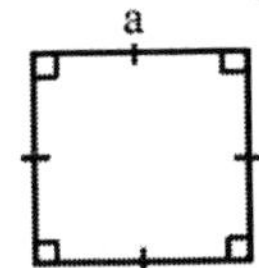

Square, Perimeter = 4a

Area = a × a = a^2

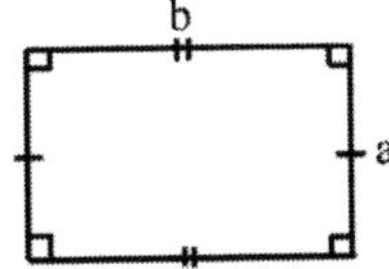

Rectangle, Perimeter = 2b + 2a

Area = a × b

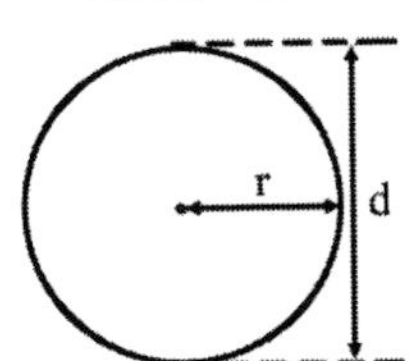

Circle, Perimeter = $2\pi r = \pi d$

Area = πr^2

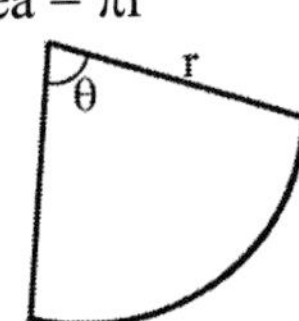

Sector, Perimeter = $\frac{\theta}{360^\circ} \times 2\pi r + 2r$

Area = $\frac{\theta}{360^\circ} \times \pi r^2$

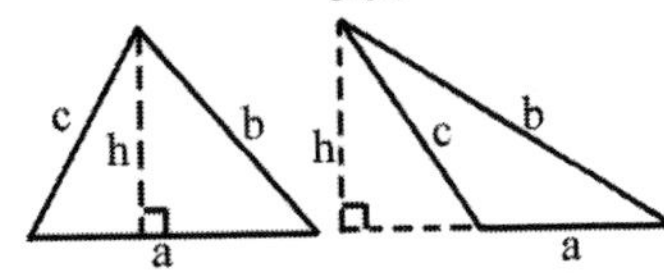

Triangle, Perimeter = a + b + c

Area = $\frac{1}{2} \times a \times h$

Heron's Formula:
(To find area of non-right angle triangles)

Semi perimeter, $s = \frac{a+b+c}{2}$

Area = $\sqrt{s(s-a)(s-b)(s-c)}$

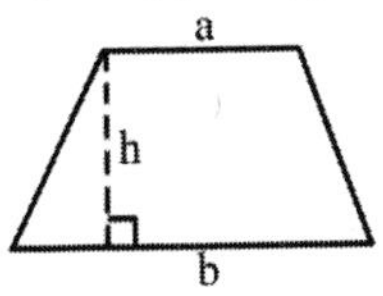

Trapezoid, Area = $\frac{1}{2} \times h \times (a+b)$

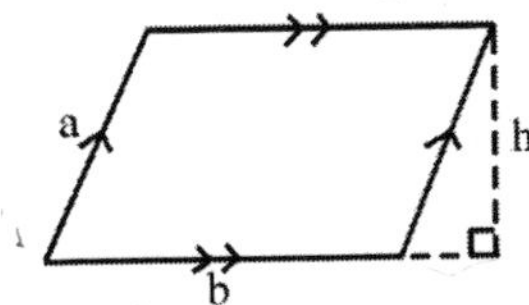

Parallelogram, Perimeter = 2a + 2b

Area = b × h

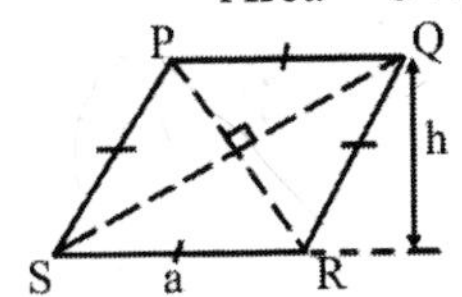

Rhombus, Perimeter = 4a

Area = $\frac{1}{2} \times PR \times QS = a \times h$

For 30°–60°–90° triangle, ratio of sides $x : x\sqrt{3} : 2x$ and $\frac{x}{2} : x\frac{\sqrt{3}}{2} : x$ are the same.

1. In Figure 6.1, △ABF and △DEF are right triangles, while BCDF is a square. Hence find the perimeter of the diagram ABCDEF.

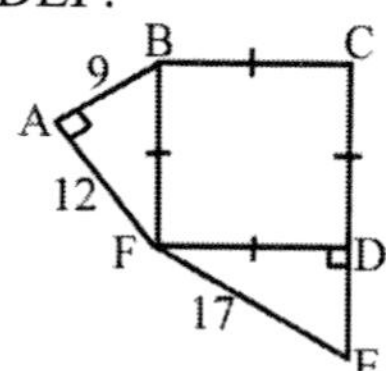

Figure 6. 1

Answer:
Given AB = 9
AF = 12
Using Pythagorean Theorem:
$BF^2 = AB^2 + AF^2$
$= 9^2 + 12^2$
$= 81 + 144$
$= 225$
$BF = \sqrt{225}$
= 15 units

Given BCDF = square
=> All sides are equilateral
BF = BC = CD = DF = 15
Also given EF = 17
Using Pythagorean Theorem:
$EF^2 = DE^2 + DF^2$
$17^2 = DE^2 + 15^2$
$289 = DE^2 + 225$
$DE^2 = 289 - 225 = 64$
$DE = \sqrt{64} = 8$ units

∴ Perimeter of diagram ABCDEF:
= AB + BC + CD + DE + EF + AF
= 9 + 15 + 15 + 8 + 17 + 12
= 76 units //

2. Figure 6.2, shows a square, ABCD. $\overline{BD}$ is the diagonal and BD = $\sqrt{32}$ in. Find the perimeter of the square.

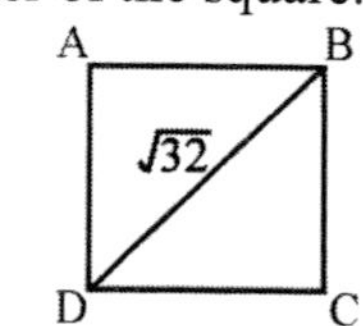

Figure 6. 2

Answer:
Given BD = $\sqrt{32}$
Also given ABCD = square
Let x = AB
=> AB = BC = CD = AD = x
Using Pythagorean Theorem:
$BD^2 = AB^2 + AD^2$
$(\sqrt{32})^2 = x^2 + x^2$
$32 = 2x^2$
$x^2 = 16$
x = 4 inches

Perimeter of square, ABCD:
$= 4 \times x$
$= 4 \times 4$
= 16 inches //

Alternatively:
Since diagonal $\overline{BD}$, of a square is an angle bisector:
=> △ABD and △BCD = 45°–45°–90° triangles:

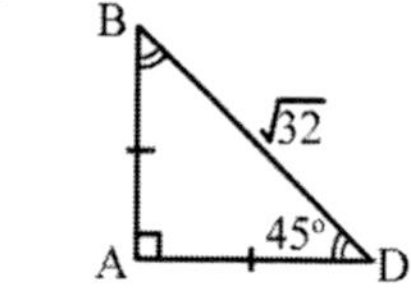

Let x = AB = AD

$BD = x\sqrt{2}$

$\sqrt{32} = x\sqrt{2}$

$x = \dfrac{\sqrt{32}}{\sqrt{2}} = \sqrt{\dfrac{32}{2}}$

$= \sqrt{16} = 4$ units

$\therefore$ Perimeter of ABCD = 4 × 4 = 16 units //

3. In Figure 6.3, ABDC is an isosceles trapezoid. Find the height of the trapezoid if its bases are 10 and 18 units. Subsequently find the perimeter of trapezoid, ABDC.

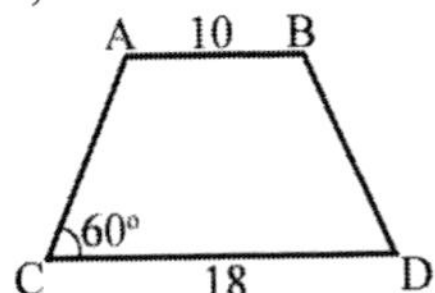

Figure 6. 3

Answers:

Let h = height of ABDC

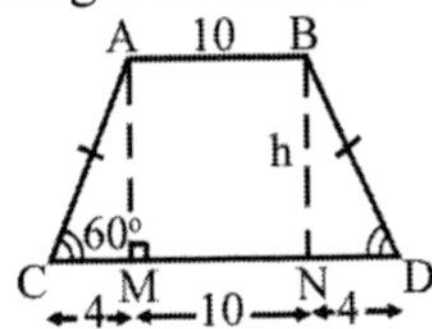

Given $\angle ACD = 60°$

Also given ABDC = isosceles trapezoid

=> AC = BD

=> $\triangle ACM = \triangle BDN = 30°–60°–90° \triangle$

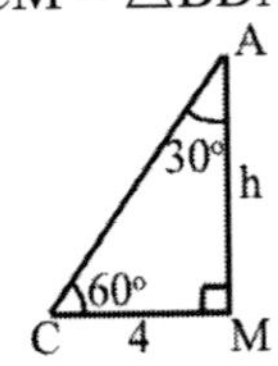

Since CM = 4

Let a = hypotenuse, AC

$CM = \dfrac{a}{2}$

$4 = \dfrac{a}{2}$

a = 8 ⇦ Since AC = a and BD = AC, BD = 8

$AM = \dfrac{\sqrt{3}}{2} a$

$h = \dfrac{\sqrt{3}}{2} \times 8$

$= 4\sqrt{3}$ units or 6.9282 units //

Perimeter of trapezoid, ABDC:

= AB + BD + CD + AC

= 10 + 8 + 18 + 8

= 44 units //

4. In Figure 6.4, PQRS is a rhombus. $\overline{PT}$ is the altitude and PT = $\sqrt{12}$ cm. Find the perimeter of PQRS.

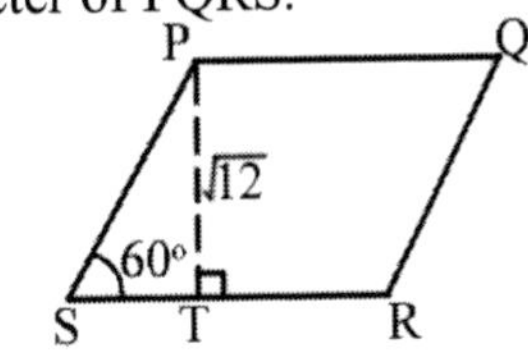

Figure 6. 4

Answer:

Given PQRS = rhombus

and $\angle PSR = 60°$

=> $\triangle PST = 30°–60°–90°$ triangle

Let PS = a

$PT = \dfrac{\sqrt{3}}{2} \times a$

$\sqrt{12} = \dfrac{\sqrt{3}}{2} \times a$

$a = \sqrt{12} \times \dfrac{2}{\sqrt{3}}$

$= \sqrt{\dfrac{12}{3}} \times 2$

$= \sqrt{4} \times 2$

$= 2 \times 2 = 4$ units

$\therefore$ Perimeter of rhombus, PQRS:

$= 4 \times a$

$= 4 \times 4$
$= 16$ units //

5. Find the circumference of circle O if its:
a) Diameter is 8 yards
b) Radius is 3 inches

Answers:
a)
Given diameter = 8
Circumference:
$= \pi \times \text{diameter}$
$= \frac{22}{7} \times 8$
$= 25.1429$ yards //

b)
Given radius, r = 3
Circumference:
$= \pi \times 2r$
$= \frac{22}{7} \times 2 \times 3$
$= 18.8571$ inches //

6. Given circle O, whose radius is 5 units. What is the arc length whose degree measure is 35°?

Answer:
Given radius, r = 5
And central angle, $\theta = 35°$
Length of arc:
$= \frac{\theta}{360°} \times 2\pi r$
$= \frac{35°}{360°} \times 2 \times \frac{22}{7} \times 5$
$= 3.0556$ units //

7. In a circle whose radius is 15 inches, find the length of arc whose sector measure is 2 rad.

Answer:
Given $\theta = 2$ rad
And radius, r = 15
Length of arc:
$= \theta \times r$
$= 2 \times 15$
$= 30$ inches //

Alternatively:
Changing radians to degrees:
Sector's angle $= \theta \times \frac{180°}{\pi}$
$= 2 \times \frac{180°}{\frac{22}{7}}$
$= 114.55°$
Thus,
Length of arc:
$= \frac{\text{Sector's angle}}{360°} \times 2\pi r$
$= \frac{114.55°}{360°} \times 2 \times \pi \times 15$
$= 30$ inches //

8. In a circle whose circumference is 21 inches, what is the length of arc whose degree measure is 120°?

Answer:
Let r = radius
Given circumference, $2\pi r = 21$ ...(1)
Degree measure of arc, $\theta = 120°$ ...(2)
Length of arc:
$= \frac{\theta}{360°} \times 2\pi r$
Substitute (1) and (2):

$= \frac{120°}{360°} \times 21$
= 7 inches //

9. A circular lens is inscribed in a rectangular metal disc as shown in Figure 6.5. The diagonal of the rectangular disc is 25 cm and its width is 24 cm. What is the circumference of the circular lens?

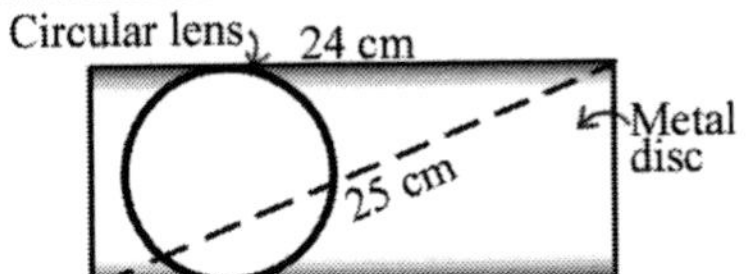

Figure 6. 5

Answer:
Let h = height of metal disc
Given width, w = 24
And diagonal, d = 25
Using Pythagorean Theorem:
$d^2 = w^2 + h^2$
$25^2 = 24^2 + h^2$
$625 = 576 + h^2$
$h^2 = 625 - 576 = 49$
$h = \sqrt{49}$
= 7 cm
Since circular lens is inscribed in rectangular disc
=> Diameter of circular lens = h
Thus,
Circumference of circular lens:
$= \pi \times h$
$= \frac{22}{7} \times 7$
= 22 cm
∴ Circumference of circular lens is 22 cm.//

10. What is the circumference of a pizza that is baked using a 12 ″ pan?

Answer:
Given diameter of pan, d = 12
=> Diameter of pizza baked = 12
Circumference of pizza:
$= \pi \times d$
$= \frac{22}{7} \times 12$
= 37. 7143 ″
∴ Circumference of pizza is 37.7143 ″. //

11. The largest bowl made out of poplar wood has a diameter of 10 feet. What is its circumference?

Answer:
Given diameter, d = 10
Circumference of bowl:
$= \pi \times d$
$= \frac{22}{7} \times 10$
= 31.4286 feet
∴ Circumference of bowl is 31.4286 feet. //

12. Determine the length of the fence around the children's playground shown in Figure 6.6.

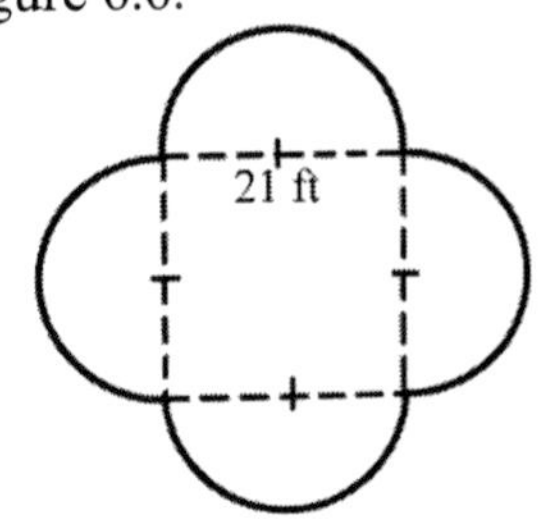

Figure 6. 6

Answer:
There are 2 identical circles
=> 2 circumferences
Given diameter of semicircle = 21
Perimeter of playground:
$= 2 \times \pi d$
$= 2 \times \frac{22}{7} \times 21$
= 132 feet
∴ Length of fence around the children's playground is 132 feet. //

13. Figure 6.7 shows the arrangement of U.S quarter dollars to form the outer frame of a photo frame. Find the perimeter of the picture frame, if the diameter of each quarter is 1 inch.

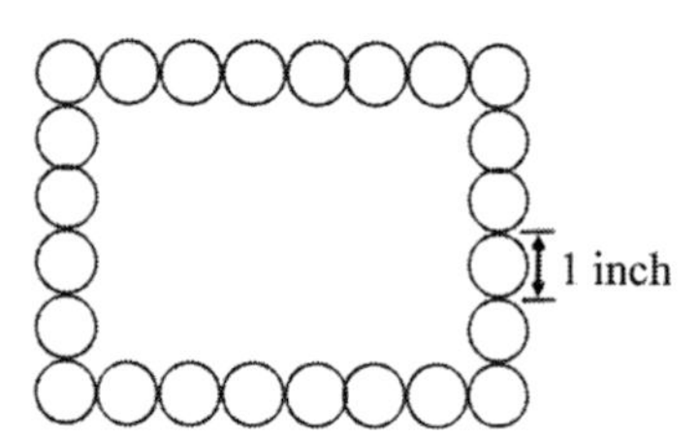

Figure 6. 7

Answer:
Given diameter of quarter = 1

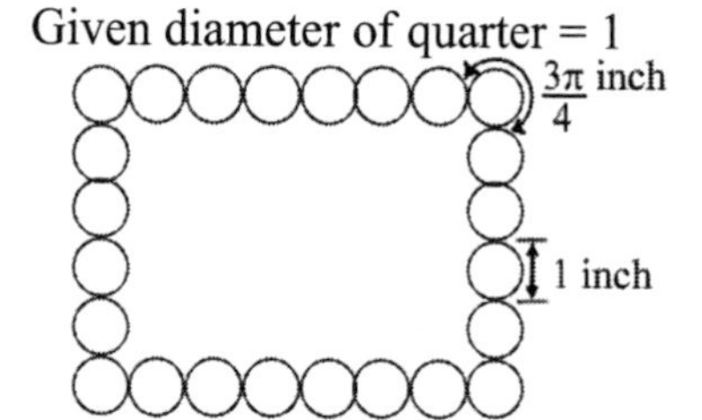

Perimeter of photo frame:

$= 20 \times \frac{1}{2} \times \pi \times 1 + 4 \times \frac{3}{4} \times \pi \times 1$

$= 10\pi + 3\pi$

$= 13 \times \frac{22}{7}$

$= 40.8571$ inches //

14. Figure 6.8, shows two overlapping regular polygons, A and B. If the sides of both polygons are 2 units, what is the perimeter of the shaded area?

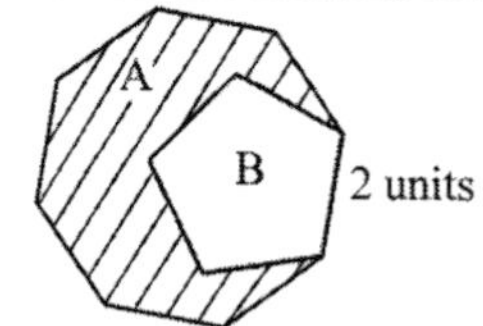

Figure 6. 8

Answer:
Let N = number of sides
Given A = regular octagon
=> $N_A = 8$
Also given B = regular pentagon
=> $N_B = 5$
Since sides of polygons are 2 units:
Perimeter of shaded area:
$= (7 + 4) \times 2$
$= 22$ units //

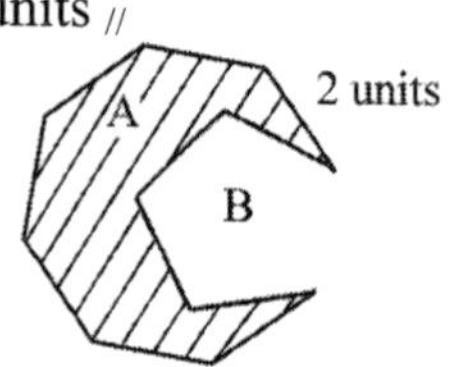

15. In Figure 6.9, ABCD is a rhombus while $\triangle CDE$ and $\triangle DEF$ are right triangles. Hence find the perimeter of Figure 6.9.

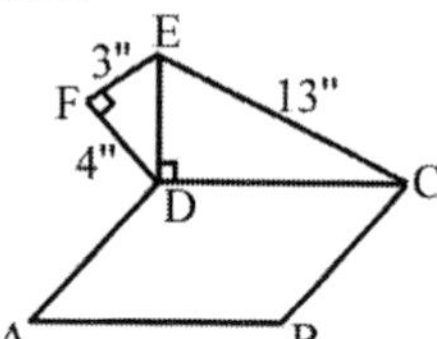

Figure 6. 9

Answer:
Given EF = 3
DF = 4
Using Pythagorean Theorem:

$DE^2 = EF^2 + DF^2$
$= 3^2 + 4^2$
$= 9 + 16$
$= 25$

$DE = \sqrt{25}$
$= 5''$

Also given CE = 13
Using Pythagorean Theorem:

$CE^2 = CD^2 + DE^2$
$13^2 = CD^2 + 5^2$
$CD^2 = 169 - 25$
$= 144$
$CD = \sqrt{144} = 12''$

Since ABCD = rhombus
=> CD = BC = AB = AD = 12
Perimeter of ABCEFD:
= AB + BC + CE + EF + DF + AD
= 12 + 12 + 13 + 3 + 4 + 12

$= 56''$ //

16. Georgina wraps a gift for her mom's birthday. The gift is a rectangular box as shown in Figure 6.10. If the knot and bow use 30″ of the ribbon and there is a 2″ overlap at the top and bottom of the box, what is the length of the ribbon used?

Figure 6. 10

Answer:
Given ribbon length:
Let a = Knot and bow = 30
b = Overlap top of box = 2
c = Overlap bottom of box = 2
Length of ribbon used:
$= 4 \times \text{height} + 2 \times \text{length} + 2 \times \text{width} + a + b + c$
$= 4 \times 15 + 2 \times 20 + 2 \times 10 + 30 + 2 + 2$
$= 60 + 40 + 20 + 34$
$= 154$ in //
∴ Length of ribbon used is 154 inches. //

17. Figure 6.11 is an isosceles trapezoid whose bases are 20″ and 30″. If the legs of the trapezoid are 13″, find the median and altitude of the trapezoid. Subsequently find the area of the top half of the trapezoid.

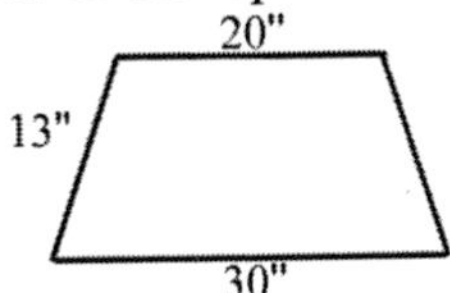

Figure 6. 11

Answers:
Let h = altitude
m = median

$$\text{Median} = \frac{1}{2} \times (\text{sum of bases})$$
$$= \frac{1}{2} \times (20 + 30)$$
$$= \frac{1}{2} \times 50$$
$$= 25''$$

∴ Median of trapezoid is 25″. //

Using Pythagorean Theorem:
$13^2 = h^2 + 5^2$
$169 = h^2 + 25$
$h^2 = 169 - 25$
$= 144$
$h = \sqrt{144}$
$= 12''$
∴ Altitude of trapezoid is 12″. //

Area of top half of trapezoid:
$$= \frac{1}{2} \times (\text{top base} + \text{median}) \times (\frac{1}{2} \times \text{altitude})$$
$$= \frac{1}{2} \times (20 + 25) \times (\frac{1}{2} \times 12)$$
$$= 135 \text{ in}^2$$
∴ Area of top half of trapezoid is 135 in². //

18. Figure 6.12, shows circle O whose radius is 5 units. Rectangle PQRS is inscribed in circle O. It is known that QR = 6 units, what is the area of rectangle PQRS?

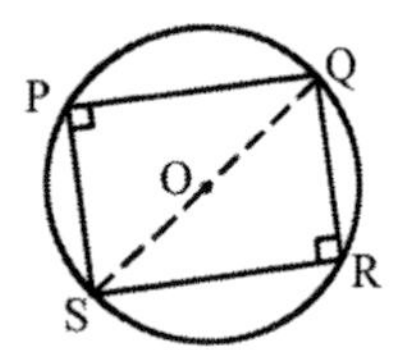

Figure 6. 12

Answer:
Given circle O's radius, r = 5
=> OS = OQ = 5
=> Diameter = QS
QS = OS + OQ
= 5 + 5 = 10
Given PQRS = inscribed rectangle
=> ∠QPS = ∠QRS = 90°
Using Pythagorean Theorem:
$QS^2 = QR^2 + RS^2$
$10^2 = 6^2 + RS^2$
$100 = 36 + RS^2$
$RS^2 = 100 - 36 = 64$
$RS = \sqrt{64} = 8$ units
Area of rectangle PQRS:
= RS × QR
= 8 × 6
= 48 $units^2$ //

19. In Figure 6.13, PQRS represents a parallelogram. Given that the square units of PQRS is 24 in^2 and the base, $\overline{RS}$ is 6 in, find the altitude of the parallelogram.

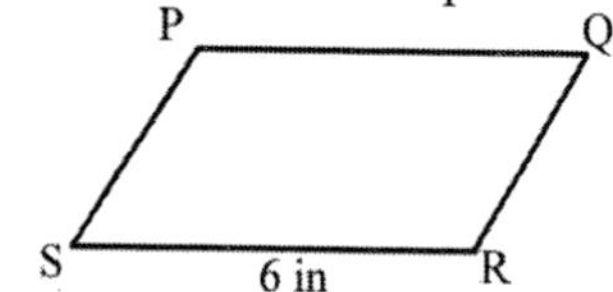

Figure 6. 13

Answer:
Given area of PQRS = 24, and RS = 6
Let h = altitude of PQRS

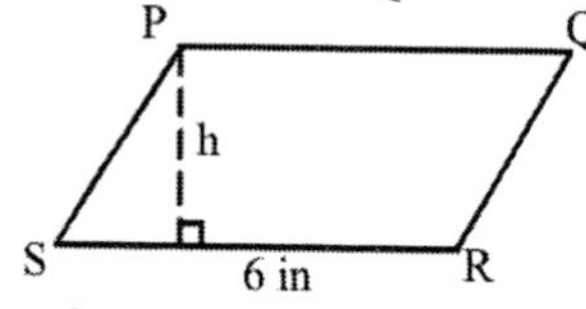

Area of parallelogram PQRS:
RS × h = 24
6 × h = 24
h = 4 in
∴ Altitude of PQRS = 4 in. //

20. In Figure 6.14, PQRS is a parallelogram. It is further given that RS = 12 yards, PS = 8 yards and ∠PSR = 60°. What is the area of the parallelogram PQRS?

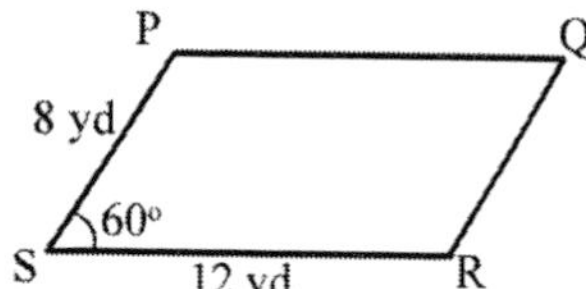

Figure 6. 14

Answer:
Given RS = 12
PS = 8
∠PSR = 60°
Let h = altitude of parallelogram
Using 30°–60°–90° triangle side ratio:

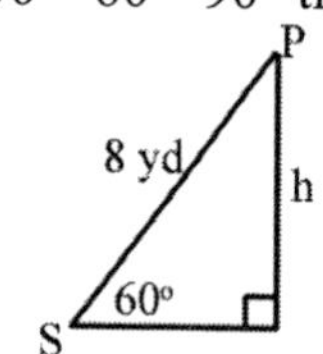

$h = \frac{\sqrt{3}}{2} \times PS$
$= \frac{\sqrt{3}}{2} \times 8$
$= 4\sqrt{3}$ or 6.9282 yd
Area of parallelogram PQRS:
= RS × h
= 12 × 6.9282
= 83.1384 yd^2 //

21. Figure 6.15 shows $\triangle ABC$, a scalene triangle. It is also given that $\angle ACB = 120°$, BC = 8 ″ and AC = 6 ″. What is the area of $\triangle ABC$?

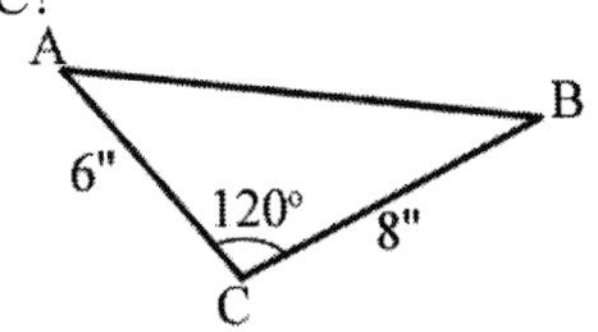

Figure 6. 15

Answer:
Given AC = 6 and BC = 8
Figure 6.15 is reproduced for labeling:

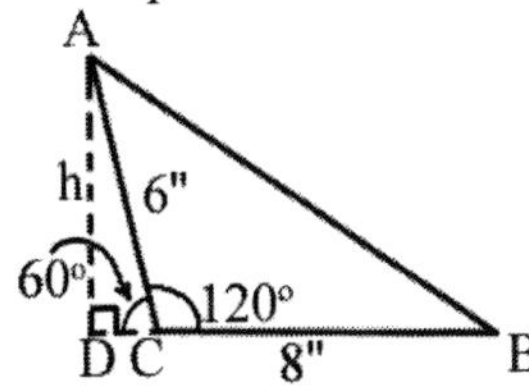

Let h = altitude of $\triangle ABC$

=> $\overline{AD} \perp \overline{BD}$

=> $\angle ADB = 90°$

Since $\angle ACB = 120°$

Sum of straight line angles = 180°

$\angle ACD + \angle ACB = 180°$

$\angle ACD = 180° - 120°$

$= 60°$

Using 30°–60°–90° triangle side ratio:

$$h = \frac{\sqrt{3}}{2} \times AC$$

$$= \frac{\sqrt{3}}{2} \times 6$$

$$= 3\sqrt{3} \text{ or } 5.1962\,''$$

Area of $\triangle ABC$:

$$= \frac{1}{2} \times h \times BC$$

$$= \frac{1}{2} \times 5.1962 \times 8$$

$$= 20.7848 \text{ inches}^2 \ //$$

22. Figure 6.16 is a trapezoid. It is further given that the altitude is 6 inches and the length of the upper base is also 6 inches. If the ratio of the upper base to the lower base is $2:3$, find the area of the trapezoid.

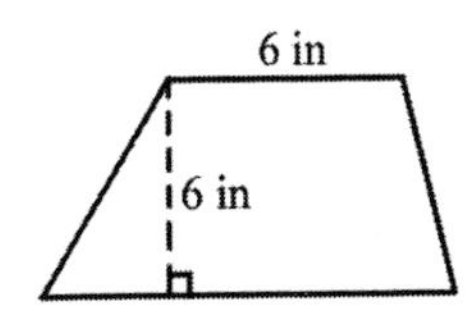

Figure 6. 16

Answer:
Given length of upper base = 6
Also given ratio,
Upper base : Lower base = $2:3$

$$\frac{\text{Upper base}}{\text{Lower base}} = \frac{2}{3}$$

$$\frac{6}{\text{Lower base}} = \frac{2}{3}$$

$$\text{Lower base} = \frac{3}{2} \times 6$$

$$= 9 \text{ inches}$$

Area of trapezoid:

$$= \frac{1}{2} \times \text{altitude} \times (\text{Upper base} + \text{Lower base})$$

$$= \frac{1}{2} \times 6 \times (6 + 9)$$

$$= 45 \text{ inches}^2 \ //$$

23. Figure 6.17 is an equilateral triangle whose sides are 8 units. Find the area of the triangle.

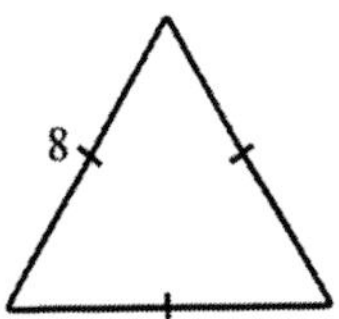

Figure 6. 17

Answer:
Given equilateral triangle
=> all sides are ≅
=> all angles are ≅
Sum of interior angles = 180°
=> each angle = 180° ÷ 3 = 60°
Let h = altitude

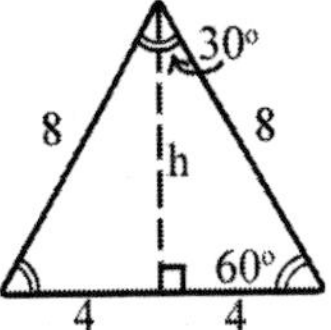

Using 30°–60°–90° triangle side ratio:

$h = \frac{\sqrt{3}}{2} \times \text{hypotenuse side}$

$= \frac{\sqrt{3}}{2} \times 8$

$= 4\sqrt{3}$ or 6.9282 units

Area of equilateral triangle:

$= \frac{1}{2} \times h \times \text{base}$

$= \frac{1}{2} \times 6.9282 \times 8$

$= 27.7128 \text{ units}^2$ //

24. The diagonals of a rhombus are 6 yd and 5 yd. What is the area of the rhombus?

Answer:
Given length of diagonals = 6 and 5
Area of rhombus, A:

$A = \frac{1}{2}(\text{diagonal} \times \text{diagonal})$

$= \frac{1}{2}(6 \times 5) = 15 \text{ yd}^2$

$\therefore$ Area of rhombus is 15 yd^2. //

25. The ratio of the diagonals in a rhombus is $1:2$. If the area of the rhombus is 9 in^2, find the length of the diagonals.

Answers:
Given ratio of diagonals = $1:2$
Let diagonals = x and 2x

Also given area of rhombus, A = 9
Thus, area of rhombus:

$\frac{1}{2}(\text{diagonal} \times \text{diagonal}) = A$

$\frac{1}{2}(x \times 2x) = 9$

$\frac{1}{2} \times 2x^2 = 9$

$x^2 = 9$

$x = \sqrt{9} = 3$

$\therefore$ Length of diagonals x = 3 in //

2x = 2 × 3 = 6 in //

26. Find the area of the shaded region in Figure 6.18.

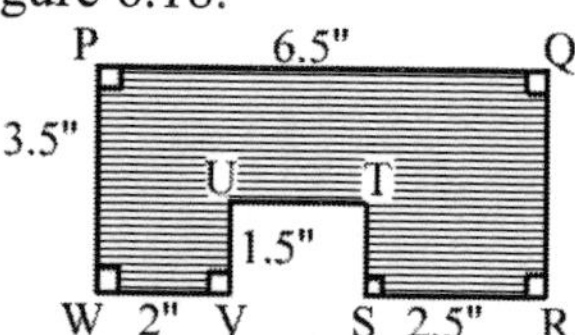

Figure 6. 18

Answer:
Using subtraction of parts method:

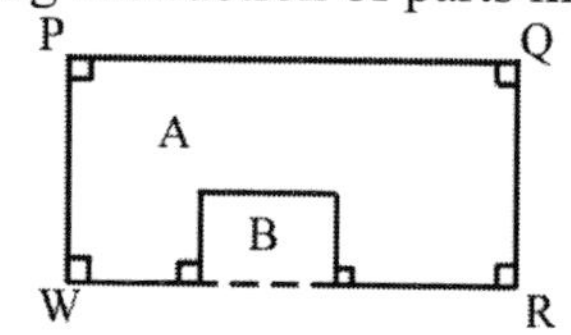

Shaded area:
= rectangle PQRW – rectangle STUV
= A – B
= 6.5 × 3.5 – 1.5 × (6.5 – 2 – 2.5)
= 22.75 – 1.5 × 2
= 22.75 – 3
= 19.75 in^2 //

Alternatively:
Using addition of parts method:

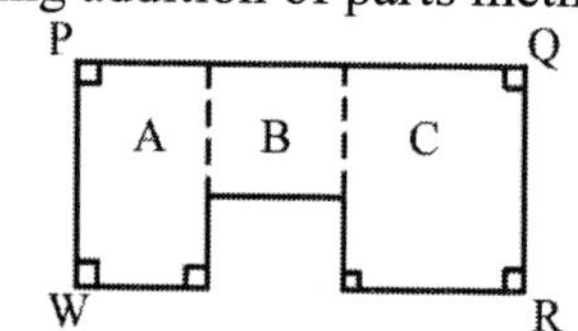

Shaded area:
= A + B + C
= 3.5 × 2 + 2 × 2 + 3.5 × 2.5
= 7 + 4 + 8.75
= 19.75 in^2 //

27. A seamstress cuts a cloth for the pocket of a pair of jeans she is sewing (see Figure 6.19). It is known that $\overline{AB}$ and $\overline{EC}$ are parallel lines, while CD = DE. The perpendicular distance from point D to $\overline{AB}$ is 13 cm. Determine the area of the pocket she is sewing. What is the minimum area of cloth she will need to make 3 pairs of the same pocket?

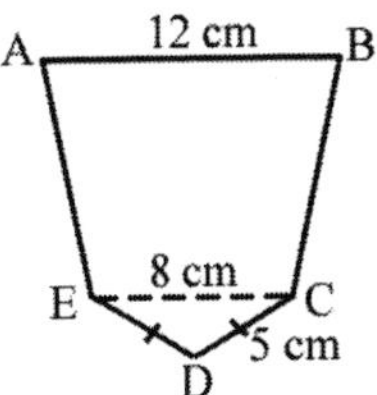

Figure 6. 19

Answers:
Figure 6.19 reproduced for labeling:

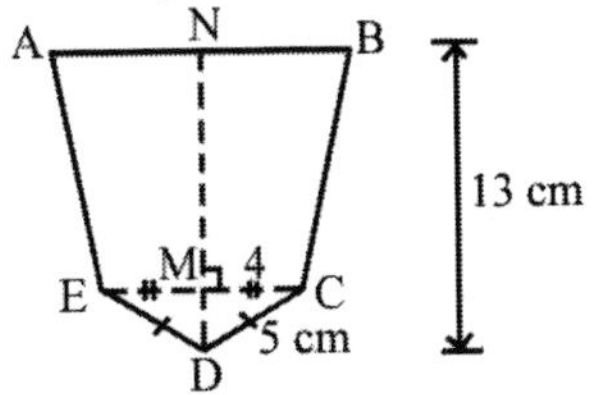

Using Pythagorean Theorem:

$CD^2 = DM^2 + CM^2$

$5^2 = DM^2 + 4^2$

$25 = DM^2 + 16$

$DM^2 = 25 - 16$

$= 9$

$DM = \sqrt{9} = 3$ cm

Area of $\triangle$CDE:

$= \frac{1}{2} \times DM \times CE$

$= \frac{1}{2} \times 3 \times 8$

$= 12 \text{ cm}^2$

Area of trapezoid ABCE:

$= \frac{1}{2} \times MN \times (AB + CE)$

$= \frac{1}{2} \times (13 - 3) \times (12 + 8)$

$= 100 \text{ cm}^2$

Area of pocket:

= Area of $\triangle$CDE + Area of ABCE

$= 12 + 100$

$= 112 \text{ cm}^2$ //

For 3 pairs of pockets, the minimum area of cloth needed:

$= 2 \times 3 \times 112$

$= 672 \text{ cm}^2$ //

28. Find the area of a regular hexagon whose sides are 6 ″ in length.

Answer:
Let N = number of sides
Given regular hexagon, N = 6
Also given length of side, s = 6
Thus,
Area of regular hexagon:

$$= \frac{N\left(\frac{s}{2}\right)^2}{\tan\left(\frac{180^\circ}{N}\right)}$$

$$= \frac{6\left(\frac{6}{2}\right)^2}{\tan\left(\frac{180^\circ}{6}\right)}$$

$$= \frac{6 \times 9}{\tan 30^\circ}$$

$$= \frac{54}{0.5774}$$

$= 93.5227 \text{ in}^2$ //

29. The diameter of a circular cake is 12″. If the cake is to be divided equally between 8 friends with no left-over, what is the area of each slice of cake?

Answer:
Given cake's diameter, d = 12

$=> \text{Radius, } r = \frac{1}{2} \times d$

$= \frac{1}{2} \times 12$

$= 6''$

Area of cake (undivided):

$= \pi r^2$

$= \frac{22}{7} \times 6^2$

$= 113.1429 \text{ in}^2$

Cake is divided into 8 equal slices.
Thus, area of each slice:

= Area of undivided cake ÷ 8

= 113.1429 ÷ 8

$= 14.1429 \text{ in}^2$ //

Alternatively:

Cake is divided into 8 equal slices

=> Angle of each sector = 360° ÷ 8
= 45°

Area of 1 slice of cake:

$= \frac{\theta}{360°} \times \pi r^2$

$= \frac{45°}{360°} \times \frac{22}{7} \times 6^2$

$= 14.1429 \text{ in}^2$ //

30. In Figure 6.20, ABC is a straight line whose length is 17 units. ABGH is a right trapezoid and BCDE is a rhombus. $\overline{AC}$ is parallel to $\overline{GF}$. Find the perimeter and area of the diagram.

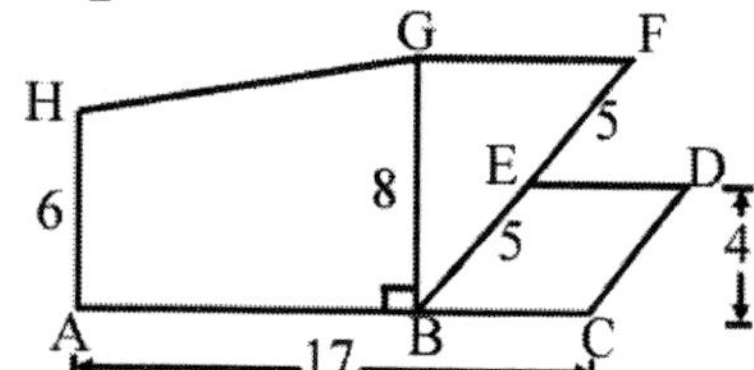

Figure 6. 20

Answers:

Given $\overline{AC} \parallel \overline{GF}$

Alternate interior angles are ≅:

$\angle ABG \cong \angle BGF = 90°$

=> △BFG = right triangle

Using Pythagorean Theorem:

$BF^2 = GF^2 + BG^2$

$(5 + 5)^2 = GF^2 + 8^2$

$100 = GF^2 + 64$

$GF^2 = 100 - 64$

$= 36$

$GF = \sqrt{36}$

= 6 units

Given BCDE = rhombus

=> All sides are equilateral

=> BC = CD = DE = BE = 5 units

Let $\overline{HI} \perp \overline{BG}$

Figure 6.20 is reproduced for labeling:

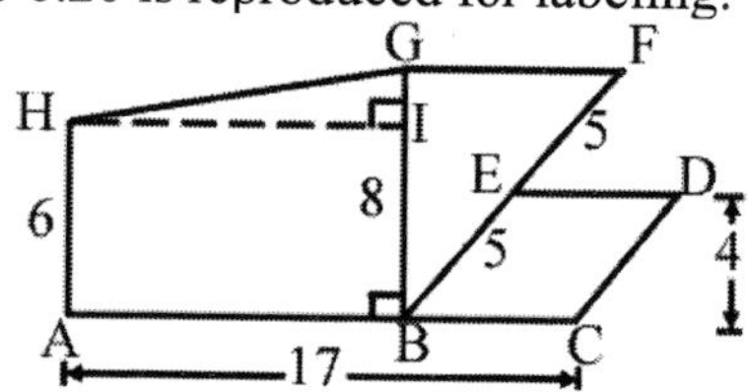

Using Pythagorean Theorem:

$GH^2 = GI^2 + HI^2$

$= (8 - 6)^2 + (17 - 5)^2$

$= 2^2 + 12^2$

$= 4 + 144$

$= 148$

$GH = \sqrt{148}$ or 12.1655 units

∴Perimeter of ACDEFGH:

= AC + CD + DE + EF + GF + GH + AH

= 17 + 5 + 5 + 5 + 6 + 12.1655 + 6

= 56.1655 units //

Area of ACDEFGH:

Using addition of parts method:

= trapezoid ABGH + △BFG + rhombus BCDE

$= \frac{1}{2} \times AB \times (AH + BG) + \frac{1}{2} \times BG \times GF +$ height × BC

$= \frac{1}{2} \times 12 \times (6 + 8) + \frac{1}{2} \times 8 \times 6 + 4 \times 5$

= 84 + 24 + 20

$= 128 \text{ units}^2$ //

31. In Figure 6.21, ABDF is a quadrilateral in the shape of a 'kite', where AB = AF and BD = DF. $\widehat{BCD}$ and $\widehat{DEF}$ are two congruent semicircles whose diameters are 5 inches. The midpoint of $\overline{BF}$ divides $\overline{DA}$ in the ratio of 1:2. What is the perimeter and area of ABCDEF?

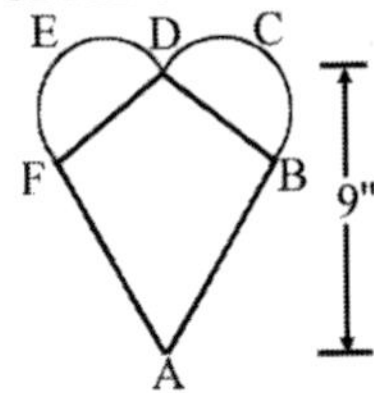

Figure 6. 21

Answers:

Let M = midpoint of $\overline{BF}$

Given $DM:MA = 1:2$

$\Rightarrow \dfrac{DM}{MA} = \dfrac{1}{2}$

$2DM = MA$ ...*

Also given DA = 9

=> DM + MA = DA

Substitute (*):

DM + 2DM = 9

3DM = 9

DM = 3 ″

=> MA = 2DM

= 2 × 3

= 6 ″

Figure 6.21 is reproduced for labeling:

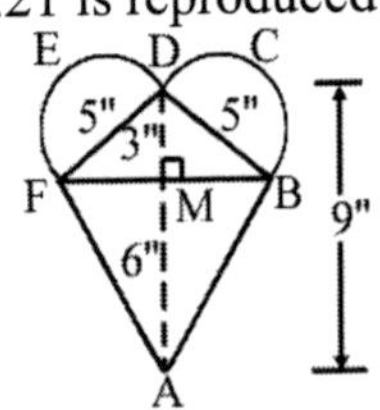

Given diameter of semicircles, d = 5

=> BD = DF = 5

Using Pythagorean Theorem:

$BD^2 = BM^2 + DM^2$

$5^2 = BM^2 + 3^2$

$25 = BM^2 + 9$

$BM^2 = 25 - 9$

$= 16$

$BM = \sqrt{16}$

$= 4''$

Again using Pythagorean Theorem:

$AB^2 = BM^2 + MA^2$

$= 4^2 + 6^2$

$= 16 + 36 = 52$

$AB = \sqrt{52}$ or 7.2111 ″

Given AB = AF

=> AF = 7.2111 ″

Since $\widehat{BCD}$ and $\widehat{DEF}$ are semicircles

=> 2 × semicircle = 1 circle

Length of $\widehat{BCD}$ + Length of $\widehat{DEF}$

$= \pi d$

$= \dfrac{22}{7} \times 5$

= 15.7143 ″

∴ Perimeter of ABCDEF:

= AB + AF + length of $\widehat{BCD}$ + length of $\widehat{DEF}$

= 7.2111 + 7.2111 + 15.7143

= 30.1365 ″ //

∴ Area of ABCDEF:

= area of 'kite' ABDF + area of circle BCDEF

$= \dfrac{1}{2} \times BF \times DA + \pi\left(\dfrac{d}{2}\right)^2$

$= \dfrac{1}{2} \times (4+4) \times 9 + \dfrac{22}{7} \times \left(\dfrac{5}{2}\right)^2$

= 36 + 19.6429

= 55.6429 in^2 //

32. In Figure 6.22, CDEF is a parallelogram and EFGH is a rhombus. What is the perimeter and area of the diagram, ABCFGHED?

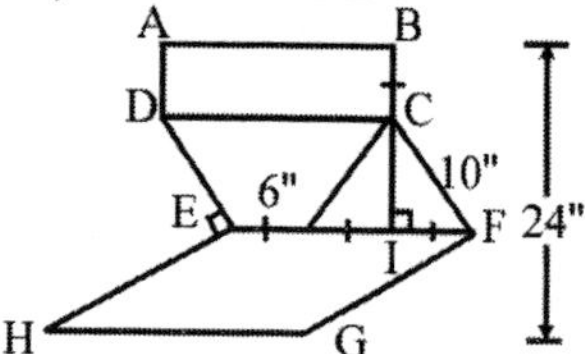

Figure 6. 22

Answers:
Given CDEF = parallelogram
=> CD = EF = 6 × 3 = 18
Given FI = 6
Given CF = 10
Using Pythagorean Theorem:
$CF^2 = CI^2 + IF^2$
$10^2 = CI^2 + 6^2$
$100 = CI^2 + 36$
$CI^2 = 100 - 36$
$= 64$
$CI = \sqrt{64}$ or 8 ″
Also given EFGH = rhombus
=> EF = FG = GH = EH = 18

∴ Perimeter of ABCFGHED:
= GH + FG + EH + CF + DE + BC + AD + AB
= 18 + 18 + 18 + 10 + 10 + 6 + 6 + 18
= 104 ″ //

∴ Area of ABCFGHED:
Using addition of parts method:
= rectangle ABCD + parallelogram CDEF + rhombus EFGH
= AB × BC + CD × CI + EF × (24 – BC – CI)
= 18 × 6 + 18 × 8 + 18 × (24 – 6 – 8)
= 108 + 144 + 180
= 432 in^2 //

33. In Figure 6.23, ABCD is a geometrical tile. ABCD and PQRS are squares and the ratio of area ABCD to area PQRS is $4:1$. It is further known that points U, V, W and X are midpoints. If the area ABCD is 144 in^2, what is the area of the shaded region?

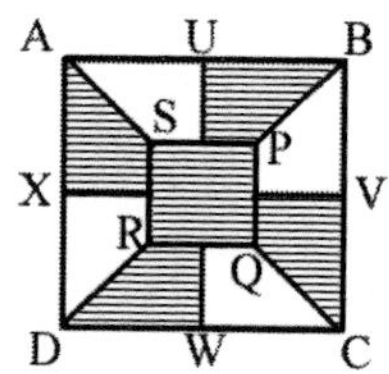

Figure 6. 23

Answer:
Given ABCD = square
Let x = length of ABCD's side
Given area of ABCD = 144
=> x × x = 144
$x^2 = 144$
$x = \sqrt{144}$
x = 12 in
Also given
Area of ABCD : Area of PQRS = $4:1$

$$\frac{\text{Area of ABCD}}{\text{Area of PQRS}} = \frac{4}{1}$$

Area of ABCD = 4 × Area of PQRS
144 = 4 × Area of PQRS
Area of PQRS = 144 ÷ 4
= 36 in^2
Since PQRS = square
Let y = length of PQRS's side
Area of PQRS:
y × y = 36
$y^2 = 36$
$y = \sqrt{36}$
= 6 in
Area of 1 shaded trapezoid, UBPS:

$$= \frac{1}{2} \times \text{height} \times (\text{short base} + \text{long base})$$

$$= \frac{1}{2} \times \left(\frac{1}{2} \times (12-6)\right) \times \left(\frac{1}{2} \times 6 + \frac{1}{2} \times 12\right)$$

$$= \frac{1}{2} \times 3 \times (3+6)$$

= 13.5 in^2
Since there are 4 shaded trapezoids:
= 4 × area of 1 trapezoid
= 4 × 13.5
= 54 in^2
Area of shaded region:
= Area of PQRS + Area of 4 shaded trapezoids
= 36 + 54
= 90 in^2 //

Alternatively:

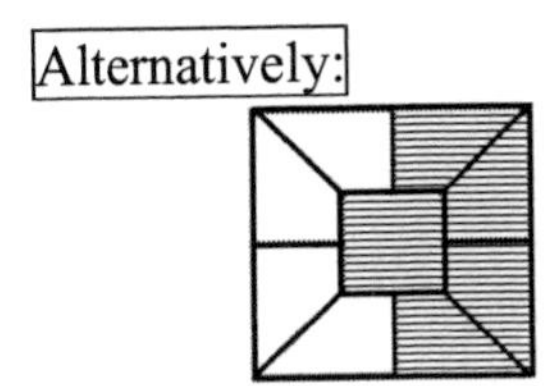

Using addition of parts method:

$= \frac{1}{2} \times$ Area ABCD + $\frac{1}{2} \times$Area PQRS

$= \frac{1}{2} \times 144 + \frac{1}{2} \times 36$

$= 72 + 18$

$= 90$ in^2 //

34. In Figure 6.24, ABCD and CDEG are parallelograms. What is the area of the shaded region?

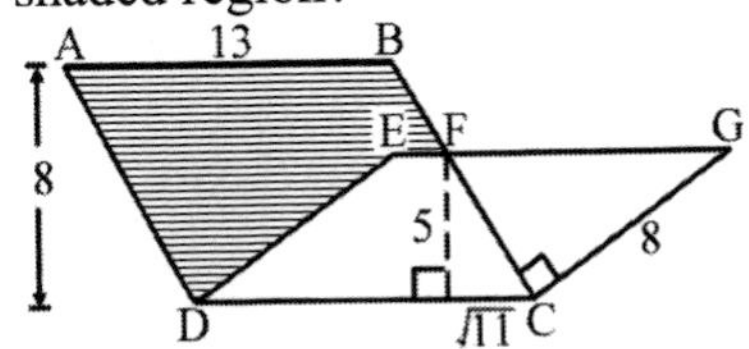

Figure 6. 24

Answer:
Using Pythagorean Theorem:

$FC^2 = 5^2 + \sqrt{11}^2$
$= 25 + 11$
$= 36$

$FC = \sqrt{36} = 6$ units

Again from Pythagorean Theorem:

$FG^2 = FC^2 + 8^2$
$= 36 + 64$
$= 100$

$FG = \sqrt{100}$
$= 10$ units

Given ABCD = parallelogram
=> Opposite sides are ≅:
AB ≅ DC = 13 units
Since CDEG = parallelogram
=> DC ≅ EG = 13 units
Thus,
EG = EF + FG
13 = EF + 10
EF = 3 units
Area of shaded region:
Using subtraction of parts method:
= area of parallelogram ABCD – area of trapezoid CDEF

$= 13 \times 8 - \frac{1}{2} \times 5 \times (3 + 13)$

$= 104 - 40$

$= 64$ units^2 //

35. In Figure 6.25, ABCDEF is a regular polygon while BCQP is an isosceles trapezoid. If the perpendicular distance from B to $\overline{PQ}$ and $\overline{AC}$ are equal, find the perimeter and area of the shaded region, ABPQ.

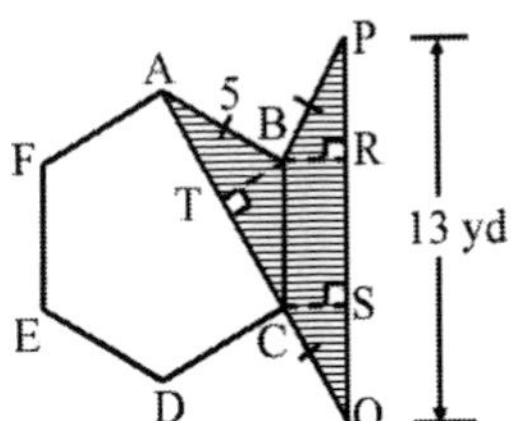

Figure 6. 25

Answers:
Given BCQP = isosceles trapezoid
=> △BPR ≅ △CQS
=> PR ≅ QS ...(1)
Figure 6.25 is reproduced for labeling:

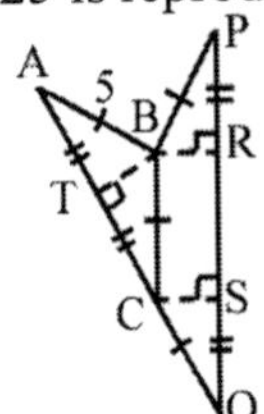

Given PQ = 13
AB = BC = RS = 5
PR + RS + QS = PQ

Substitute (1):
PR + 5 + PR = 13
2PR = 13 – 5 = 8
PR = 4 yd
Using Pythagorean Theorem:
$BP^2 = PR^2 + BR^2$
$5^2 = 4^2 + BR^2$
$25 = 16 + BR^2$
$BR^2 = 9$
$BR = \sqrt{9}$
$= 3$ yd
Given BR = BT
Using Pythagorean Theorem:
$AB^2 = AT^2 + BT^2$
$5^2 = AT^2 + 3^2$
$25 = AT^2 + 9$
$AT^2 = 25 - 9$
$= 16$
$AT = \sqrt{16}$
$= 4$ yd
Since AT = CT
=> CT = 4 yd
AC = AT + CT
= 4 + 4 = 8

∴ Perimeter of shaded region:
= PQ + BP + AB + AC + CQ
= 13 + 5 + 5 + 8 + 5
= 36 yd //

∴ Area of shaded region:
= trapezoid BCQP + △ABC
$= \frac{1}{2} \times BR \times (BC + PQ) + \frac{1}{2} \times AC \times BT$
$= \frac{1}{2} \times 3 \times (5 + 13) + \frac{1}{2} \times 8 \times 3$
= 27 + 12
$= 39$ yd^2 //

36. In Figure 6.26, $\overline{AB}$ is the common tangent of circle O and circle P. $\overline{OA}$ passes through the center points O and P. The radius of circle O and circle P are 7 inches and 4 inches respectively. It is further given that $\overline{AB}$ is 24 inches and points Q and R are on the circumference of circle P while Q is on circle O. Hence find the perimeter and area of the shaded region.

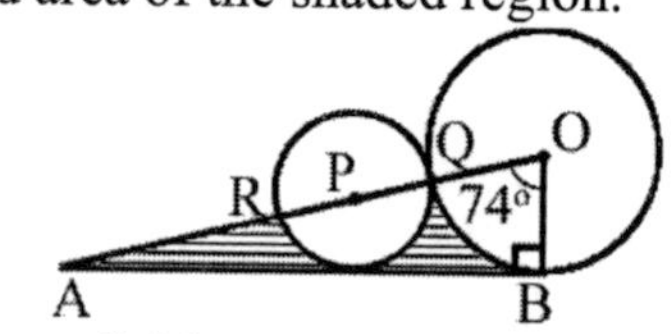

Figure 6. 26

Answers:
Given $\overline{AB}$ = common tangent
=> ∠ABO = 90°
=> △ABO = right triangle
Using Pythagorean Theorem:
$OA^2 = OB^2 + AB^2$
$= 7^2 + 24^2$
$= 49 + 576$
$= 625$
$OA = \sqrt{625}$
= 25 inches ...(1)
Area of △ABO:
$= \frac{1}{2} \times AB \times OB$
$= \frac{1}{2} \times 24 \times 7$
$= 84$ in^2 ...(2)

Given ∠BOQ = 74°
Arc length of $\widehat{BQ}$:
$= \frac{\angle BOQ}{360°} \times \pi \times 2 \times OB$
$= \frac{74°}{360°} \times \frac{22}{7} \times 2 \times 7$
= 9.0444 in ...(3)
Area of sector BOQ:
$= \frac{\angle BOQ}{360°} \times \pi \times OB^2$
$= \frac{74°}{360°} \times \frac{22}{7} \times 7^2$
$= 31.6556$ in^2 ...(4)

Given P = center point of circle P
Since points Q and R are on the circumference of circle P,
$\overline{QR}$ = diameter of circle P
=> Sector QPR = semicircle
Arc length of semicircle, $\widehat{QR}$:

$= \frac{1}{2} \times \pi \times 2 \times PQ$

$= \frac{1}{2} \times \frac{22}{7} \times 2 \times 4$

$= 12.5714$ in ...(5)

Area of semicircle QPR:

$= \frac{1}{2} \times \pi \times PQ^2$

$= \frac{1}{2} \times \frac{22}{7} \times 4^2$

$= 25.1429\ in^2$ ...(6)

∴ Perimeter of shaded region:

= AB + arc length of $\overset{\frown}{BQ}$ + arc length of $\overset{\frown}{QR}$ + AR

Substitute (1), (3), and (5):

= 24 + 9.0444 + 12.5714 + (OA – QR – OQ)

= 45.6158 + (25 – 8 – 7)

= 55.6158 in //

∴ Area of shaded region:

= △ABO – sector BOQ – semicircle QPR

Substitute (2), (4), and (6):

= 84 – 31.6556 – 25.1429

$= 27.2015\ in^2$ //

37. Figure 6.27 shows a scalene triangle, ABC. What is the area of △ABC?

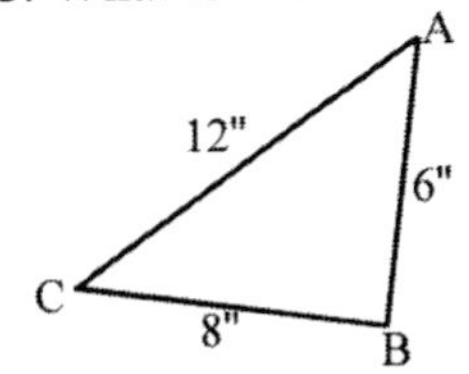

Figure 6. 27

Answer:

Given ABC = scalene triangle

Semi perimeter, $s = \frac{AB + AC + BC}{2}$

$= \frac{6 + 12 + 8}{2}$

$= 13$

Area of △ABC:

Using Heron's formula:

$= \sqrt{s(s - AB)(s - AC)(s - BC)}$

$= \sqrt{13(13 - 6)(13 - 12)(13 - 8)}$

$= \sqrt{13(7)(1)(5)}$

$= \sqrt{455}$

$= 21.3307\ in^2$ //

38. Figure 6.28 shows ABCD a trapezium. $\overline{BD}$ is a diagonal. What is the area of trapezium ABCD?

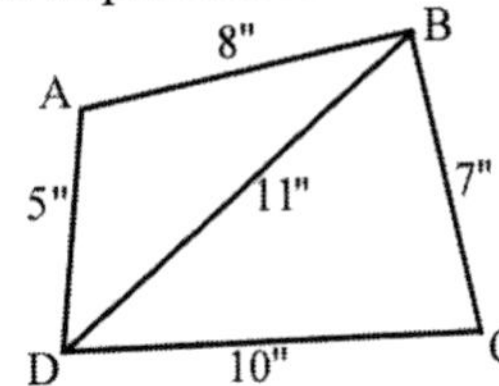

Figure 6. 28

Answer:

Given △ABD and △BCD = scalene △s

Using Heron's Formula:

For △ABD:

$s_1 = \frac{AB + BD + AD}{2}$

$= \frac{8 + 11 + 5}{2} = 12$

Area of △ABD:

$= \sqrt{s_1(s_1 - AB)(s_1 - BD)(s_1 - AD)}$

$= \sqrt{12(12 - 8)(12 - 11)(12 - 5)}$

$= \sqrt{12(4)(1)(7)}$

$= \sqrt{336}$

$= 18.3303\ in^2$

For $\triangle BCD$:

$s_2 = \dfrac{BC + CD + BD}{2}$

$s_2 = \dfrac{7 + 10 + 11}{2}$

$= 14$

Area of $\triangle BCD$:

$= \sqrt{s_2(s_2 - BC)(s_2 - CD)(s_2 - BD)}$

$= \sqrt{14(14-7)(14-10)(14-11)}$

$= \sqrt{14(7)(4)(3)}$

$= \sqrt{1176}$

$= 34.2929\ in^2$

$\therefore$ Area of trapezium, ABCD:

= Area of $\triangle ABD$ + Area of $\triangle BCD$

$= 18.3303 + 34.2929$

$= 52.6232\ in^2$ //

39. Figure 6.29 shows ABCD a square. Find the area of the shaded sector, BCE. (Let $\pi = \dfrac{22}{7}$)

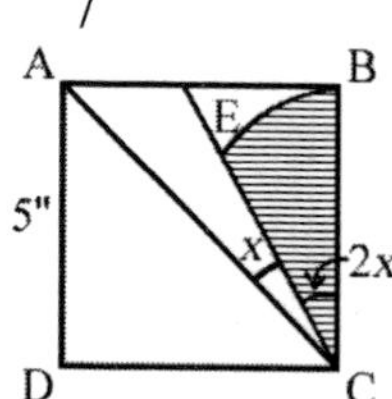

Figure 6. 29

Answer:

Given ABCD = square

=> $\overline{AC}$ = diagonal

=> $\overline{AC}$ = angle bisector

=> $\angle ACB = \angle BCD \div 2$

$= 90° \div 2$

$= 45°$

$\angle BCE + \angle ACE = \angle ACB$

$2x + x = 45°$

$3x = 45°$

$x = 15°$

Area of sector BCE:

$= \dfrac{\angle BCE}{360°} \times \pi \times BC^2$

$= \dfrac{2(15°)}{360°} \times \dfrac{22}{7} \times 5^2$

$= 6.5476\ in^2$ //

40. Figure 6.30 shows POR a sector whose radius is 17 ″ and $\angle POR = 40°$. Find the area of the shaded region, PQR.

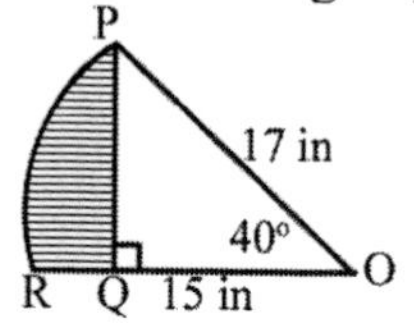

Figure 6. 30

Answer:

Given radius, r = 17

Using Pythagorean Theorem:

$OP^2 = PQ^2 + OQ^2$

$17^2 = PQ^2 + 15^2$

$289 = PQ^2 + 225$

$PQ^2 = 64$

$PQ = \sqrt{64} = 8$

Area of sector POR:

$= \dfrac{40°}{360°} \times \dfrac{22}{7} \times 17^2$

$= 100.9206\ in^2$

Area of $\triangle OPQ$:

$= \dfrac{1}{2} \times PQ \times OQ$

$= \dfrac{1}{2} \times 8 \times 15$

$= 60\ in^2$

$\therefore$ Area of shaded region, PQR:

= area of sector POR – area of $\triangle OPQ$

$= 100.9206 - 60$

$= 40.9206\ in^2$ //

Chapter 7
Coordinate Geometry

Cartesian coordinate system – in 2 dimensions is represented by two perpendicular axes, x–axis (**abscissa**) and y–axis (**ordinate**).

Points on the Cartesian plane are **ordered pair**.

(0, 0) – **Origin**

Coordinate plane can be divided into 4 quadrants:

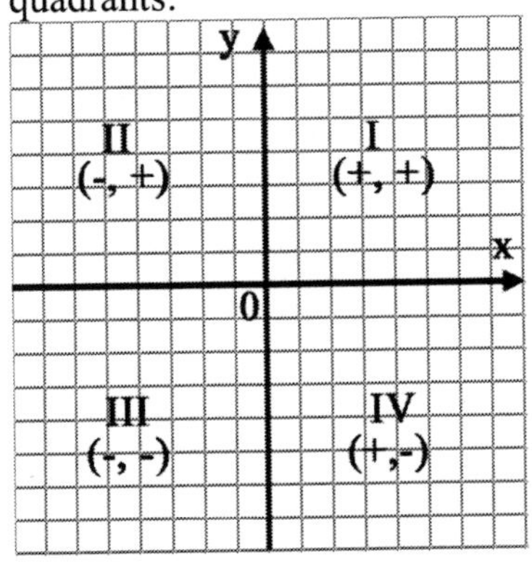

Slope/gradient, $m = \dfrac{y_2 - y_1}{x_2 - x_1}$

Equation of a **straight line**:

General form:

$Ax + By + C = 0$

Point slope form:

$y - y_1 = m(x - x_1)$

Slope intercept form:

$y = mx + c$

If two lines, L_1 and L_2, are **parallel**, their slopes:

$m_1 = m_2$

If two lines, L_1 and L_2, are **perpendicular**, their slopes:

$m_1 \times m_2 = -1$

Midpoint, $M = \left(\dfrac{x_1 + x_2}{2}, \dfrac{y_1 + y_2}{2} \right)$

Distance, $d = \sqrt{(x_2 - x_1)^2 + (y_2 - y_1)^2}$

Equation of a **circle**, whose center is (x_1, y_1) and radius is r:

Standard form:

$(x - x_1)^2 + (y - y_1)^2 = r^2$

General form:

$Ax^2 + Bx + Cy^2 + Dy + E = 0$

Finding area using coordinates:

y
B(x_2, y_2)
A(x_1, y_1)
C(x_3, y_3)
x

(Notice that the coordinates of the vertices are placed **counterclockwise** in the matrix)

Area of △ABC:

$$= \frac{1}{2} \left| \begin{matrix} x_1 & x_2 & x_3 & x_1 \\ y_1 & y_2 & y_3 & y_1 \end{matrix} \right|$$

$$= \frac{1}{2}(x_1 y_2 + x_2 y_3 + x_3 y_1 - y_1 x_2 - y_2 x_3 - y_3 x_1)$$

Distance-time graph:

1. Gradient represents speed $\left(\dfrac{\text{Distance}}{\text{Time}}\right)$
2. Average speed:
   $= \dfrac{\text{Total distance traveled}}{\text{Total time taken}}$

Speed-time graph:

1. Gradient represents acceleration
   $= \dfrac{\text{Speed}}{\text{Time}}$
2. Area under graph represents distance traveled.

1. For each of the following pairs of coordinates, determine the slope of each line and sketch the line segment.

a) A(3, −2) and B(−3, 4)
b) C(6, 7) and D(0, 1)
c) E(3, − 1) and F(3, 9)
d) G(−2, 2) and H(6, 2)

Answers:

a) Given A(3, −2) and B(−3, 4)

$$\text{Slope, } m_{AB} = \frac{y_2 - y_1}{x_2 - x_1}$$
$$= \frac{4-(-2)}{-3-3}$$
$$= \frac{4+2}{-6}$$
$$= \frac{6}{-6}$$
$$= -1_{//}$$

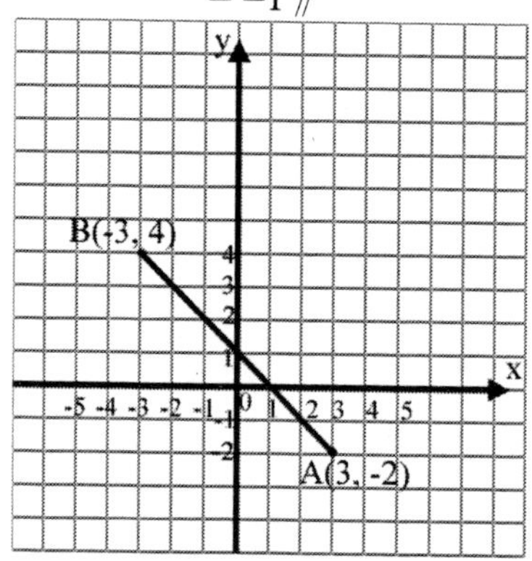

b) Given C(6, 7) and D(0, 1)

$$\text{Slope } m_{CD} = \frac{1-7}{0-6}$$
$$= \frac{-6}{-6} = 1_{//}$$

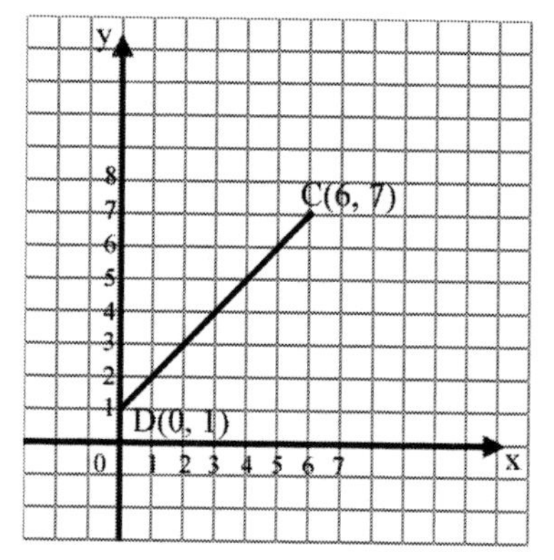

c) Given E(3, −1) and F(3, 9)

$$\text{Slope } m_{EF} = \frac{9-(-1)}{3-3}$$
$$= \frac{9+1}{0}$$
$$= \frac{10}{0}$$
= undefined slope //

∴ Line EF is a vertical line.

d) Given G(−2, 2) and H(6, 2)

$$\text{Slope, } m_{GH} = \frac{2-2}{6-(-2)}$$

$= \frac{0}{6+2}$

$= \frac{0}{8}$

$= 0$ //

$\therefore$ Line GH is a horizontal line.

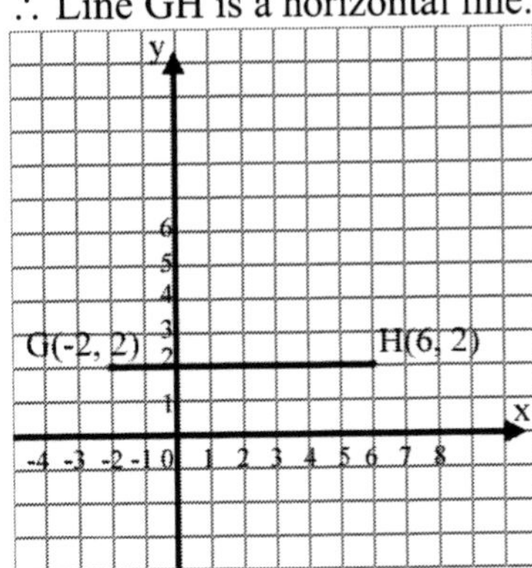

2. Figure 7.1 shows a Cartesian plane where $\overline{AB}$ and $\overleftrightarrow{CD}$ are two straight lines.

a) Find the x- and y-intercepts of each line.

b) What is the gradient of each line?

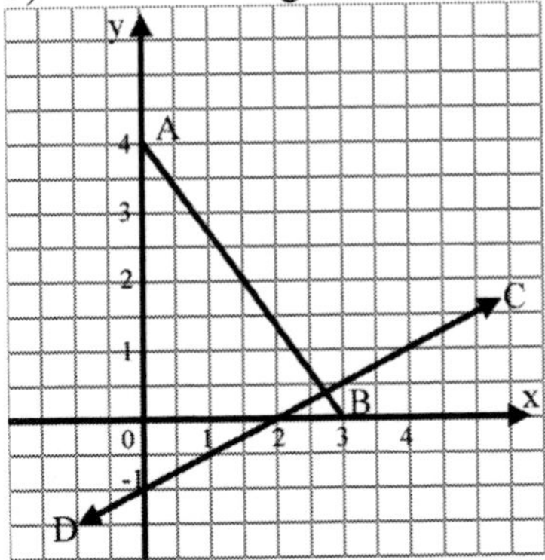

Figure 7. 1

Answers:

a) From the graph in Figure 7.1:

For $\overline{AB}$:

x- intercept = (3, 0) //

y-intercept = (0, 4) //

For $\overleftrightarrow{CD}$:

x-intercept = (2, 0) //

y-intercept = (0, −1) //

b) For $\overline{AB}$:

Gradient, $m_{AB} = \frac{y_2 - y_1}{x_2 - x_1}$

$= \frac{4-0}{0-3}$

$= -\frac{4}{3}$ //

Gradient, $m_{CD} = \frac{-1-0}{0-2}$

$= \frac{-1}{-2}$

$= \frac{1}{2}$ //

3. What is the value of k in Figure 7.2, if the slope of the straight line is −3?

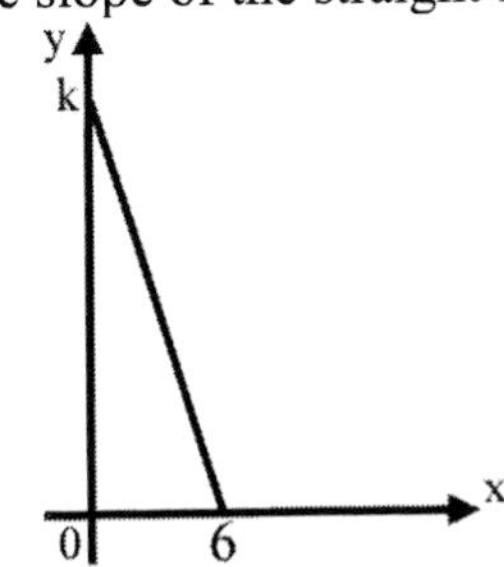

Figure 7. 2

Answer:

Given slope, m = −3

y-intercept = (0, k)

x-intercept = (6, 0)

Slope:

$\frac{k-0}{0-6} = -3$

$\frac{k}{-6} = -3$

$k = -3 \times (-6)$

$k = 18$

$\therefore k = 18$ //

4. Find the gradient of the straight line whose equation is $\frac{x}{2} - \frac{y}{5} = 4$.

Answer:

Given $\frac{x}{2} - \frac{y}{5} = 4$ ⇦ General form

Rearrange equation from general form to slope intercept form:

$$\frac{x}{2}-\frac{y}{5}=4$$

$$-\frac{y}{5}=-\frac{x}{2}+4$$

$$-\frac{y}{5}\times(-5)=-\frac{x}{2}\times(-5)+4\times(-5)$$

$$y=\frac{5}{2}x-20$$ ⇦ Slope intercept form

$\therefore$ Gradient is $\frac{5}{2}$ //

5. Given $2y - 3x + 5 = 0$ is a straight line. What is the slope of the line and its x- and y-intercepts?

Answers:

Given $2y - 3x + 5 = 0$ (*)⇦ General form

Rearrange equation from general form to slope intercept form:

$2y = 3x - 5$

$$y=\left(\frac{1}{2}\right)3x-\left(\frac{1}{2}\right)5$$ ⇦ Divide by 2

$$y=\frac{3}{2}x-\frac{5}{2}$$

In slope intercept form ($y = mx + c$), slope $= m$:

$\therefore$ Slope of line $=\frac{3}{2}$ //

For x-intercept, where $y = 0$:

Substitute into (*):

$2(0) - 3x + 5 = 0$

$-3x + 5 = 0$

$-3x = -5$

$$x=\frac{-5}{-3}=\frac{5}{3}$$

$\therefore$ x-intercept is $(\frac{5}{3}, 0)$ //

For y-intercept, where $x = 0$:

Substitute into (*):

$2y - 3(0) + 5 = 0$

$2y + 5 = 0$

$2y = -5$

$$y=\frac{-5}{2}$$

$\therefore$ y-intercept is $(0, \frac{-5}{2})$ //

6. In Figure 7.3, ABCD is a rectangle. It is further known that $5AE = CD$, $DF = BC$ and $BC = 3AE$. If $CD = 10$ units, find the gradients of $\overline{BE}$ and $\overline{EF}$.

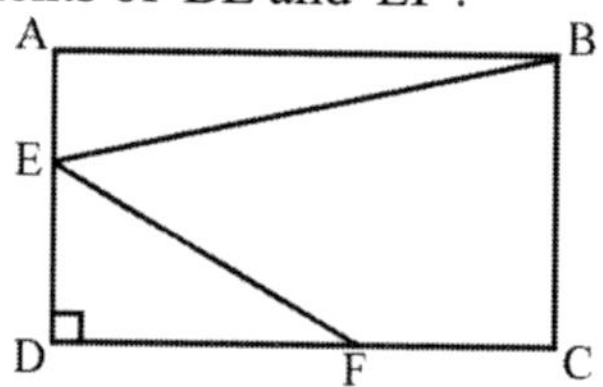

Figure 7. 3

Answers:

Given ABCD = rectangle

=> AB = CD

Also given $CD = 10$ ...(1)

$5AE = CD$ ...(2)

Substitute (1) into (2):

$5AE = 10$

$$AE=\frac{10}{5}$$

$= 2$ units

Gradient of $\overline{BE}$:

$$m_{BE}=\frac{AE}{AB}$$

$$=\frac{2}{10}$$ ⇦ AB = CD = 10

$$=\frac{1}{5}$$ //

Given DF = BC

Also given $BC = 3AE$ ...(3)

=> DF = 3AE ...(4) ⇦ transitive property
Since ABCD = rectangle
=> Opposite sides are congruent
=> $AD \cong BC$
Thus,
BC = AD
BC = AE + ED
ED = BC − AE
Substitute (3):
ED = 3AE − AE
= 2AE ...(5)
Gradient of $\overline{EF}$:

$$m_{EF} = -\frac{ED}{DF}$$

Substitute (4) and (5):

$$m_{EF} = -\frac{2AE}{3AE}$$

$$= -\frac{2\cancel{AE}}{3\cancel{AE}}$$

$$= -\frac{2}{3} \text{ //}$$

7. Find the equation of the line that passes through (1, 2) and (7, 5).

Answer:
Given points: (1, 2) and (7, 5)

Slope, $m = \frac{y_2 - y_1}{x_2 - x_1}$

$$m = \frac{5-2}{7-1}$$

$$m = \frac{3}{6}$$

$$= \frac{1}{2}$$

Using slope intercept form:
y = mx + c

Substitute (1, 2) and $m = \frac{1}{2}$:

$$2 = \frac{1}{2}(1) + c$$

$$2 = \frac{1}{2} + c$$

$$c = 2 - \frac{1}{2}$$

$$= \frac{4}{2} - \frac{1}{2} = \frac{3}{2}$$

∴ Equation of straight line is $y = \frac{1}{2}x + \frac{3}{2}$

or 2y = x + 3 //

Alternatively:
Using point slope form:
$y - y_1 = m(x - x_1)$

Substitute (1, 2) and $m = \frac{1}{2}$:

$$y - 2 = \frac{1}{2}(x - 1)$$

2y − 4 = x − 1
2y = x + 3 //

8. Determine if P(3, 5) lies on the line whose equation is 4x − y = 6.

Answer:
Given 4x − y = 6 ...(1)
Also given P(3, 5)
=> x-coordinate = 3 and y-coordinate = 5
Substitute x = 3 into (1):
=> 4(3) − y = 6
12 − y = 6
y = 12 − 6
= 6
But y-coordinate of P is 5, therefore P(3, 5) does not lie on the line 4x − y = 6. //

9. Find the value of k if P(3, 6) lies on the line whose equation is $2x - 3y = k$.

Answer:
Given P(3, 6)
Also given $2x - 3y = k$ ...(1)
Since P lies on line,
Substitute $x = 3$ and $y = 6$ into (1):
=> $2(3) - 3(6) = k$
$6 - 18 = k$
$-12 = k$
$\therefore k = -12$ //

10. Find the value of q if P(q, 2) lies on the line whose equation is $5x - 2y = q$.

Answer:
Given P(q, 2)
Also given $5x - 2y = q$ ...(1)
Since P lies on line,
Substitute $x = q$ and $y = 2$ into (1):
=> $5(q) - 2(2) = q$
$5q - 4 = q$
$5q - q = 4$
$4q = 4$
$\therefore q = \frac{4}{4} = 1$ //

11. In Figure 7.4, $\triangle ABC$ is an isosceles triangle. If the equation for $\overline{AC}$ is $4y = -5x - 20$, find the coordinate for point C. Hence what is the equation (expressed in general form) for $\overline{BC}$?

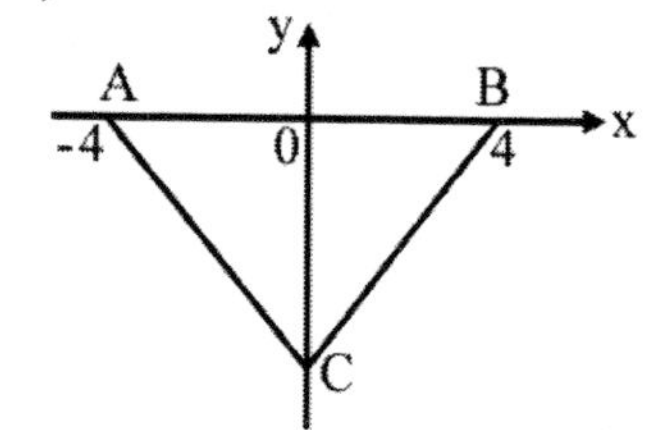

Figure 7. 4

Answers:
Given $\overline{AC}$: $4y = -5x - 20$ ...(1)
For y-intercept:
Substitute $x = 0$ into (1):
$4y = -5(0) - 20$
$4y = -20$

$y = \frac{-20}{4} = -5$

Since $\overline{AC}$ cuts y-axis at point C therefore C(0, –5) //

Also given B(4, 0)
From C(0, –5)

$$\text{Slope, } m_{BC} = \frac{y_2 - y_1}{x_2 - x_1} = \frac{0-(-5)}{4-0} = \frac{5}{4}$$

Using point slope form:
$y - y_B = m_{BC}(x - x_B)$
$y - 0 = \frac{5}{4}(x - 4)$
$4y = 5x - 20$
$-5x + 4y + 20 = 0$ ⇦ General form
$\therefore$ Equation of $\overline{BC}$ in general form is $-5x + 4y + 20 = 0$ //

Alternatively:
Using points C(0, –5) and B(4, 0):
$\frac{y - y_1}{x - x_1} = \frac{y_2 - y_1}{x_2 - x_1}$
$\frac{y-(-5)}{x-0} = \frac{0-(-5)}{4-0}$
$\frac{y+5}{x} = \frac{5}{4}$
$4(y + 5) = 5x$
$4y + 20 = 5x$
$-5x + 4y + 20 = 0$ //

12. Find the equation of a straight line that is parallel to y-axis and passes through point (3, 2).

Answer:
Given: line parallel to y-axis
=> slope, m undefined
Also given: line passes through (3, 2)
=> x-intercept is (3, 0)
$\therefore$ Equation of line is $x = 3$. //

13. Given two straight lines, $y = mx + 3$ and $3x - 2y = 4$, are parallel. What is the value of m?

Answer:
Given, $y = mx + 3$ ...(1)
Also given, $3x - 2y = 4$
=> $-2y = -3x + 4$

$y = \frac{1}{-2}(-3x) + \frac{1}{-2}(4)$ ⇦Divide by −2

$y = \frac{3}{2}x - 2$ ...(2)

Parallel lines have same slope:
Compare (1) with (2):

$\therefore m = \frac{3}{2}$ //

14. Find the intersection point for the two straight lines, $y = 2x + 3$ and $2y = 4x + 8$.

Answer:
Given $y = 2x + 3$ ...(1)
And $2y = 4x + 8$
=> $y = 2x + 4$ ...(2) ⇦ divide equation by 2
From slope intersecting form:
$y = mx + c$
where, m = slope or gradient
Compare (1) and (2), equations have same slopes, m = 2
$\therefore$ Equations (1) and (2) are parallel lines.
Since parallel lines do not intersect, hence there is no intersection point. //

Further test:
(1) – (2):
$y - y = 2x + 3 - (2x + 4)$
$0 = 2x + 3 - 2x - 4$
$0 \neq -1$ (which is not true)
$\therefore$ We have shown that the lines do not intersect. //

15. Determine if the lines, $y = 2x + 5$ and $4y + 2x = 3$ are perpendicular.

Answer:
Let, m = slope
Given, $y = 2x + 5$ ...(1)
=> $m_1 = 2$
Also given, $4y + 2x = 3$ ...(2)
=> $4y = -2x + 3$

$y = \frac{1}{4}(-2x) + \frac{1}{4}(3)$ ⇦Divide by 4

$y = -\frac{1}{2}x + \frac{3}{4}$

=> $m_2 = -\frac{1}{2}$

Perpendicular lines: product of slopes = −1
=> $m_1 \times m_2$

$= 2 \times \left(-\frac{1}{2}\right)$

$= -1$

$\therefore$ The lines are perpendicular. //

16. Given two straight lines, $py = 2x - 1$ and $4x - 3y - 5 = 0$, are perpendicular. What is the value of p?

Answer:
Let m = slope
Given, Line 1: $py = 2x - 1$

=> $y = \frac{2}{p}x - \frac{1}{p}$

$m_1 = \frac{2}{p}$

Also given, Line 2: $4x - 3y - 5 = 0$
=> $-3y = -4x + 5$

$y = \frac{-4}{-3}x + \left(-\frac{1}{3}\right)(5)$ ⇦Divide by −3

$= \frac{4}{3}x - \frac{5}{3}$

$m_2 = \frac{4}{3}$

Given Line 1 and Line 2 are perpendicular:

$\Rightarrow m_1 \times m_2 = -1$

$\frac{2}{p} \times \frac{4}{3} = -1$

$\frac{8}{3p} = -1$

$8 = -3p$

$\therefore p = -\frac{8}{3}$ //

17. Find the distance between the following two endpoints:
a) A(3, 2) and B(3, 12)
b) C(–4, 2) and D(8, 2)
c) E(–2, 3) and F(5, 5)

Answers:

a) Given A(3, 2) and B(3, 12)
Since the x-coordinates (abscissas) are the same, A and B are on the same vertical line.
Distance, AB:
= 12 – 2
= 10 units //

b) Given C(–4, 2) and D(8, 2)
Since the y-coordinates (ordinates) are the same, C and D are on the same horizontal line.
Distance, CD:
= 8 – (–4)
= 8 + 4
= 12 units //

c) Given E(–2, 3) and F(5, 5)
The x-coordinates and y-coordinates are different.
Use distance formula.
Distance, EF:

$= \sqrt{(x_2 - x_1)^2 + (y_2 - y_1)^2}$

$= \sqrt{(5-(-2))^2 + (5-3)^2}$

$= \sqrt{7^2 + 2^2}$

$= \sqrt{49+4} = \sqrt{53}$ units //

18. Find the distance from P(3, 6) to the straight line y = 4x – 12.

Answer:
Given P(3, 6)
$y = 4x - 12$
$\Rightarrow -4x + y + 12 = 0$ ⇦ General form
Let d = distance from P to straight line

$d = \frac{|Ax_1 + By_1 + C|}{\sqrt{A^2 + B^2}}$

$= \frac{|-4(3) + (1)(6) + 12|}{\sqrt{(-4)^2 + (1)^2}}$

$= \frac{|-12 + 6 + 12|}{\sqrt{16+1}}$

$= \frac{|6|}{\sqrt{17}}$

= 1.4552 units //

19. Find the distance between the parallel lines:

$4x + 3y = 4$
$3y = -4x + 9$

Answer:
Given parallel lines:
$4x + 3y = 4$ ...(1)
$3y = -4x + 9$ ...(2)
$\Rightarrow 4x + 3y - 9 = 0$ ⇦ General form
A = 4, B = 3, C = –9
For line (1), when y = 0

[We chose y = 0 for simplicity, you can choose any real value for y]:
$4x + 3(0) = 4$
$4x = 4$
$x = 1$
Point on line (1) is (1, 0)
Distance from (1, 0) to line (2):

$= \frac{|Ax_1 + By_1 + C|}{\sqrt{A^2 + B^2}}$

$= \frac{|4(1) + 3(0) - 9|}{\sqrt{4^2 + 3^2}}$

$= \frac{|-5|}{\sqrt{16+9}}$

$= \frac{5}{\sqrt{25}}$ ⇦ Absolute value is always positive

$= \frac{5}{5}$

= 1 unit //

20. In Figure 7.5, $\overline{RQ}$ is parallel to x-axis and PQ is 5 units. Find the coordinate for point Q.

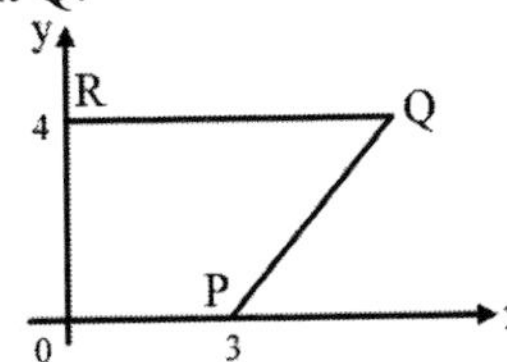

Figure 7. 5

Answer:
Given $\overline{RQ}$ parallel to x-axis
=> Q's y-coordinate = 4
Let a = Q's x-coordinate
Also given P(3, 0)
Since PQ = 5:

$\sqrt{(a-3)^2 + (4-0)^2} = 5$
$a^2 - 6a + 9 + 16 = 5^2$
$a^2 - 6a + 25 = 25$
$a^2 - 6a = 0$
$a(a - 6) = 0$
∴ Critical values of a = 0 and 6
From Figure 7.5, a must be 6.
=> Q(6, 4) //

Note: a = 0 refers to coordinate P.

21. Find the midpoint of the line segment whose endpoints are (4, 5) and (–2, 7).

Answer:
Given endpoints: (4, 5) and (–2, 7)
Midpoint:

$= \left(\frac{x_1 + x_2}{2}, \frac{y_1 + y_2}{2}\right)$

$= \left(\frac{4 + (-2)}{2}, \frac{5+7}{2}\right)$

$= \left(\frac{2}{2}, \frac{12}{2}\right)$

= (1, 6)
∴ Midpoint is (1, 6). //

22. Given B(–3, 4) is the midpoint of chord $\overline{AC}$. If A(7, –2), find the coordinate of C.

Answer:
Given midpoint, B = (–3, 4)
Endpoint, A = (7, –2)
Let C = (x_C, y_C)

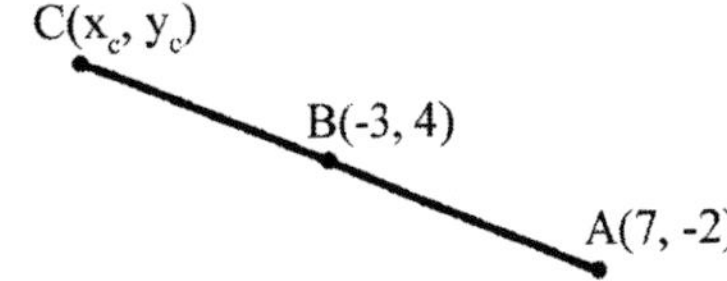

Midpoint B:

$(-3, 4) = \left(\frac{x_1 + x_2}{2}, \frac{y_1 + y_2}{2}\right)$

$= \left(\frac{7 + x_C}{2}, \frac{-2 + y_C}{2}\right)$

Compare corresponding components:
x-coordinate:

$-3 = \frac{7 + x_C}{2}$
$-6 = 7 + x_C$
$x_C = -6 - 7 = -13$

y-coordinate:

$4 = \frac{-2 + y_C}{2}$
$8 = -2 + y_C$
$y_C = 8 + 2 = 10$
∴ C(–13, 10) //

23. In Figure 7.6, OPQR is a parallelogram. What is the coordinate of point Q?

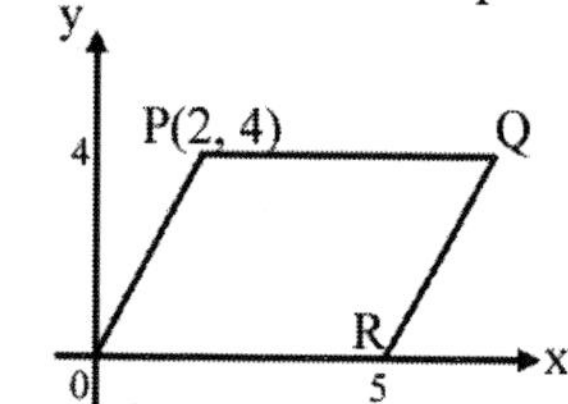

Figure 7. 6

Answer:
Given P = (2, 4) and R = (5, 0)
Also given OPQR = parallelogram
=> $\overline{PR}$ and $\overline{OQ}$ = diagonals
=> Diagonals in a parallelogram intersect at midpoint.

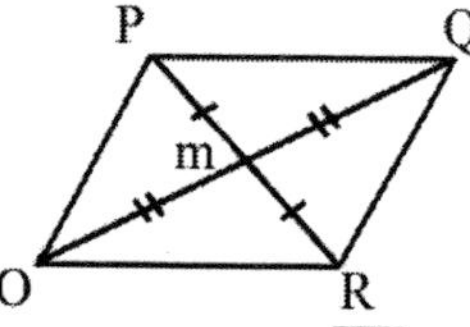

Let m = midpoints of $\overline{PR}$ and $\overline{OQ}$

Midpoint of $\overline{PR}$:

$$m = \left(\frac{x_P + x_R}{2}, \frac{y_P + y_R}{2}\right)$$

$$= \left(\frac{2+5}{2}, \frac{4+0}{2}\right)$$

$$= \left(\frac{7}{2}, 2\right)$$

Let Q = (x_Q, y_Q)

Midpoint of $\overline{OQ}$:

$$m = \left(\frac{x_O + x_Q}{2}, \frac{y_O + y_Q}{2}\right)$$

$$\left(\frac{7}{2}, 2\right) = \left(\frac{0 + x_Q}{2}, \frac{0 + y_Q}{2}\right)$$

Compare like components:
x-coordinate:

$$\frac{7}{2} = \frac{x_Q}{2}$$

$x_Q = 7$

y-coordinate:

$$2 = \frac{y_Q}{2}$$

$y_Q = 4$

$\therefore$ Q(7, 4) //

24. Given P(−2, 1) and R(8, 6) are endpoints of a line segment $\overline{PR}$. If Q(2, 3) is located on $\overline{PR}$, what is the ratio of PQ : QR ?

Answer:
Let m = PQ
n = QR
Given,

m n
P(-2, 1) Q(2, 3) R(8, 6)

Thus point Q:

$$(2, 3) = \left(\frac{x_P n + x_R m}{m+n}, \frac{y_P n + y_R m}{m+n}\right)$$

$$= \left(\frac{-2n + 8m}{n+m}, \frac{(1)n + 6m}{n+m}\right)$$

Comparing like terms:
x-coordinate:

$$2 = \frac{-2n + 8m}{n+m}$$

2(n + m) = −2n + 8m
2n + 2m = −2n + 8m
8m − 2m = 2n + 2n
6m = 4n
3m = 2n

$$\frac{m}{n} = \frac{2}{3}$$

m : n = 2 : 3
Substitute m = PQ and n = QR:
$\therefore$ PQ : QR = 2 : 3 //

Alternatively:
Length PQ:

$= \sqrt{(2-(-2))^2 + (3-1)^2}$
$= \sqrt{4^2 + 2^2}$
$= \sqrt{16+4}$
$= \sqrt{20}$ ⇦ Note: $\sqrt{20} = \sqrt{4 \times 5}$
$= 2\sqrt{5}$

Length QR:

$= \sqrt{(8-2)^2 + (6-3)^2}$
$= \sqrt{6^2 + 3^2}$
$= \sqrt{36+9}$
$= \sqrt{45}$ ⇦ Note: $\sqrt{45} = \sqrt{9 \times 5}$
$= 3\sqrt{5}$

Thus,

$PQ:QR = 2\sqrt{5}:3\sqrt{5}$

$= 2:3$ //

25. Given P(2, 4) and Q(8, 10) are endpoints of a line segment. If point R divides $\overline{PQ}$ in the ratio $1:2$, find the coordinate of point R.

Answer:

Given, P = (2, 4)

Q = (8, 10)

Also given R divides $\overline{PQ}$ in ratio $1:2$

Let m = 1

n = 2

1 2
P R Q

Coordinate R:

$= \left(\frac{x_P n + x_Q m}{n+m}, \frac{y_P n + y_Q m}{n+m} \right)$

$= \left(\frac{2(2)+8(1)}{2+1}, \frac{4(2)+10(1)}{2+1} \right)$

$= \left(\frac{4+8}{3}, \frac{8+10}{3} \right)$

$= \left(\frac{12}{3}, \frac{18}{3} \right)$

= (4, 6)

∴ R(4, 6) //

26. It is known that B divides line segment, $\overline{AC}$ in the ratio $2:3$. If B(5, 6) and C(11, 9), find the coordinate of point A.

Answer:

Given B(5, 6) and C(11, 9)

Let m = 2

n = 3

A = (x_A, y_A)

A B C
2 3

Coordinate B:

$(5, 6) = \left(\frac{3x_A + 2(11)}{3+2}, \frac{3y_A + 2(9)}{3+2} \right)$

Comparing like terms:

x-coordinate:

$5 = \frac{3x_A + 22}{5}$

$25 = 3x_A + 22$

$3x_A = 25 - 22$

$= 3$

$x_A = 1$

y-coordinate:

$6 = \frac{3y_A + 18}{5}$

$30 = 3y_A + 18$

$3y_A = 30 - 18$

$= 12$

$y_A = 4$

∴ A(1, 4) //

27. In Figure 7.7, $\overline{AE}$ and $\overline{BD}$ are straight lines. If it is further known that C is the midpoint of $\overline{AE}$ and C divides $\overline{BD}$ in the ratio $2:1$, find:

a) Coordinate C
b) Coordinate B

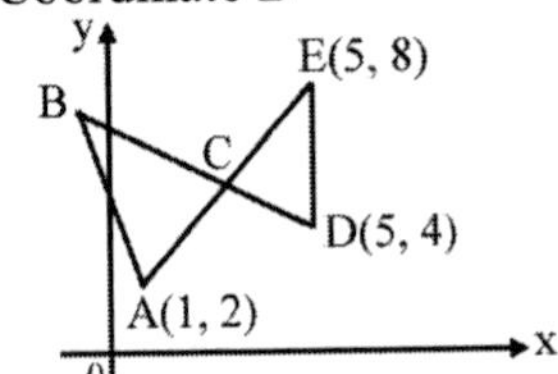

Figure 7. 7

Answers:

a) Given A = (1, 2)

E = (5, 8)

Since C = midpoint $\overline{AE}$

Let $C = (x_C, y_C)$
Thus, coordinate C:

$$(x_C, y_C) = \left(\frac{1+5}{2}, \frac{2+8}{2}\right)$$
$$= \left(\frac{6}{2}, \frac{10}{2}\right)$$
$$= (3, 5)$$

$\therefore$ C(3, 5) //

b) Given D = (5, 4)
From (a) we found C = (3, 5)
Since C divides $\overline{BD}$ in the ratio 2:1
Let $B = (x_B, y_B)$
Coordinate C:

$$(3, 5) = \left(\frac{(1)x_B + 2(5)}{1+2}, \frac{1(y_B) + 2(4)}{1+2}\right)$$
$$= \left(\frac{x_B + 10}{3}, \frac{y_B + 8}{3}\right)$$

Compare like components:
x-coordinate:

$$3 = \frac{x_B + 10}{3}$$
$$9 = x_B + 10$$
$$x_B = 9 - 10$$
$$= -1$$

y-coordinate:

$$5 = \frac{y_B + 8}{3}$$
$$15 = y_B + 8$$
$$y_B = 15 - 8$$
$$= 7$$

$\therefore$ B(−1, 7) //

28. Find the shortest distance from P(3, 8) to the y-axis.

Answer:
Given P = (3, 8)
Shortest distance from P to y-axis is the perpendicular distance between P and the y-coordinate, (0, 8).

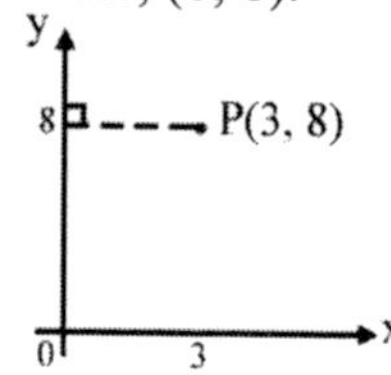

Thus,
The shortest distance from P to y-axis is:
= 3 − 0
= 3 units //

29. Sketch $y = x^2 + 2x + 1$ on a Cartesian plane.

Answer:
Given $y = x^2 + 2x + 1$
In the form $y = ax^2 + bx + c$
a = 1, b = 2, c = 1
<u>To find shape of graph</u>:
Since a > 0:
=> Graph has the shape ∪
<u>To find y-intercept, let x = 0:</u>

$$y = (0)^2 + 2(0) + 1$$
$$= 1$$

$\therefore$ y-intercept is (0, 1)
<u>To find x-intercept, let y = 0:</u>

$$0 = x^2 + 2x + 1$$
$$= (x + 1)(x + 1)$$
$$x = -1$$

$\therefore$ x-intercept is (−1, 0)
<u>To find vertex, use complete the squares:</u>

$$y = x^2 + 2x + 1$$
$$= x^2 + 2x + \frac{2^2}{(2(1))^2} - \frac{2^2}{(2(1))^2} + 1$$
$$= \left(x + \frac{2}{2}\right)^2 - \frac{2^2}{2^2} + 1$$
$$= (x + 1)^2 - 1 + 1$$
$$= (x + 1)^2 + 0$$

$\therefore$ Vertex is (−1, 0)

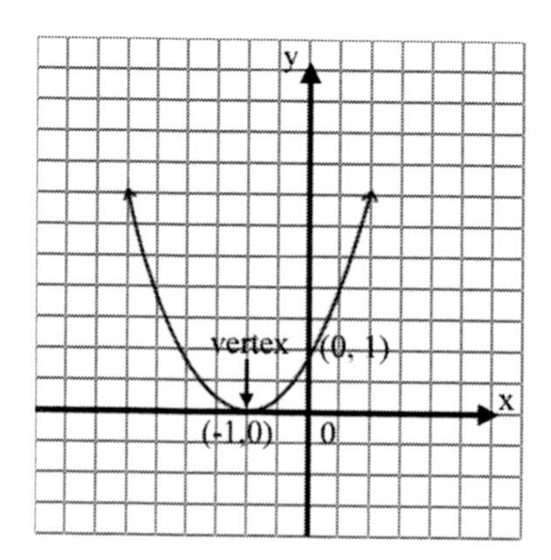

30. Sketch the quadratic curves, $y = x^2$ and $y = -x^2$ on the same Cartesian plane.

Answers:
Given $y = x^2$
In the form $y = ax^2 + bx + c$
$a = 1$, $b = 0$ and $c = 0$
=> Since $a = 1$ thus $a > 0$
Graph has the shape ∪
=> Vertex has a minimum point
To find y-intercept, let $x = 0$:
$y = (0)^2$
$= 0$
∴ y-intercept is (0, 0)
To find x-intercept, let $y = 0$:
$0 = x^2$
$x = 0$
∴ x-intercept is (0, 0)
To find vertex, use complete the squares:
$y = x^2$
$= (x + 0)^2 + 0$
∴ Vertex is (0, 0)

Also given $y = -x^2$
In the form $y = ax^2 + bx + c$
$a = -1$, $b = 0$, $c = 0$
=> Since $a = -1$ thus $a < 0$
Graph has the shape ∩
=> Vertex has maximum point
To find y-intercept, let x= 0:
$y = -(0)^2$
$= 0$
∴ y-intercept is (0, 0)
To find x-intercept, let $y = 0$:
$0 = -x^2$
$x = 0$
∴ x-intercept is (0, 0)
Vertex of graph $y = -x^2$ is also (0, 0)

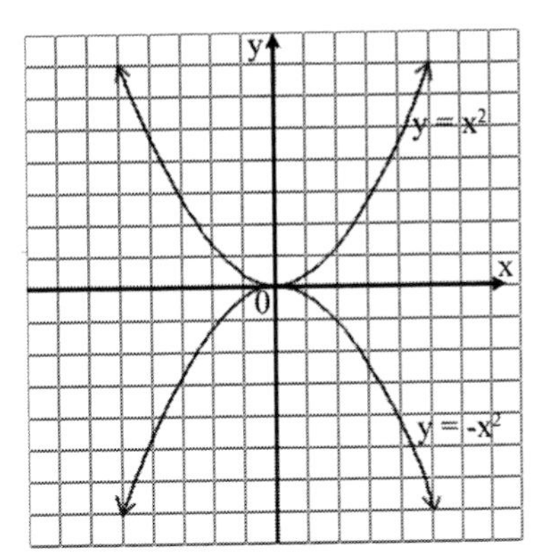

31. Sketch the graphs $y = ax^2 - 2$, where
a) $a = 2$
b) $a = 8$
c) $a = \frac{1}{2}$
on the same Cartesian plane.

Answers:
Given $y = ax^2 - 2$
a) For $a = 2$
=> $y = 2x^2 - 2$
In the form $y = ax^2 + bx + c$
$a = 2$, $b = 0$, $c = -2$
Since $a = 2$, $a > 0$
Graph has the shape ∪
To find y-intercept, let $x = 0$:
$y = 2(0)^2 - 2$

$y = -2$

$\therefore$ y-intercept is (0, –2)

To find x-intercept, let y = 0:

$0 = 2x^2 - 2$

$0 = 2(x^2 - 1)$

$0 = 2(x + 1)(x - 1)$

$x = -1$ or 1

$\therefore$ x-intercepts are (–1, 0) and (1, 0)

To find vertex, use complete the squares:

$y = 2x^2 - 2$

$y = 2(x)^2 - 2$ ⇦already in the form $y = a(x + A)^2 + B$

Vertex = (–A, B)

$\therefore$ Vertex is (0, –2) …(*)

b) Given a = 8

=> $y = 8x^2 - 2$

Since a = 8, a > 0

Graph has the shape ∪

To find y-intercept, let x = 0:

$y = 8(0)^2 - 2$

$y = -2$

$\therefore$ y-intercept is (0, –2)

To find x-intercept, let y = 0:

$0 = 8x^2 - 2$

$0 = 4x^2 - 1$ ⇦Divide by 2

$0 = (2x + 1)(2x - 1)$

$x = \pm\frac{1}{2}$

$\therefore$ x-intercepts are $(-\frac{1}{2}, 0)$ and $(\frac{1}{2}, 0)$

To find vertex, use complete the squares:

$y = 8x^2 - 2$

$y = 8(x)^2 - 2$ [refer to (*)]

$\therefore$ Vertex is (0, –2)

c) For $a = \frac{1}{2}$

=> $y = \frac{1}{2}x^2 - 2$

Since $a = \frac{1}{2}$, a > 0

Graph has the shape ∪

To find y-intercept, let x = 0:

$y = \frac{1}{2}(0) - 2$

$y = -2$

$\therefore$ y-intercept is (0, –2)

To find x-intercept, let y = 0:

$0 = \frac{1}{2}x^2 - 2$

$0 = x^2 - 4$ ⇦× 2 to remove fraction

$0 = (x + 2)(x - 2)$

$x = -2, 2$

$\therefore$ x-intercepts are (–2, 0) and (2, 0)

To find vertex, use complete the squares:

$y = \frac{1}{2}x^2 - 2$

$y = \frac{1}{2}(x)^2 - 2$ [refer to (*)]

$\therefore$ Vertex is (0, –2)

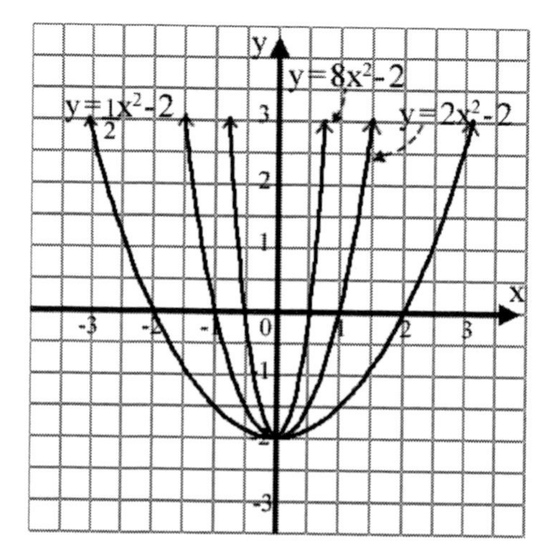

32. Given the quadratic curve, $y = 2x^2 + 4x - 6$. Determine the vertex and sketch the graph.

Answers:
Given $y = 2x^2 + 4x - 6$
In the form $y = ax^2 + bx + c$
$a = 2, b = 4, c = -6$
To find vertex, use complete the squares:
$y = 2x^2 + 4x - 6$

$$= 2\left(x^2 + \frac{4}{2}x + \frac{4^2}{(2(2))^2} - \frac{4^2}{(2(2))^2} - \frac{6}{2}\right)$$

$$= 2\left(\left(x + \frac{4}{2(2)}\right)^2 - \frac{4^2}{(4)^2} - \frac{6}{2}\right)$$

$$= 2\left((x+1)^2 - 1 - 3\right)$$

$$= 2\left((x+1)^2 - 4\right)$$

$$= 2(x+1)^2 - 8$$

$\therefore$ Vertex is $(-1, -8)$ //

Since $a = 2$, $a > 0$
Graph is in the shape ∪
=> Vertex has a minimum point
To find y-intercept, let $x = 0$:
$y = 2(0)^2 + 4(0) - 6$
$= -6$
$\therefore$ y-intercept is $(0, -6)$
To find x-intercept, let $y = 0$:
$0 = 2x^2 + 4x - 6$
$0 = x^2 + 2x - 3$ ⇦Divide by 2
$0 = (x + 3)(x - 1)$
$x = -3, 1$
$\therefore$ x-intercepts are $(-3, 0)$ and $(1, 0)$

Therefore on Cartesian plane:

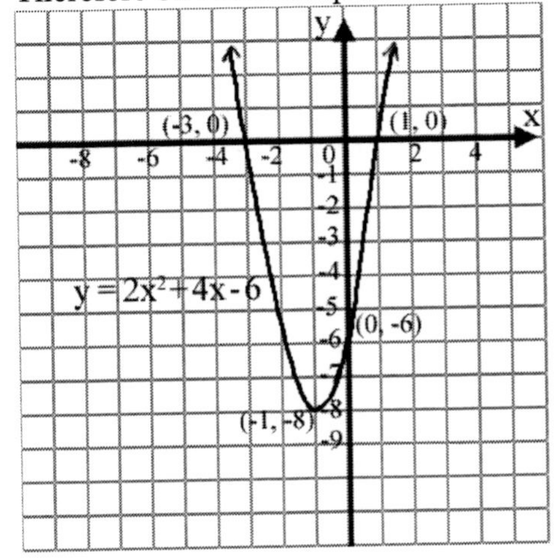

33. What is the equation of the circle in Figure 7.8, if its radius is 3 units and the center point is (3, 3)?

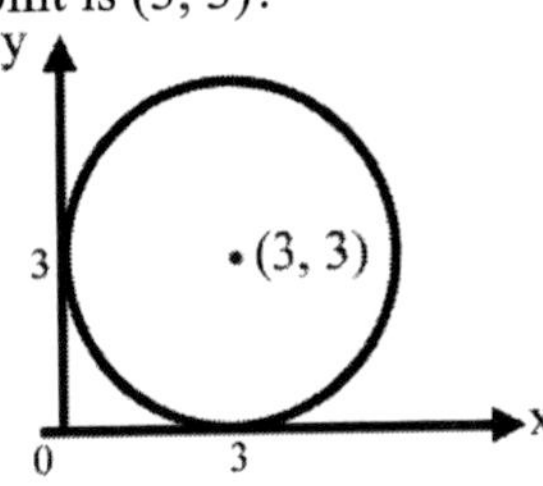

Figure 7. 8

Answer:
Given radius, $r = 3$
Center point = (3, 3)
Equation of circle:
$(x - x_1)^2 + (y - y_1)^2 = r^2$
$(x - 3)^2 + (y - 3)^2 = 3^2$
$(x - 3)^2 + (y - 3)^2 = 9$ //

34. In a Cartesian plane, sketch the graph of the equation $(x - 1)^2 + (y + 2)^2 = 4$.

Answer:
Given equation: $(x - 1)^2 + (y + 2)^2 = 4$
=> $(x - 1)^2 + (y + 2)^2 = 2^2$ ...(*)
Since equation of a circle takes the form:
$(x - x_1)^2 + (y - y_1)^2 = r^2$
Equation in (*) is a circle
=> $x_1 = 1, y_1 = -2, r = 2$
=> Center point, $(x_1, y_1) = (1, -2)$
=> Radius, $r = 2$

Thus, on Cartesian plane:

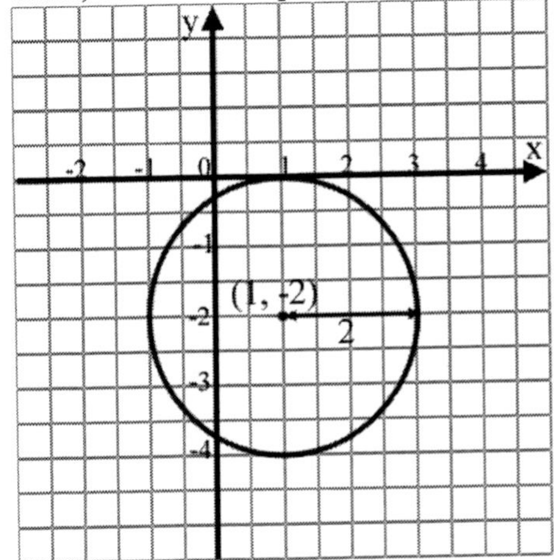

35. Determine the equation of the following circles:

a) center (2, –3) and radius 7 units in length
b) center at the origin and radius 3 units in length

Answers:
a) Given center = (2, –3)
radius, r = 7
Equation of circle:
$(x - x_1)^2 + (y - y_1)^2 = r^2$
$(x - 2)^2 + (y + 3)^2 = 7^2$
$(x - 2)^2 + (y + 3)^2 = 49$ //

b) Given center = origin
= (0, 0)
radius, r = 3
Equation of circle:
$(x - x_1)^2 + (y - y_1)^2 = r^2$
$(x - 0)^2 + (y - 0)^2 = 3^2$
$x^2 + y^2 = 9$ //

36. Circle O has center point O(2, 3) and radius, 5 units. Find the equation of circle O. Hence find a point on its circumference.

Answers:
Given center point = (2, 3)
Radius, r = 5
Equation of circle O:
$(x - x_1)^2 + (y - y_1)^2 = r^2$
$(x - 2)^2 + (y - 3)^2 = 5^2$
$(x - 2)^2 + (y - 3)^2 = 25$ // ...(*)

A point on the circumference:
Circumference of a circle is made up of many points.
Let $y_2 = 3$,
Substitute $x = x_2$ and $y_2 = 3$ into (*):
$(x_2 - 2)^2 + (3 - 3)^2 = 25$
$(x_2 - 2)^2 = 25$
$x_2 - 2 = \pm\sqrt{25}$
$= \pm 5$
$x_2 = -5 + 2, 5 + 2$
$= -3, 7$
∴ Point on the circumference is (–3, 3) or (7, 3). //

37. Find the center point and radius of the following circles:

a) $x^2 + y^2 - 4x + 2y - 3 = 0$
b) $2x^2 + 2y^2 + 2x - 4y - 7 = 0$

Answers:
a) Given equation of circle:
$x^2 + y^2 - 4x + 2y - 3 = 0$
To convert to standard form, use complete the squares:
=> $x^2 + y^2 - 4x + 2y - 3 = 0$
$(x^2 - 4x) + (y^2 + 2y) = 3$

$$\left[\left(x + \frac{-4}{2}\right)^2 - \frac{(-4)^2}{4}\right] + \left[\left(y + \frac{2}{2}\right)^2 - \frac{2^2}{4}\right] = 3$$

$(x - 2)^2 - 4 + (y + 1)^2 - 1 = 3$
$(x - 2)^2 + (y + 1)^2 = 8$
Compare to the standard form of equation of circle:
$(x - x_1)^2 + (y - y_1)^2 = r^2$
=> Center point, $(x_1, y_1) = (2, -1)$ //
=> Radius, $r = \sqrt{8}$ units //

b) Given equation of circle:
$2x^2 + 2y^2 + 2x - 4y - 7 = 0$
To convert to standard form, use complete the squares:
=> $2x^2 + 2y^2 + 2x - 4y - 7 = 0$
$x^2 + y^2 + x - 2y - \frac{7}{2} = 0$ ⇦Divide by 2
$(x^2 + x) + (y^2 - 2y) = \frac{7}{2}$

$$\left[\left(x+\frac{1}{2}\right)^2-\frac{1}{4}\right]+\left[\left(y+\frac{-2}{2}\right)^2-\frac{(-2)^2}{4}\right]=\frac{7}{2}$$

$$\left(x+\frac{1}{2}\right)^2-\frac{1}{4}+(y-1)^2-1=\frac{7}{2}$$

$$\left(x+\frac{1}{2}\right)^2+(y-1)^2=\frac{19}{4}$$

Compare to the standard form of equation of circle: $(x-x_1)^2+(y-y_1)^2=r^2$

=> Center point, $(x_1, y_1) = (-\frac{1}{2}, 1)$ //

=> Radius, $r=\sqrt{\frac{19}{4}}=\frac{1}{2}\sqrt{19}$ units //

38. Figure 7.9 represents a circle with radius 2 units and center point, (3, 2). A straight line, $y = 2x - 8$ passes through the circumference of the circle. Find the coordinates where the straight line intersects the circle.

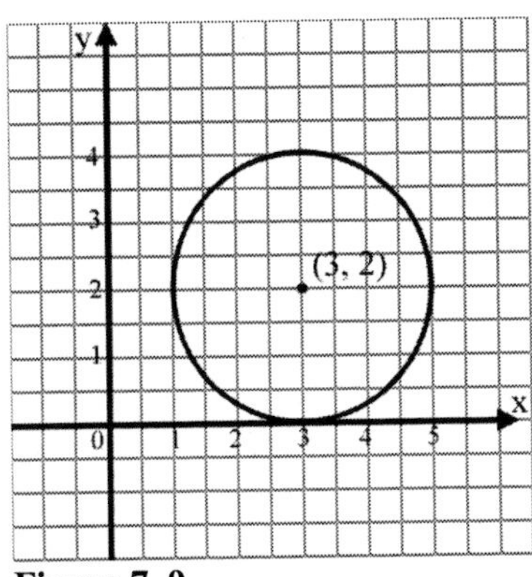

Figure 7. 9

Answers:
Given radius, $r = 2$
Center point = (3, 2)
Equation of circle:
$(x-x_1)^2+(y-y_1)^2=r^2$
$(x-3)^2+(y-2)^2=2^2$
$(x-3)^2+(y-2)^2=4$ ...(1)
Also given $y = 2x - 8$ ...(2)
Substitute (2) into (1):
$(x-3)^2+(2x-8-2)^2=4$
$x^2-6x+9+(2x-10)^2=4$
$x^2-6x+5+4x^2-40x+100=0$
$5x^2-46x+105=0$ ...(3)
Solve equation (3) using quadratic formula:

$$x=\frac{-b\pm\sqrt{b^2-4ac}}{2a}$$

$$x=\frac{46\pm\sqrt{(-46)^2-4(5)(105)}}{2(5)}$$

$$=\frac{46\pm\sqrt{2116-2100}}{10}$$

$$=\frac{46\pm\sqrt{16}}{10}$$

$$=\frac{46\pm 4}{10}$$

$= 5$ or 4.2
Substitute $x = 5$ into (2):
$y = 2(5) - 8$
$= 10 - 8$
$= 2$
∴ Intersection point is (5, 2)
Substitute $x = 4.2$ into (2):
$y = 2(4.2) - 8$
$= 8.4 - 8$
$= 0.4$
∴ Intersection point is (4.2, 0.4)
∴ Line cuts the circle at (5, 2) and (4.2, 0.4). //

39. Circle O has the equation $(x-2)^2+(y-1)^2=3^2$. Determine if point, M(2, 5) is located inside, outside or on the circumference of circle O.

Answer:
Given M = (2, 5)
Equation of circle O:

$(x-2)^2 + (y-1)^2 = 3^2$
=> radius, r = 3
=> center of circle O = (2, 1)
Distance from O to M:

$$OM = \sqrt{(x_2 - x_1)^2 + (y_2 - y_1)^2}$$
$$= \sqrt{(2-2)^2 + (5-1)^2}$$
$$= \sqrt{0^2 + 4^2}$$
$$= \sqrt{4^2}$$
$$= 4 \text{ units}$$

Since radius of circle O is 3 units, while distance from point O to point M is 4 units, point M lies *outside* of circle O. //

40. Circle O has a center point at O(3, 6). P(6, 10) is a point on the circumference of circle O. What is the diameter of circle O?

Answer:
Given center point of circle O = (3, 6)
Point on circumference, P = (6, 10)
Length of radius, r = OP

$$OP = \sqrt{(x_2 - x_1)^2 + (y_2 - y_1)^2}$$
$$= \sqrt{(6-3)^2 + (10-6)^2}$$
$$= \sqrt{3^2 + 4^2}$$
$$= \sqrt{9+16}$$
$$= \sqrt{25}$$
$$= 5 \text{ units}$$

Diameter = 2 × radius
= 2 × 5
= 10 units //

41. Find the equation of the circle, whose endpoints of its diameter are P(0, 4) and Q(6, 2).

Answer:
Given P = (0, 4)
Q = (6, 2)
$\overline{PQ}$ = diameter
Center of circle:
= midpoint of $\overline{PQ}$

$$= \left(\frac{x_1 + x_2}{2}, \frac{y_1 + y_2}{2}\right)$$
$$= \left(\frac{0+6}{2}, \frac{4+2}{2}\right)$$
$$= (3, 3) \quad \ldots(1)$$

Length of diameter, PQ:

$$= \sqrt{(x_2 - x_1)^2 + (y_2 - y_1)^2}$$
$$= \sqrt{(6-0)^2 + (2-4)^2}$$
$$= \sqrt{6^2 + (-2)^2}$$
$$= \sqrt{36+4}$$
$$= \sqrt{40} \quad \Leftarrow \text{Note: } \sqrt{40} = \sqrt{4 \times 10}$$
$$= 2\sqrt{10} \text{ units}$$

Radius, r = diameter ÷ 2
$= 2\sqrt{10} \div 2$
$= \sqrt{10}$ units ...(2)

Equation of circle:
=> $(x - x_1)^2 + (y - y_1)^2 = r^2$

$(x-3)^2 + (y-3)^2 = \sqrt{10}^2$ ⇦From (1) & (2)
$(x-3)^2 + (y-3)^2 = 10$ //

42. P(3, 4) and Q(7, 12) are on the circumference of a circle whose diameter is $\overline{PQ}$. What is the equation of the circle?

Answer:
Given P = (3, 4)
Q = (7, 12)
$\overline{PQ}$ = diameter
Center point of circle:
= midpoint of $\overline{PQ}$

$$= \left(\frac{x_1 + x_2}{2}, \frac{y_1 + y_2}{2}\right)$$

$= \left(\frac{3+7}{2}, \frac{4+12}{2} \right)$

$= \left(\frac{10}{2}, \frac{16}{2} \right)$

$= (5, 8)$ ...(1)

Length of diameter PQ:

$= \sqrt{(x_2 - x_1)^2 + (y_2 - y_1)^2}$

$= \sqrt{(7-3)^2 + (12-4)^2}$

$= \sqrt{4^2 + 8^2}$

$= \sqrt{16 + 64}$

$= \sqrt{80}$ ⇦ Note: $\sqrt{80} = \sqrt{4 \times 20}$

$= 2\sqrt{20}$ units

Radius, r = diameter ÷ 2

$= 2\sqrt{20} \div 2$

$= \sqrt{20}$ units ...(2)

Equation of circle:

=> $(x - x_1)^2 + (y - y_1)^2 = r^2$

$(x - 5)^2 + (y - 8)^2 = \sqrt{20}^2$ ⇦From (1) & (2)

$(x - 5)^2 + (y - 8)^2 = 20$ //

43. Find the points where the circles, $(x - 2)^2 + (y + 1)^2 = 5$ and $(x + 1)^2 + (y + 4)^2 = 5$ meet.

Answers:

Given, $(x - 2)^2 + (y + 1)^2 = 5$ ...(1)

$x^2 - 4x + 4 + y^2 + 2y + 1 = 5$

$x^2 - 4x + y^2 + 2y = 0$ ...(2)

Also given, $(x + 1)^2 + (y + 4)^2 = 5$

$x^2 + 2x + 1 + y^2 + 8y + 16 = 5$

$x^2 + 2x + y^2 + 8y + 12 = 0$ ...(3)

Let (3) – (2):

$6x + 6y + 12 = 0$

$x + y + 2 = 0$

$y = -x - 2$ ...(4)

Substitute (4) into (1):

$(x - 2)^2 + (-x - 2 + 1)^2 = 5$

$x^2 - 4x + 4 + x^2 + 2x + 1 = 5$

$2x^2 - 2x = 0$

$2x(x - 1) = 0$

$x = 0$ or 1

For x = 0, substitute into (4): ...**

$y = -0 - 2$

$y = -2$

∴ Point is (0, –2)

For x = 1, substitute into (4):

$y = -1 - 2$

$y = -3$

∴ Point is (1, –3)

∴ The two circles meet at (0, –2) and (1, –3). //

**Note: If you had substituted x = 0 and x = 1 into equation (1), you will find 4 points which is not true because 2 circles that intersect will meet at most at 2 points (it is possible they meet at 1 point or not meet at all!).

44. Figure 7.10 is a circle whose center point is (6, 2) and radius, 4 units. M is a straight line with equation, y = –2. Show that M is a tangent to the circle.

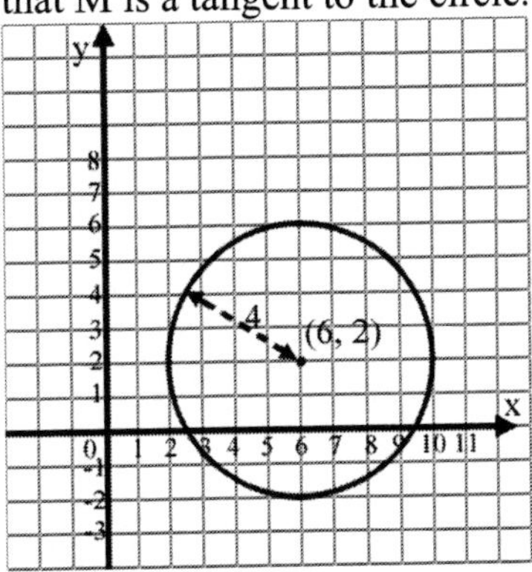

Figure 7. 10

Answer:

Given radius, r = 4

Center point = (6, 2)

Equation of circle:

$(x - x_1)^2 + (y - y_1)^2 = r^2$

$(x - 6)^2 + (y - 2)^2 = 4^2$ ...(1)

Equation of straight line, M:

$y = -2$ ...(2)

Substitute (2) into (1):

$(x - 6)^2 + (-2 - 2)^2 = 4^2$

$(x - 6)^2 + (-4)^2 = 16$

$(x - 6)^2 + 16 = 16$

$(x - 6)^2 = 0$

$x = 6$

Since line M touches the circumference of the circle at *one point*, (6, –2), thus line M is a tangent to the circle. //

45. Figure 7.11 shows a distance-time graph for Corey's travel to his grandma's home. Find:
a) The duration Corey had stopped for rest.
b) Corey's speed during the first 9 minutes
c) Corey's average speed for the entire journey.

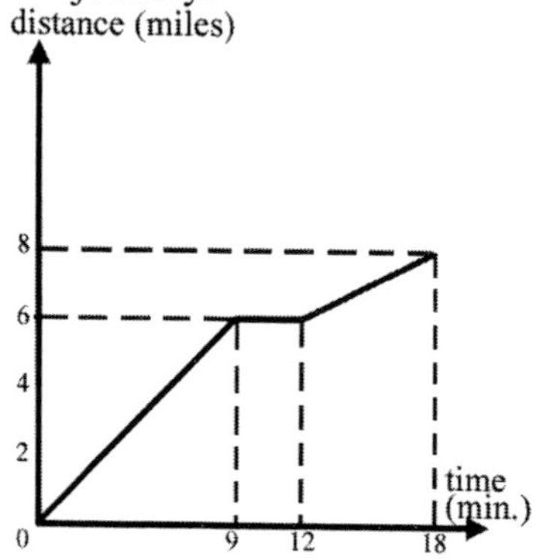

Figure 7. 11

Answers:
a) Duration Corey had stopped:
= duration where slope is 0
= 12 – 9
= 3 minutes //

b) $\text{Slope} = \dfrac{\text{Distance traveled}}{\text{Time}}$

$= \dfrac{6-0}{9-0}$

$= \dfrac{2}{3}$ miles per minute. //

c) Average speed:

$= \dfrac{\text{Total distance traveled}}{\text{Total time taken}}$

$= \dfrac{8}{18}$

$= \dfrac{4}{9}$ miles per minute. //

46. The distance-time graph shown in Figure 7.12 represents Juan's drive from point A to B and B to A. Find Juan's speed from A to B and B to A. Hence find Juan's average speed.

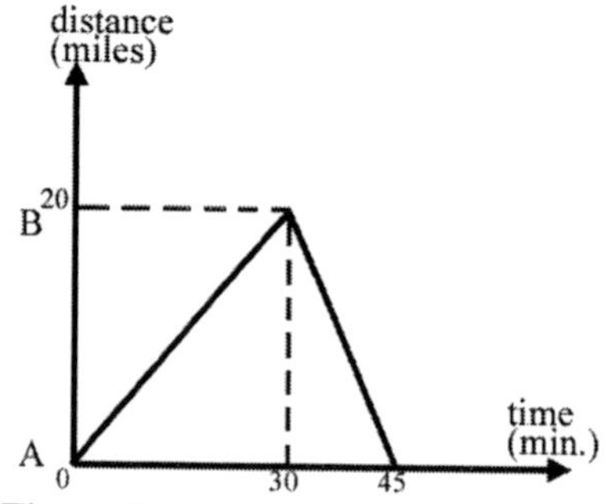

Figure 7. 12

Answers:
Juan's speed from A to B:

$\text{Slope} = \dfrac{\text{Distance traveled}}{\text{Time}}$

$= \dfrac{20-0}{30-0}$

$= \dfrac{2}{3}$ miles per minute. //

Juan's speed from B to A:

$\text{Slope} = \dfrac{\text{Distance traveled}}{\text{Time}}$

$= \dfrac{0-20}{45-30}$

$= -\dfrac{20}{15}$

$= -\dfrac{4}{3}$ miles per minute. //

Average speed:

$= \dfrac{\text{Total distance}}{\text{Total time}}$

$= \dfrac{20+20}{45}$

$= \dfrac{40}{45}$

$= \dfrac{8}{9}$ miles per minute. //

47. The speed-time graph in Figure 7.13 shows the truck's motion over a distance in 10 minutes. Find its

a) Acceleration
b) Deceleration

speed (miles per minute)

0.75 B C
0.50 A
D time (min.)
0 3 5 10

Figure 7. 13

Answers:

a) Acceleration:

$= \text{gradient of line } \overline{AB}$

$= \dfrac{0.75-0.5}{3-0}$

$= \dfrac{0.25}{3}$

$= \dfrac{2.5}{30}$

$= \dfrac{0.5}{6}$

$= 0.08333 \text{ miles minute}^{-2}$ //

b) Deceleration:

$= \text{gradient of line } \overline{CD}$

$= \dfrac{0-0.75}{10-5}$

$= \dfrac{-0.75}{5}$

$= -0.15 \text{ miles minute}^{-2}$ //

48. Figure 7.14 shows the speed-time graph for a car in 45 minutes. Find the:

a) Distance when the car had traveled at a constant speed.
b) Mean speed of the car for the whole 45 minutes.

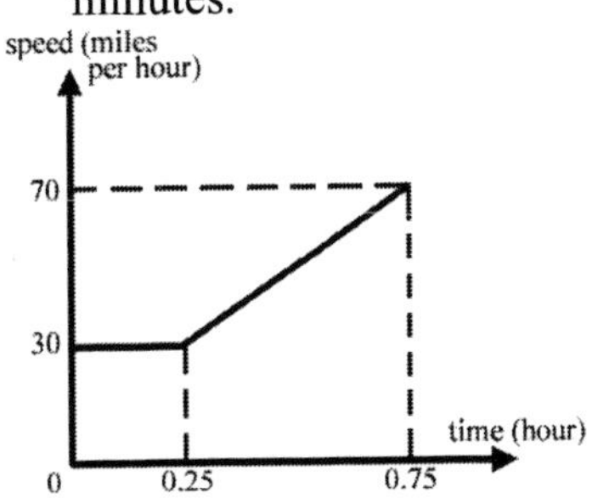

Figure 7. 14

Answers:

Graph below is a reproduction of Figure 7.14 for labeling.

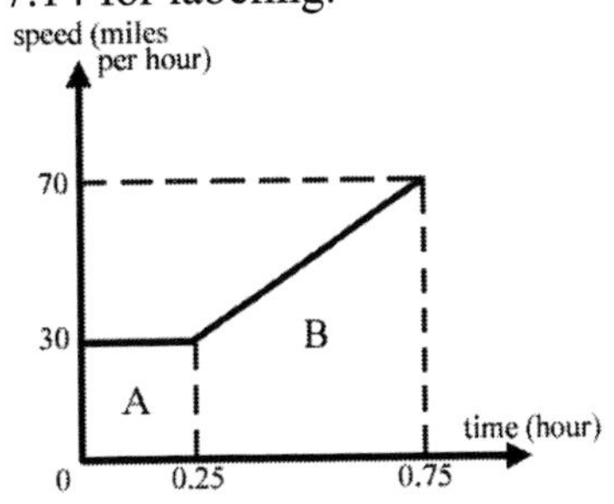

a) Distance car had traveled at constant speed:

= area of rectangle A

$= 30 \times 0.25$

$= 7.5$ miles //

b) Total distance the car had traveled in 0.75 hour (i.e. 45 minutes):

= area of rectangle A + area of trapezoid B

$= 7.5 + \dfrac{1}{2} \times (0.75-0.25) \times (70+30)$

$= 7.5 + \dfrac{1}{2} \times 0.5 \times 100$

$= 7.5 + 25$

$= 32.5$ miles

Mean speed:

$= \dfrac{\text{Total distance traveled}}{\text{Total time}}$

$= \dfrac{32.5}{0.75}$

$= 43.33$ miles per hour. //

49. Figure 7.15 shows the speed-time graph of a particle in one hour. Find the:
a) Rate of change of the particle's speed in the last 0.5 hour.
b) w, if the particle had traveled 25 miles in the first 0.5 hour.

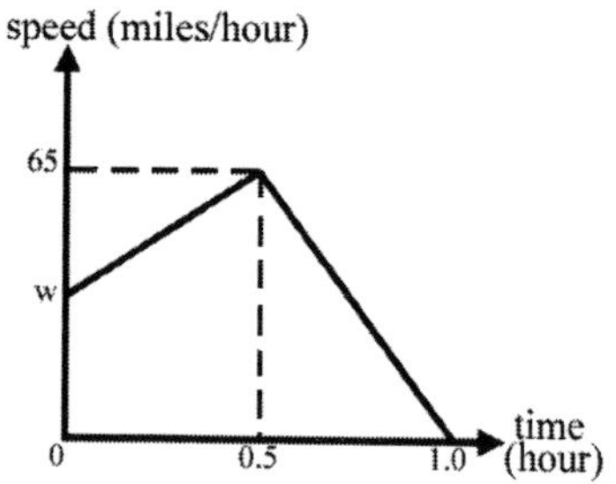

Figure 7. 15

Answers:
a) Deceleration:

$= \dfrac{0-65}{1-0.5}$

$= \dfrac{-65}{0.5}$

$= -130 \text{ miles hour}^{-2}$ //

b) Given distance traveled in the first 0.5 hours = 25
=> Area of trapezoid = 25

$\frac{1}{2} \times 0.5 \times (w+65) = 25$

$w + 65 = \frac{25}{0.5} \times 2$

$w + 65 = 100$

$\therefore\ w = 35$ miles/hour //

50. Figure 7.16 shows the speed-time graph for two vehicles, A and B for 1 hour. What is the difference of the distance traveled by vehicle A and vehicle B during the first hour?

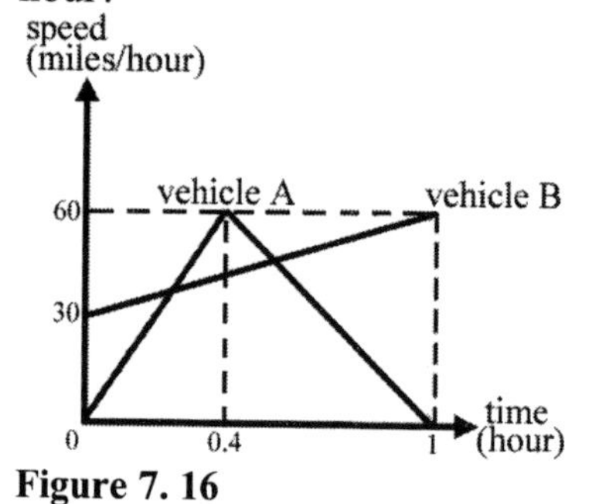

Figure 7. 16

Answer:
Let D = distance traveled
Total distance traveled by vehicle A:
D_A = area of triangle

$= \frac{1}{2} \times \text{height} \times \text{base}$

$= \frac{1}{2} \times 60 \times 1$

= 30 miles

Total distance traveled by vehicle B:
D_B = area of trapezoid

$= \frac{1}{2} \times \text{height} \times (\text{bases})$

$= \frac{1}{2} \times 1 \times (30 + 60)$

= 45 miles

Difference of total distance traveled:
$= D_B - D_A$
$= 45 - 30$
= 15 miles //

51. Given A(6, 8), B(1, 0) and C(5, 4). What is the area of the triangle ABC?

Answer:

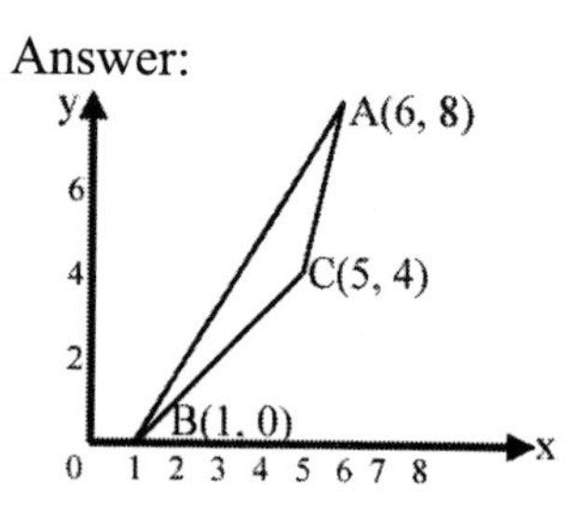

Given A(6, 8), B(1, 0) and C(5, 4)

Coordinates are placed counterclockwise in matrix.

Area of $\triangle ABC$:

$$=\frac{1}{2}\begin{vmatrix} 6 & 1 & 5 & 6 \\ 8 & 0 & 4 & 8 \end{vmatrix}$$

$$=\frac{1}{2}\left(6(0)+1(4)+5(8)-8(1)-0(5)-4(6)\right)$$

$$=\frac{1}{2}\left(4+40-8-24\right)$$

$$=\frac{1}{2}\times 12$$

$= 6 \text{ units}^2$ //

52. Given P(−2, 3), Q(4, 4) and R(1, 0). Find the area of triangle PQR.

Answer:

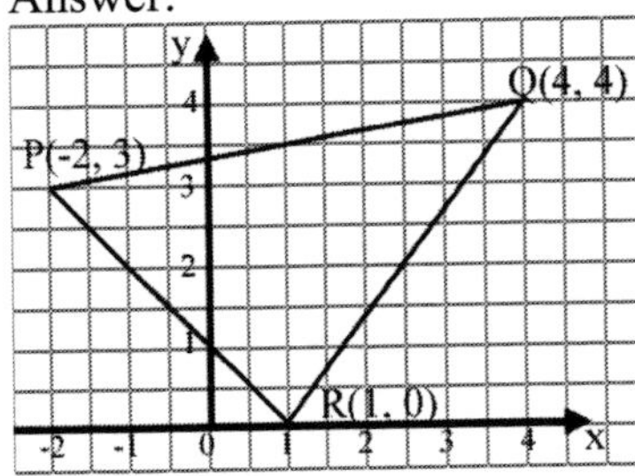

Given P(−2, 3), Q(4, 4) and R(1, 0)

Area of $\triangle PQR$:

$$=\frac{1}{2}\begin{vmatrix} -2 & 1 & 4 & -2 \\ 3 & 0 & 4 & 3 \end{vmatrix}$$

$$=\frac{1}{2}\left(-2(0)+1(4)+4(3)-3(1)-0(4)-4(-2)\right)$$

$$=\frac{1}{2}\left(4+12-3+8\right)$$

$$=\frac{1}{2}\times 21$$

$= 10.5 \text{ units}^2$ //

53. A(2, 1), B(m, 2) and C(4, 4) are vertices in a triangle. If the area of triangle ABC is 5 units^2, what are the possible values of m?

Answer:

Given area of $\triangle ABC = 5$

Since area can be positive or negative depending on the arrangement of vertices:

Thus, area of $\triangle ABC = \pm 5$

$$\frac{1}{2}\begin{vmatrix} 2 & m & 4 & 2 \\ 1 & 2 & 4 & 1 \end{vmatrix} = \pm 5$$

$$\frac{1}{2}\left(2(2)+m(4)+4-m-2(4)-4(2)\right)=\pm 5$$

$$\frac{1}{2}\left(4+4m+4-m-8-8\right)=\pm 5$$

$$\frac{1}{2}(3m-8)=\pm 5$$

$3m - 8 = \pm 10$

For +10:

$3m - 8 = 10$

$3m = 18$

$m = 6$

For −10:

$3m - 8 = -10$

$3m = -2$

$m = -\frac{2}{3}$

$\therefore$ m is 6 or $-\frac{2}{3}$ //

If the vertices in the matrix were placed counterclockwise then m = 6

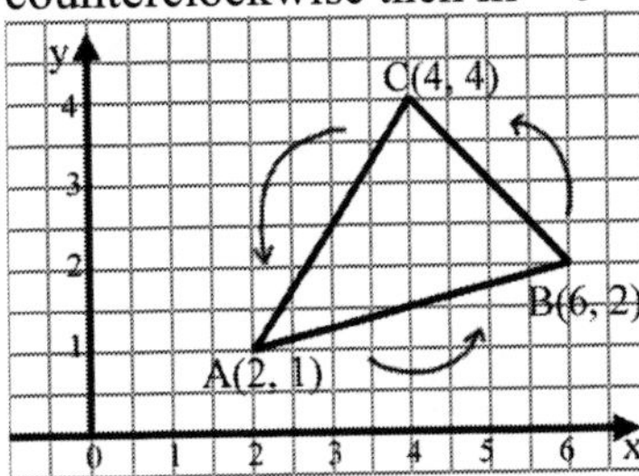

If the vertices in the matrix were placed clockwise then $m = -\frac{2}{3}$

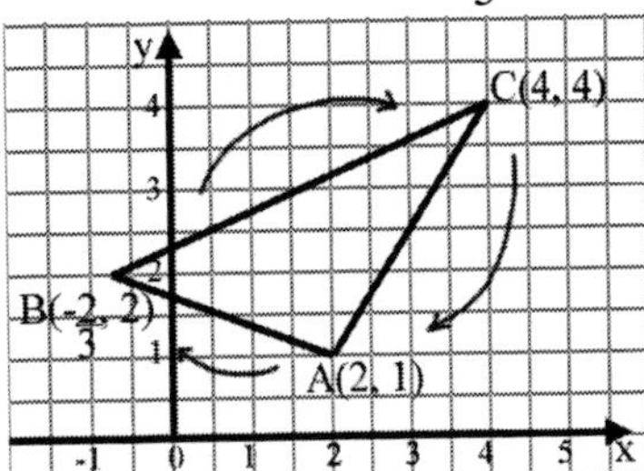

54. Given A(1, 2), B(7, 4) and C(−2, 1). Determine if ABC is a triangle.

Answer:

Given A(1, 2), B(7, 4) and C(−2, 1)

$= \frac{1}{2}\begin{vmatrix} 1 & 7 & -2 & 1 \\ 2 & 4 & 1 & 2 \end{vmatrix}$

$= \frac{1}{2}\left(1(4)+7(1)-2(2)-2(7)-4(-2)-1(1)\right)$

$= \frac{1}{2}\left(4+7-4-14+8-1\right)$

$= \frac{1}{2}\times(0)$

$= 0 \text{ units}^2$

$\therefore$ ABC is a straight line. //

55. A(2, 6), B(6, 7), C(5, 4) and D(1, 3) are vertices in a parallelogram. Find the area of parallelogram ABCD.

Answer:
Given A(2, 6), B(6, 7), C(5, 4) and D(1, 3)

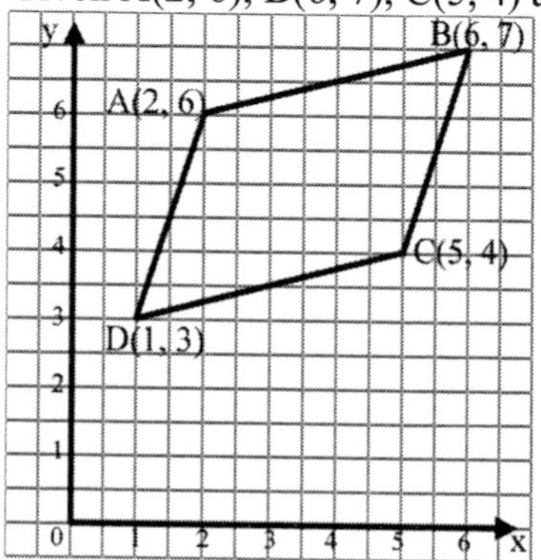

Area of parallelogram ABCD:

$= \frac{1}{2}\begin{vmatrix} 2 & 1 & 5 & 6 & 2 \\ 6 & 3 & 4 & 7 & 6 \end{vmatrix}$

$= \frac{1}{2}\left(2(3)+1(4)+5(7)+6(6)-6(1)-3(5)-4(6)-7(2)\right)$

$= \frac{1}{2}\left(6+4+35+36-6-15-24-14\right)$

$= \frac{1}{2}(22)$

$= 11 \text{ units}^2$ //

56. Figure 7.17 shows a quadrilateral PQRS. What is the area of the quadrilateral PQRS?

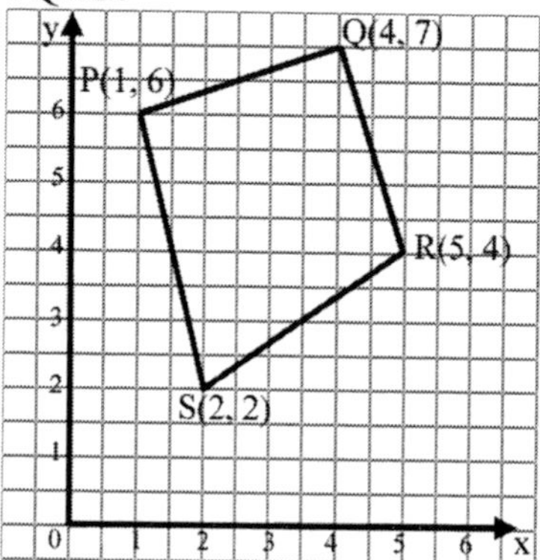

Figure 7. 17

Answer:
Given P(1, 6), Q(4, 7), R(5, 4) and S(2, 2)
Area of quadrilateral PQRS:

$= \frac{1}{2}\begin{vmatrix} 2 & 5 & 4 & 1 & 2 \\ 2 & 4 & 7 & 6 & 2 \end{vmatrix}$

$= \frac{1}{2}\left(2(4)+5(7)+4(6)+1(2)-2(5)-4(4)-7(1)-6(2)\right)$

$= \frac{1}{2}\left(8+35+24+2-10-16-7-12\right)$

$= \frac{1}{2}(24)$

$= 12 \text{ units}^2$ //

57. Figure 7.18 shows a rhombus ABCD. Find:

a) Equation $\overline{BD}$
b) Intersection point of lines, $\overline{AC}$ and $\overline{BD}$
c) Coordinate B
d) Area of rhombus ABCD

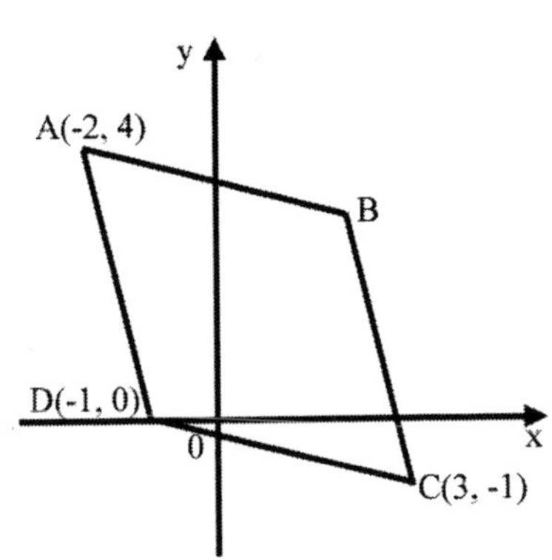

Figure 7. 18

Answers:

a) Given A(−2, 4) and C(3, −1)

Slope of $\overline{AC}$, $m_{AC} = \dfrac{y_2 - y_1}{x_2 - x_1}$

$= \dfrac{-1-4}{3-(-2)}$

$= \dfrac{-5}{5}$

$= -1$ …(1)

Diagonals of a rhombus are ⊥:

=> $\overline{AC}$ and $\overline{BD}$ are diagonals

=> $\overline{AC} \perp \overline{BD}$

Let, m_{BD} = slope of $\overline{BD}$

Product of gradient:

$m_{AC} \times m_{BD} = -1$

Substitute (1):

$-1 \times m_{BD} = -1$

$m_{BD} = 1$

Using point slope form: D(−1, 0), $m_{BD} = 1$

$y - y_1 = m(x - x_1)$

$y - (0) = 1(x - (-1))$

$y = x + 1$

∴ Equation of $\overline{BD}$: y = x + 1 // …(2)

b) To find equation of $\overline{AC}$:

Using point slope form: A(−2, 4), $m_{AC} = -1$

$y - y_1 = m(x - x_1)$

$y - 4 = -1(x - (-2))$

$y - 4 = -x - 2$

$y = -x + 2$ …(3)

Substitute (2) into (3):

$x + 1 = -x + 2$

$2x = 1$

$x = \dfrac{1}{2}$

Substitute $x = \dfrac{1}{2}$ into (2):

$y = \dfrac{1}{2} + 1$

$y = \dfrac{3}{2}$

∴ Intersection point of $\overline{AC}$ and $\overline{BD}$ is $(\dfrac{1}{2}, \dfrac{3}{2})$. //

c) From part (b), the intersection point $(\dfrac{1}{2}, \dfrac{3}{2})$ of $\overline{AC}$ and $\overline{BD}$ is the midpoint of $\overline{BD}$.

Let B = (x_B, y_B)

Using midpoint formula:

Midpoint of $\overline{BD} = \left(\dfrac{x_1 + x_2}{2}, \dfrac{y_1 + y_2}{2}\right)$

$\left(\dfrac{1}{2}, \dfrac{3}{2}\right) = \left(\dfrac{-1 + x_B}{2}, \dfrac{0 + y_B}{2}\right)$

Compare like components:

x–coordinate:

$\dfrac{1}{2} = \dfrac{-1 + x_B}{2}$

$1 = -1 + x_B$

$x_B = 1 + 1 = 2$

y–coordinate:

$\dfrac{3}{2} = \dfrac{0 + y_B}{2}$

$3 = 0 + y_B$

$y_B = 3$

∴ B(2, 3) //

d) Area of rhombus ABCD:

$= \dfrac{1}{2}\begin{vmatrix} -2 & -1 & 3 & 2 & -2 \\ 4 & 0 & -1 & 3 & 4 \end{vmatrix}$

$= \dfrac{1}{2}\big((-2)(0) + (-1)(-1) + 3(3) + 2(4)$
$\quad - 4(-1) - 0(3) - (-1)(2) - 3(-2)\big)$

$= \dfrac{1}{2}(1 + 9 + 8 + 4 + 2 + 6)$

$= \dfrac{1}{2}(30)$

= 15 units^2

$\therefore$ Area of rhombus is 15 units^2. //

58. Figure 7.19 shows two parallel lines $\overline{AB}$ and $\overline{CD}$. If the equation for $\overline{AB}$ is 2x + 3y = 6, find the value of k and the area of the shaded $\triangle$ACD.

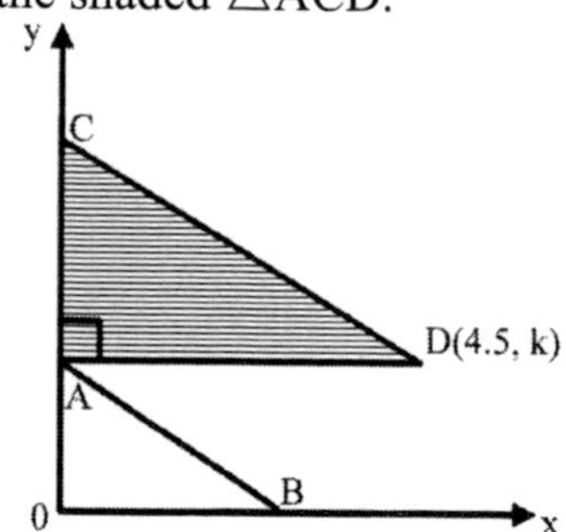

Figure 7. 19

Answers:

Given, $\overline{AB}$: 2x + 3y = 6 ...(1)

At y-axis, let x = 0:

=> 2(0) + 3y = 6

3y = 6

y = 2

Point D is on the same horizontal line as point A,

$\therefore$ k = 2 //

From (1),

2x + 3y = 6 ⇦ General form

$y = -\frac{2}{3}x + 2$ ⇦ slope intercept form

Slope of $\overline{AB}$, $m_{AB} = -\frac{2}{3}$

Since given, $\overline{AB} \parallel \overline{CD}$:

=> Slope, $m_{AB} = m_{CD} = -\frac{2}{3}$

By point slope form: D(4.5, 2), $m_{CD} = -\frac{2}{3}$

$y - 2 = -\frac{2}{3}(x - 4.5)$

3y – 6 = –2x + 9

3y + 2x = 15 ...(2)

At y-axis, substitute x = 0 into (2):

3y + 2(0) = 15

3y = 15

y = 5

$\therefore$ C(0, 5)

Thus, area of $\triangle$ACD:

$= \frac{1}{2} \times AC \times AD$

$= \frac{1}{2} \times (5-2) \times (4.5-0)$

= 6.75 units^2 //

59. Find the approximate area of the outline duck figurine shown in Figure 7.20.

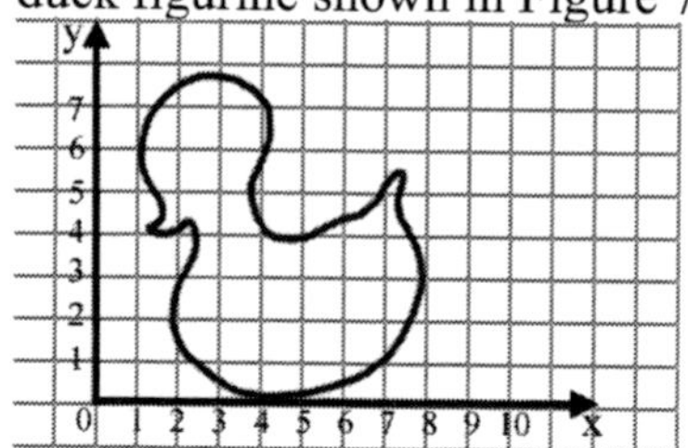

Figure 7. 20

Answer:

Question asked for approximation, thus squares that are 0.5 and more are included in our measurement:

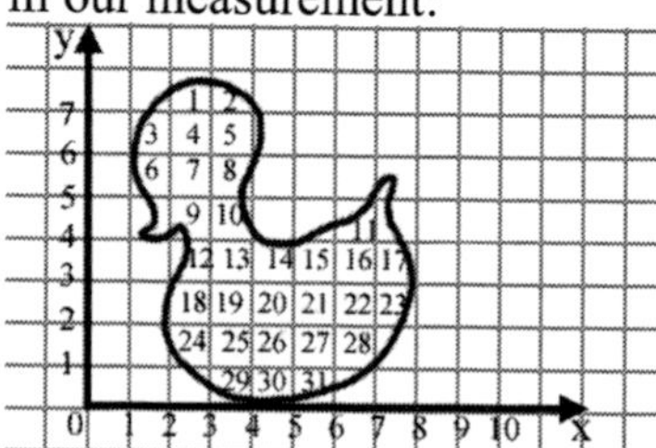

$\therefore$ The area of the outline figurine is approximately 31 units^2. //

Chapter 8
Locus

Locus (plural **loci**) – describing movement of point/points that satisfy one or more conditions

5 general conditions:

1. Fixed distance from a point – P is a point that moves such that it is always k units from a fix point, O. Thus, locus P is a **circle**.

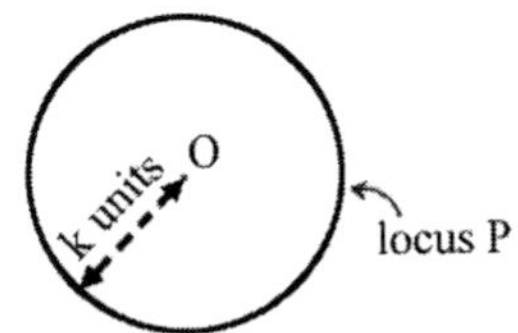

2. Fixed distance from a line – P is a point that moves such that it is always k units from line, L. Thus locus P is **two parallel lines**.

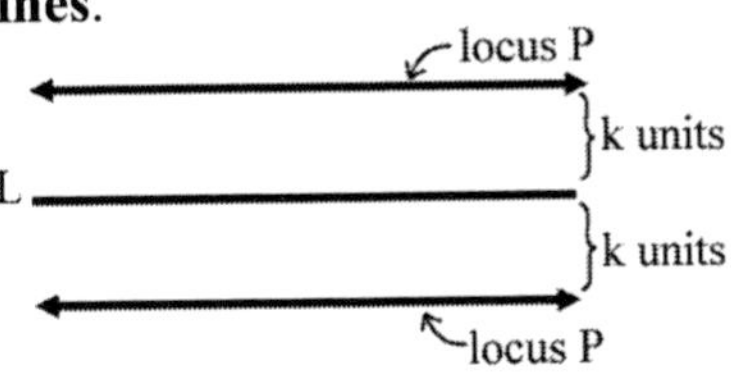

3. Equidistant from two parallel lines – P is a point that moves such that it is always equidistant between two parallel lines, L_1 and L_2. Thus P is a line parallel to and equidistant from the two lines.

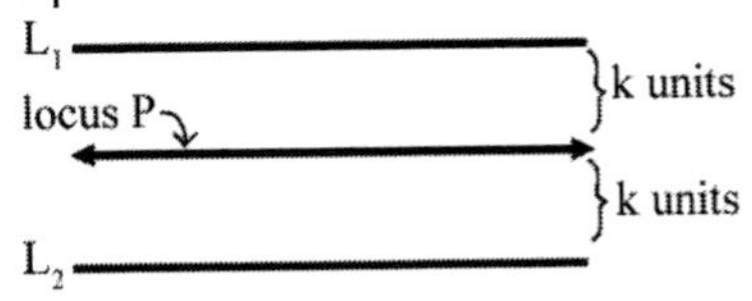

4. Equidistant from two points – P is a point that moves such that it is always equidistant from two points. Thus P is a **perpendicular bisector of the line segment**.

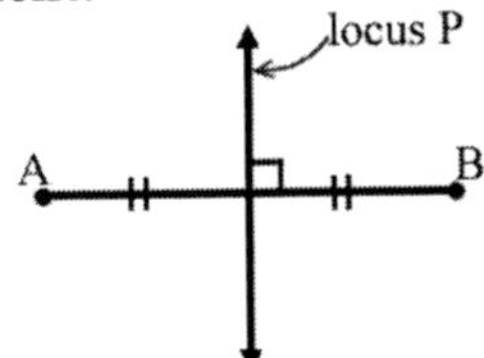

5. Equidistant from two intersecting lines – P moves such that it is always equidistant from two intersecting lines, L_3 and L_4. Thus, P is an **angle bisector**.

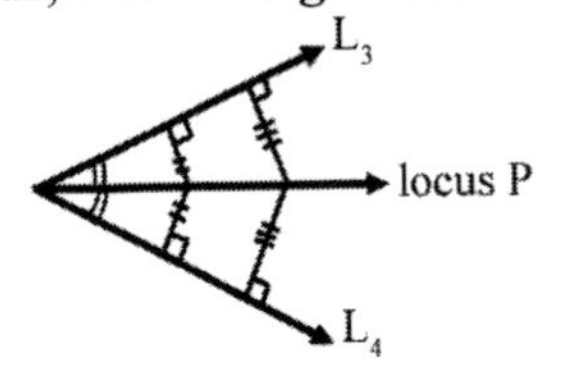

Compound locus – the number of points that satisfy all the stated conditions when two or more loci are involved

Locus

1. Circle O has radius 4 in. Determine the locus of a point, M that lies inside circle O and is always 3 in. away from the circumference of circle O.

Answer:
Given radius of circle O = 4
Conditions of locus M:

a) Inside circle O
b) 3 in. away from circumference of circle O

Thus,
Locus M is a concentric circle with circle O (i.e. circle M and circle O have the same center point, O) and has a radius of 1 in. //

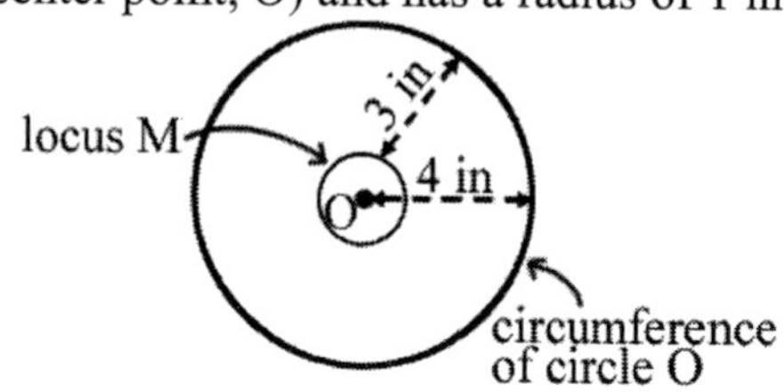

2. Point k moves such that it is always 5 units from the center point (3, 4) of a circle whose radius is 6 units. Describe locus k. What is the equation of locus k?

Answers:
Given circle's center point = (3, 4)
Circle's radius = 6
Condition of locus k:
5 units from (3, 4)
Thus,
Locus k is a circle whose radius is 5 units and is concentric with the original circle. //
Equation of locus k:

$(x-x_1)^2+(y-y_1)^2=r^2$

$(x-3)^2+(y-4)^2=5^2$

$(x-3)^2+(y-4)^2=25$ //

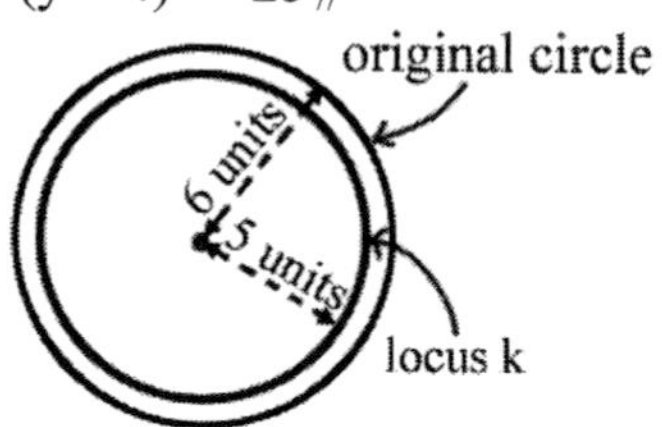

3. In circle O, $\overline{AB}$ is a diameter. Point P moves such that it is always the same distance from point B and passes through point A. Use a diagram to represent this information. Hence describe locus P.

Answers:
Given $\overline{AB}$ = diameter of circle O
Conditions for locus P:
a) Same distance from B
b) Passes through point A
Thus,

Locus P is a circle whose center point is B and radius, $\overline{BA}$. //

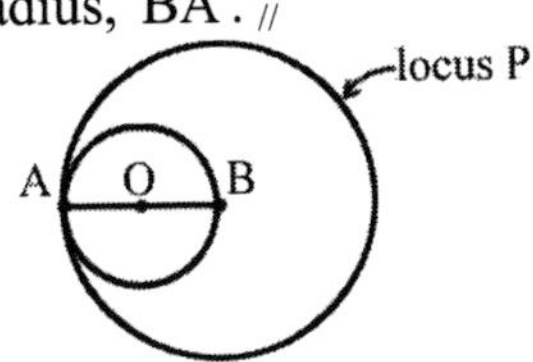

4. Figure 8.1 shows a trapezoid, ABCD where ∠BAD = 90°. Find the locus of point P such that triangle BPD is always right angle at P.
(Note: P does not necessarily lie inside the trapezoid)

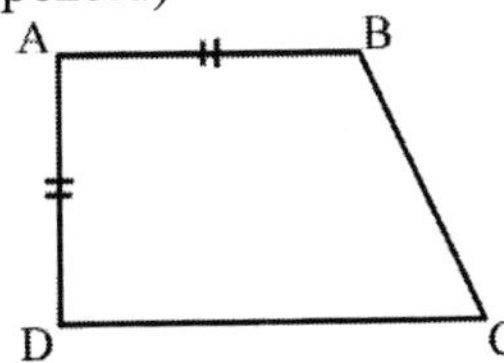

Figure 8. 1

Answer:
Given ABCD = trapezoid
Conditions for locus P:
a) △BPD is right angle triangle
b) P is always 90°

From Angles in Semicircle theorem, triangle formed from the 2 ends of a diameter of a circle and touches the circumference of the same circle is a right angle triangle.

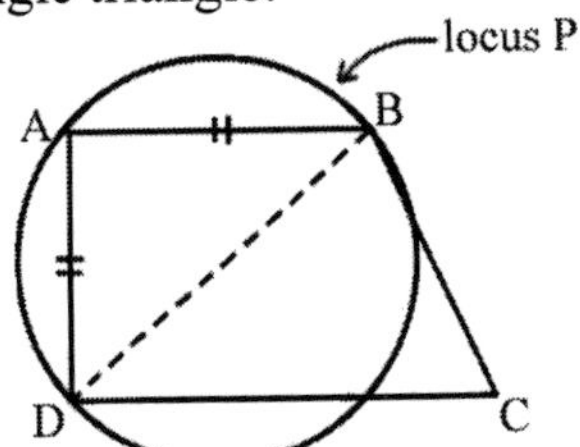

∴ For △BPD to be always 90° at point P, locus P must be the circumference of a circle whose center point is the midpoint of $\overline{BD}$ and diameter is $\overline{BD}$. //

5. Given Q(5, 4) lies on the circumference of circle whose center is O(2, 3). If point P moves on the circumference of circle O, what is the equation of locus P?

Answer:
Given center of circle O = (2, 3)
Also given Q = (5, 4)
Let r = radius
=> $\overline{OQ}$ = radius
Length of OQ:

$$OQ = \sqrt{(5-2)^2+(4-3)^2}$$
$$= \sqrt{3^2+1^2}$$
$$= \sqrt{10} \text{ units}$$

Since P moves on the circumference of circle O, therefore equation of locus P = circumference of circle O
Equation of locus P:

$$(x-x_1)^2+(y-y_1)^2=r^2$$
$$(x-2)^2+(y-3)^2=\sqrt{10}^2$$
$$(x-2)^2+(y-3)^2=10 \text{ //}$$

6. Circle O has a radius of 4 cm. Locus P is always 2 cm from the circumference of circle O. Use a diagram to represent these information. Subsequently describe locus P.

Answers:
Given radius of circle O = 4
Conditions for locus P:
a) 2 cm from circumference of circle O
b) Not specified that P has to be inside circle O

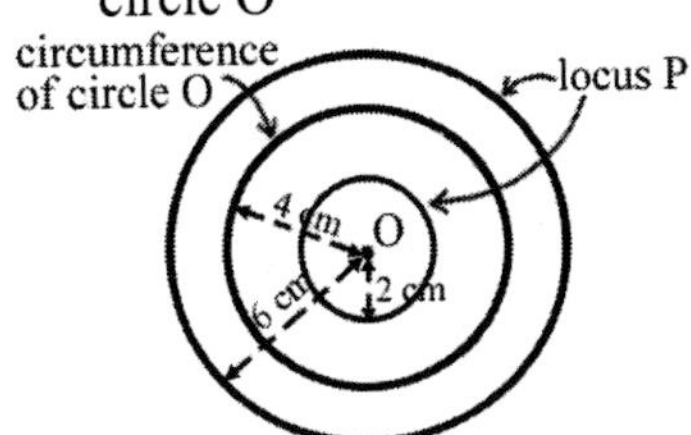

Thus,
Locus P is two circles concentric with circle O and whose radii are 6 cm and 2 cm. //

7. Determine the equation of the locus of all points whose ordinates are twice their abscissa.

Answer:
Given ordinate (y-coordinate) 2 times abscissa (x-coordinate)
Thus,
The locus is a straight line whose equation is: $y = 2x$. //

8. Determine the locus of the center point of a golf ball that rolls along a flat ground.

Answer:
Locus is a parallel line to the ground at a distance equals to the radius of the golf ball. //

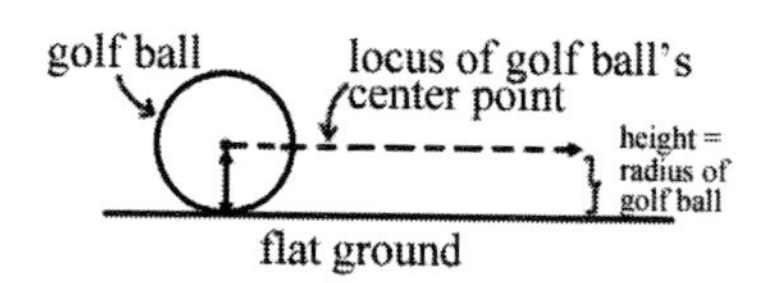

9. In a coordinate plane, point P moves such that its ordinate is always a constant 3 units from the x-axis. Use a diagram to illustrate locus P. Hence describe locus P and determine the equation or equations of locus P.

Answers:
Given condition of locus P:
3 units from x-axis

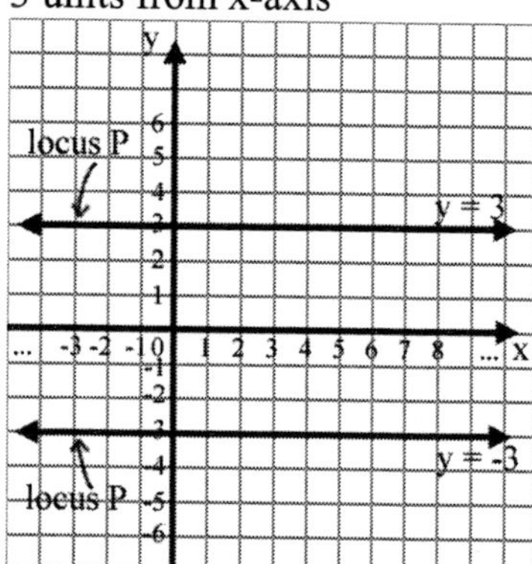

Thus,
Locus P is two parallel horizontal lines, $y = 3$ and $y = -3$ and both lines are parallel to the x-axis. //

Note: Locus P is also perpendicular to y-axis.

10. Find the locus for all points 3 units from the line $y = x$. Hence illustrate your answer in a Cartesian plane.

Answers:
Given $y = x$
Condition for locus:
3 units from the straight line $y = x$

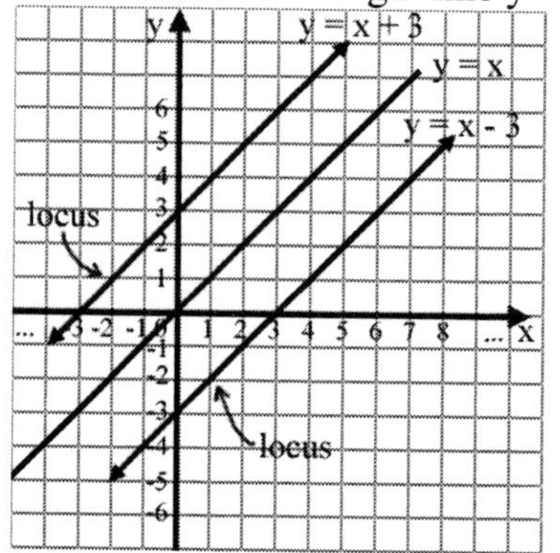

Thus,
Loci for all points 3 units from the line $y = x$, are two parallel lines, $y = x + 3$ and $y = x - 3$. //

11. Given A(4, 5) and B(8, 13) are ends of a line segment. Point P moves such that it is always equidistant from A and B. Express the information in a diagram. Hence find the equation of locus P.

Answers:
Given A(4, 5) and B(8, 13)

For locus P to be equidistant from A and B:

a) P must divide $\overline{AB}$ equally => P passes through the midpoint of $\overline{AB}$

b) Locus P $\perp$ $\overline{AB}$

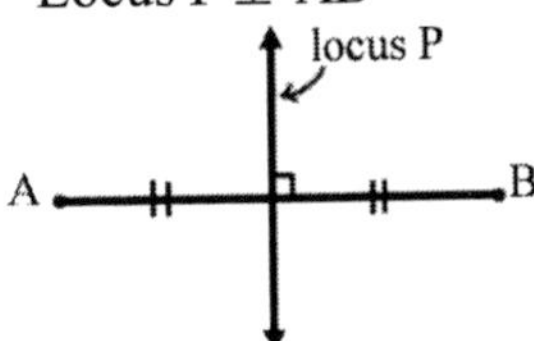

$\therefore$ Locus P is a perpendicular bisector of $\overline{AB}$. //

Let P_1 = midpoint of $\overline{AB}$

Midpoint of $\overline{AB}$:

$$P_1 = \left(\frac{x_1 + x_2}{2}, \frac{y_1 + y_2}{2}\right)$$

$$= \left(\frac{4+8}{2}, \frac{5+13}{2}\right)$$

$$= \left(\frac{12}{2}, \frac{18}{2}\right)$$

$$= (6, 9)$$

Let m_{AB} = gradient of $\overline{AB}$

m_P = gradient of locus P

Gradient of $\overline{AB}$:

$$m_{AB} = \frac{y_2 - y_1}{x_2 - x_1}$$

$$= \frac{13-5}{8-4}$$

$$= \frac{8}{4}$$

$$= 2$$

Since locus is perpendicular to $\overline{AB}$

=> $m_P \times m_{AB} = -1$

$m_P \times 2 = -1$

$$m_P = -\frac{1}{2}$$

Thus, using point slope form: $P_1(6, 9)$ and

$$m_P = -\frac{1}{2}$$

$y - y_1 = m_P(x - x_1)$

$$y - 9 = -\frac{1}{2}(x - 6)$$

$2y - 18 = -x + 6$

$x + 2y = 24$

$\therefore$ Equation of locus P is $x + 2y = 24$ //

12. Find two equations of straight lines that describe the locus of points equidistant from the line y = 3 and x = 3.

Answers:

Given, y = 3 and x = 3

Condition of locus:

Lines equidistant from y = 3 and x = 3

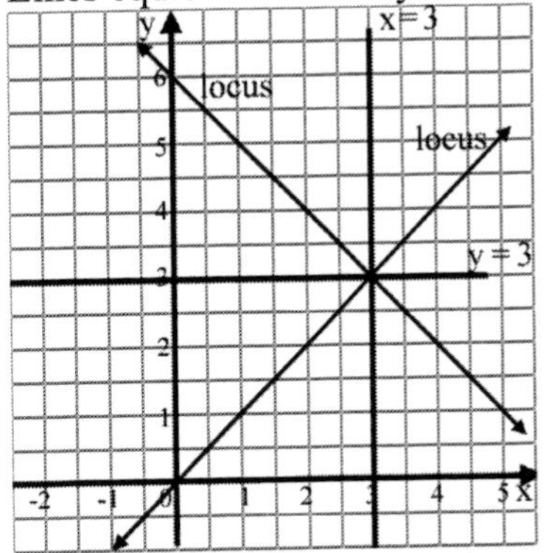

From the Cartesian plane above,

Straight line passes through (0, 0) & (3, 3):

$$\frac{y - y_1}{x - x_1} = \frac{y_2 - y_1}{x_2 - x_1}$$

$$\frac{y}{x} = \frac{3}{3} = 1$$

$y = x$

Straight line passes through (0, 6) & (3, 3):

$$\frac{y-6}{x-0} = \frac{3-6}{3-0}$$

$$\frac{y-6}{x-0} = -1$$

$x + y = 6$

$\therefore$ Equations of locus, y = x and x + y = 6, bisect the intersecting lines y = 3 and x = 3 //

13. Given P(2, 5) and Q(6, 7) are ends of a line segment. S moves such that it is always equidistant from Q and passes through the midpoint of $\overline{PQ}$. Describe and find the equation of locus S.

Answers:
Given P(2, 5) and Q(6, 7)
Conditions of locus S:
a) Equidistant from Q
b) Passes through midpoint $\overline{PQ}$

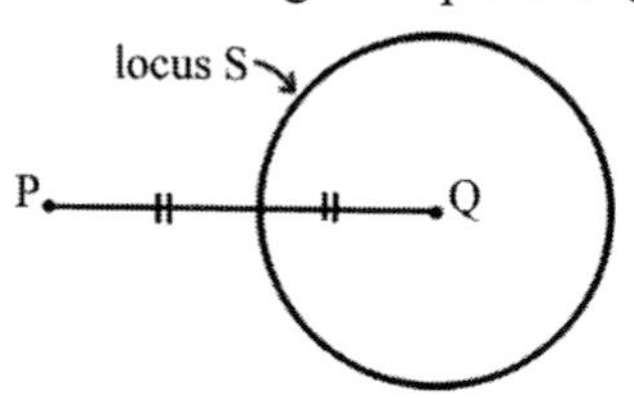

Let M = midpoint of $\overline{PQ}$
Radius of locus S, r = MQ:

$$MQ = \frac{1}{2}\times PQ$$
$$= \frac{1}{2}\times\sqrt{(x_2-x_1)^2+(y_2-y_1)^2}$$
$$= \frac{1}{2}\times\sqrt{(6-2)^2+(7-5)^2}$$
$$= \frac{1}{2}\times\sqrt{4^2+2^2}$$
$$= \frac{1}{2}\times\sqrt{20} \quad \Leftarrow \text{Note: } \sqrt{20}=\sqrt{4\times 5}$$
$$= \frac{1}{2}\times 2\times\sqrt{5}$$
$$= \sqrt{5}\text{ units}$$

Equation of locus S (equation of circle):
$$(x-x_1)^2+(y-y_1)^2=r^2$$
$$(x-6)^2+(y-7)^2=\sqrt{5}^2$$
$$(x-6)^2+(y-7)^2=5 \;//$$

∴ Locus S is a circle with center Q(6, 7) and radius, $\sqrt{5}$ units. //

14. Given, two parallel lines, $\overleftrightarrow{A}$ and $\overleftrightarrow{B}$ are located on the same plane and are 6 inches apart. Point P is located equidistant between the parallel lines. Find the number of points that are the same distance from $\overleftrightarrow{A}$ and $\overleftrightarrow{B}$, and is always 4 inches from P.

Answer:
Given $\overleftrightarrow{A}$ and $\overleftrightarrow{B}$ = parallel lines, 6″ apart
P = equidistant between parallel lines
Conditions for compound locus:
a) Same distance from $\overleftrightarrow{A}$ and $\overleftrightarrow{B}$, *and*
b) Always 4″ from P

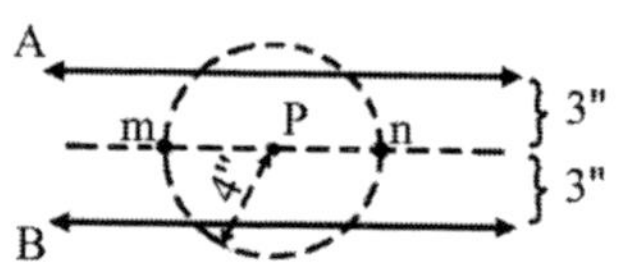

∴ There are 2 points, m and n that are equidistant from lines, $\overleftrightarrow{A}$ and $\overleftrightarrow{B}$, and are 4 inches from P. //

15. Marcy's yard has two parallel rows of flower beds, $\overline{A}$ and $\overline{B}$, that are 4 feet apart. A birdbath lies equidistant between the flower beds. Marcy had asked you to construct a path that is one foot parallel to flower bed A and 3 feet from the birdbath. The intersection points of the compound locus are where she wants you to install the lamps. How many lamps does she want to install?

Answer:
Given $\overline{A}$ & $\overline{B}$ = parallel lines, 4 feet apart
Let C = birdbath position
Given C = equidistant from $\overline{A}$ and $\overline{B}$
Conditions for locus:
a) 1 ft from $\overline{A}$, *and*
b) 3 ft from C

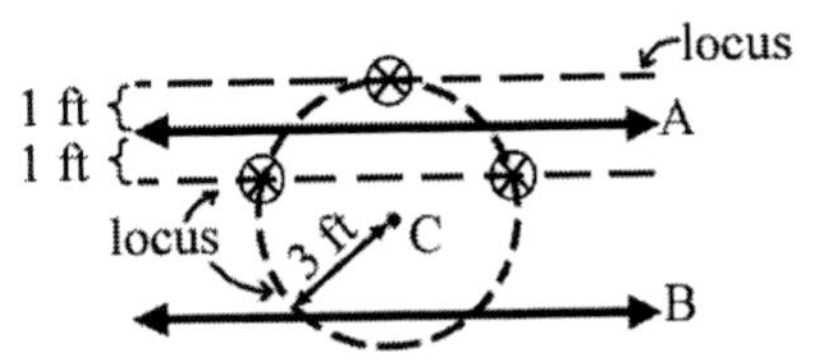

∴ The locus of points that are parallel to $\overline{A}$ and also 3 feet from birdbath, C, are 3 points (⊗) as shown in the diagram above. //

16. Mrs. Winston has invited James for tea. However she asked that James assists her in placing candle holders on her square coffee table shown in Figure 8.2. She provided James with the following information:
"Draw a locus of points that moves within ABCD such that, the perpendicular distance from the diagonals are always the same and always 0.5 ft from the center of ABCD."
How many candle holders will James need? Help James determine the position of the candle holders.

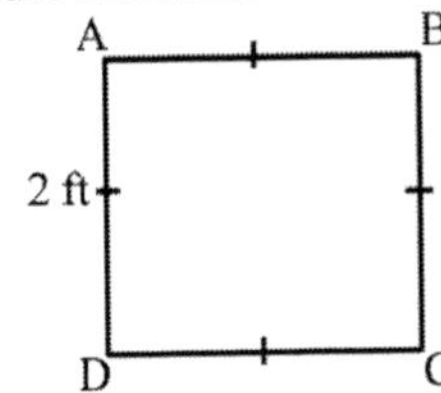

Figure 8. 2

Answers:
Given ABCD = square
=> Diagonals intersect at center point of ABCD
Conditions for locus:

a) Equal perpendicular distance from diagonals, *and*
b) 0.5 ft from the center of ABCD

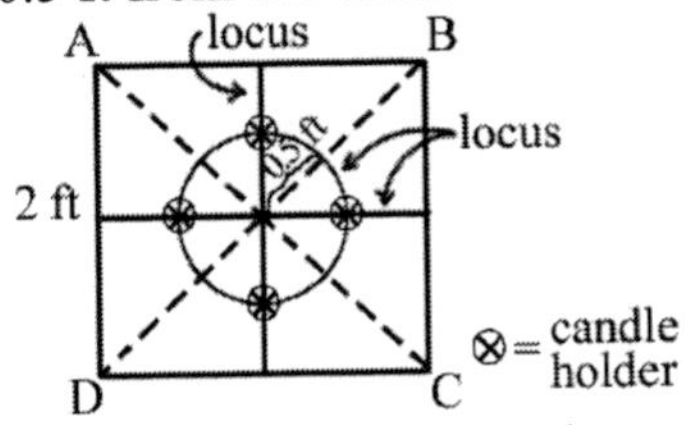

∴ The locus of points that are equal perpendicular distance from diagonals, $\overline{AC}$ and $\overline{BD}$ and always 0.5 ft from the center of ABCD are 4 points (⊗) as shown in diagram above. //

∴ James will need 4 candle holders. //

17. A farmer inherited a square plot of cornfield whose sides are p units (see Figure 8.3) and would like to install scarecrows. The farmer has asked his son, a mathematician, to prepare a blue print for the scarecrows. Here is the instruction:
"Draw a locus of points equidistant from $\overline{AD}$ and $\overline{CD}$ and r units from point A."
Determine where the scarecrows will be located given the following conditions:

a) $r = p$
b) $r < \frac{p}{2}$
c) $r \geq \frac{p}{2}$

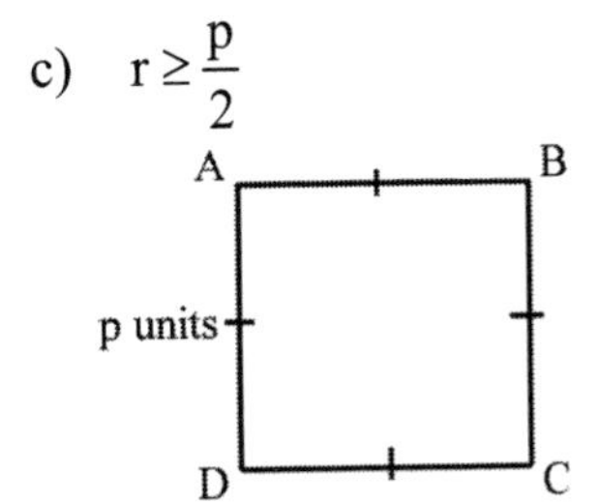

Figure 8. 3

Answers:
Given ABCD = square,
AB = BC = CD = AD = p units
Conditions of locus:

i. Equidistant from $\overline{AD}$ and $\overline{CD}$, *and*
ii. r units from A

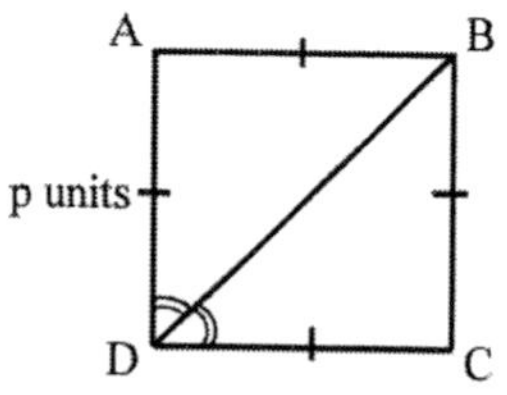

$\therefore$ locus of points is diagonal, $\overline{BD}$

a) For r = p:

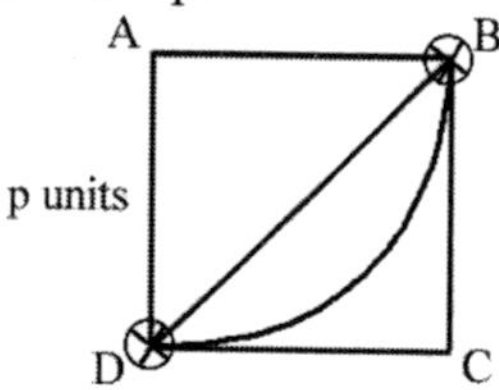

$\therefore$ The locus of points that are equidistant from $\overline{AD}$ and $\overline{CD}$ and r units from A are 2 points (⊗) as shown in the diagram above. //

b) For $r < \frac{p}{2}$

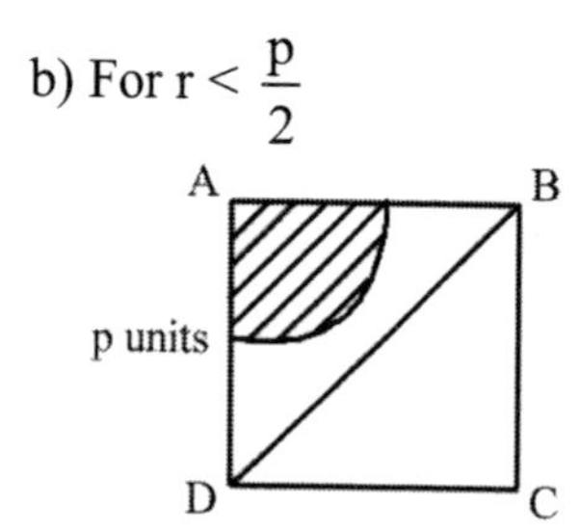

$\therefore$ The locus of points that are equidistant from $\overline{AD}$ and $\overline{CD}$ and r units from A are 0 points as shown in the diagram above. //

c) For $r \geq \frac{p}{2}$

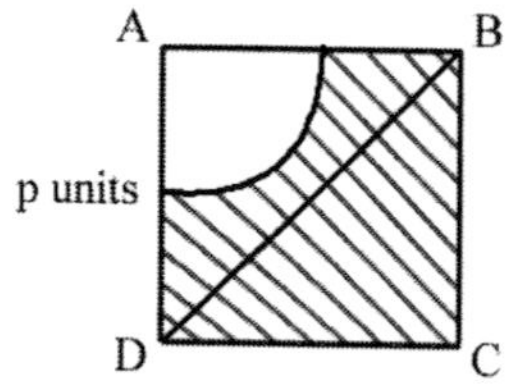

A B

p units

D C

$\therefore$ Locus of points that are equidistant from $\overline{AD}$ and $\overline{CD}$ and r units from A is 0 point or 1 point (when r touches midpoint of $\overline{BD}$) or 2 points (when r cuts diagonal $\overline{BD}$) as shown in the diagram above. //

18. At the Wagga Math Camp all campers are given a map (see Figure 8.4) and instructions upon induction. Below are the instructions for locating clean drinking fountains:

i. Points P, Q, R, and S are cherry trees. And PQRS is a square whose sides are 60 ft.

ii. Within PQRS, find locus of points, A, that is always 20 ft from $\overline{QS}$

iii. Within PQRS, find locus of points, B, that is always 25 ft from the intersection of the diagonals, $\overline{PR}$ and $\overline{QS}$

iv. The compound locus is where the clean drinking fountains are located.

Help Meg who is spending her first summer at Wagga Math Camp to locate the clean drinking fountains.

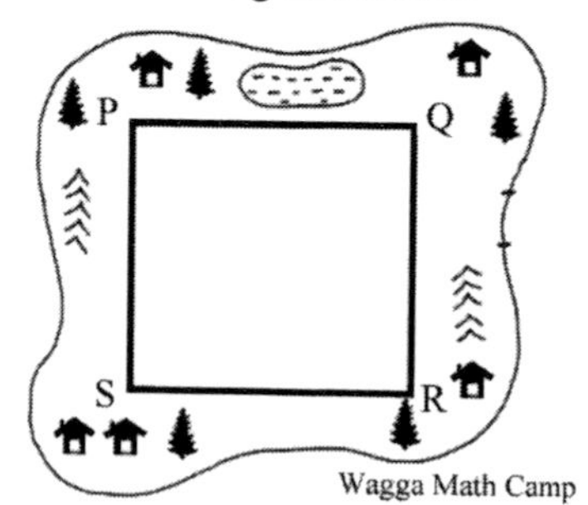

Figure 8. 4

Answer:

Given PQRS = square, side = 60

Condition for locus A:

- 20 ft from $\overline{QS}$, within PQRS

$\therefore$ Locus A is two parallel lines to $\overline{QS}$

Condition for locus B:

- 25 ft from intersection point of $\overline{PR}$ and $\overline{QS}$, within PQRS

$\therefore$ Locus B is a circle with radius 25 ft and center at intersection point of $\overline{PR}$ and $\overline{QS}$

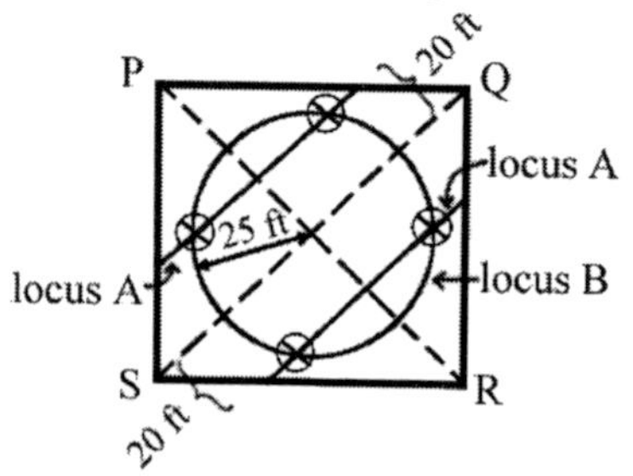

$\therefore$ The compound locus that are 20 ft from $\overline{QS}$ and 25 ft from the intersection point of $\overline{PR}$ and $\overline{QS}$ are 4 points (⊗) as shown in the diagram above. //

19. Your company's annual sale has started. As the company's customer support manager, you have received directive from Miss Sucol to station customer support personnel in a busy rotunda hall (see Figure 8.5). The diameter, $\overline{PR}$ of the hall is 50 yards. Here are her instructions:

a) Locus T moves such that PT = TR
b) Locus W is always 5 yards from the circumference of the circle
c) Locus X is 20 yards equidistant from diameter, $\overline{PR}$
d) The intersections of loci, T, W, and X are where the customer support personnel should be stationed

How many customer support personnel are required to fulfill her instructions?

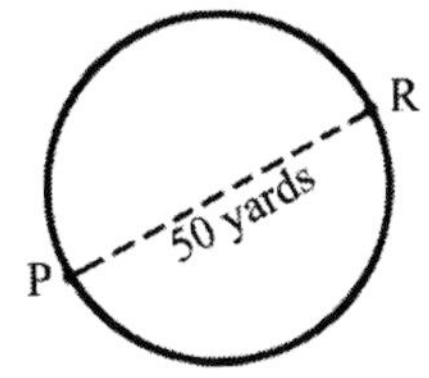

Figure 8. 5

Answer:
Given PR = diameter, 50 yards
Let O = building's center point
Condition for locus T:

- PT = TR

∴ Locus T is a perpendicular bisector of diameter, $\overline{PR}$ and passes through center, O. Locus T is also a diameter
Condition for locus W:

- 5 yards from circumference of circle

∴ Locus W is a circle with radius 20 yards and center O
Condition for locus X:

- 20 yards equidistant from $\overline{PR}$

∴ Locus X is 2 parallel lines to $\overline{PR}$ and 20 yards from $\overline{PR}$

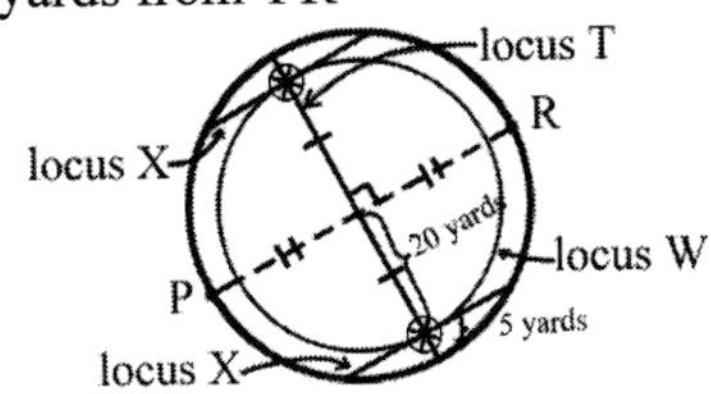

∴ Locus of points that are perpendicular bisector to $\overline{PR}$, 5 yards from circumference of circle and 20 yards equidistant from $\overline{PR}$ are 2 points (⊗) as shown in diagram above. Hence 2 customer support personnel are needed to be stationed in the location marked by ⊗. //

20. You work for the country's espionage agency. An agent has given you a map to locate the enemy's hideout with the following mathematical clues:

a) P, Q and R are city codes, while S is a radio station. △PQR is an equilateral triangle. S is located inside △PQR
b) Point M moves such that its perpendicular distance from $\overline{PQ}$ and $\overline{PR}$ is always equal, within △PQR
c) Point N moves such that its distance from S is always 2 miles
d) The intersections of the loci are where the enemy's hideouts are located.

How many hideouts have been revealed?

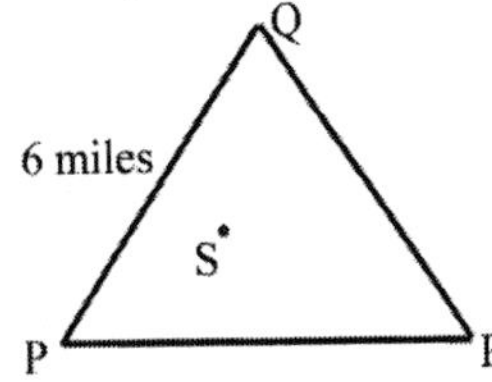

Figure 8. 6

Answer:
Given PQR = equilateral△, side = 6 miles
Condition for locus M:

- Equal perpendicular distance from $\overline{PQ}$ and $\overline{PR}$ within △PQR

$\therefore$ Locus M bisects $\angle QPR$. Since $\triangle PQR$ is an equilateral triangle, locus M is also perpendicular to $\overline{QR}$.

Condition for locus N:

- Always 2 miles from S

$\therefore$ Locus N is a circle with radius 2 miles and center, S.

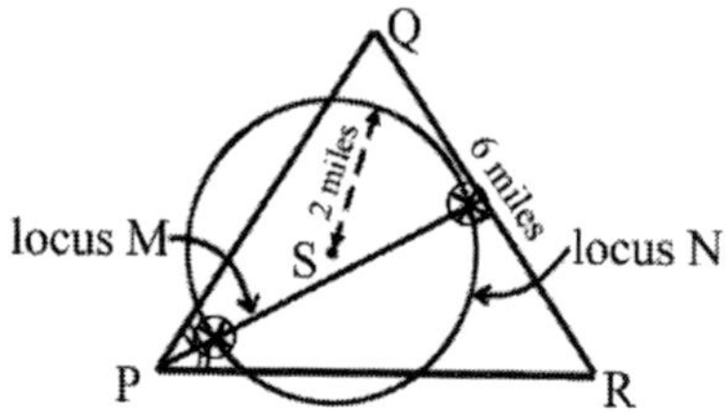

$\therefore$ Locus of points whose perpendicular distance to $\overline{PQ}$ and $\overline{PR}$ are always equal and always 2 miles from point S are 2 points (⊗) as shown in diagram above. Hence the agent has revealed 2 enemy hideouts. //

21. Mrs. Sergey is very absent-minded. Hence, her husband a math professor has hidden a second pair of house key in their lawn in case Mrs. Sergey misplaces or forgets their house key. Below are the instructions for locating the second pair of house key:

a) Points A, B, C and D represent lawn sprinklers. ABCD is a rhombus.
b) Within ABCD, locus P is always equidistant from $\overline{AD}$ and $\overline{BC}$
c) Within ABCD, locus R moves such that RD = RC
d) The intersection of loci is where the second pair of house key can be located.

Thus, describe to Mrs. Sergey where the second pair of house key can be found.

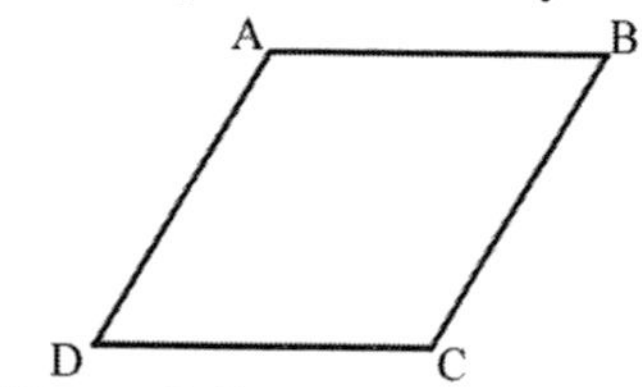

Figure 8. 7

Answer:

Given ABCD = rhombus

=> $\overline{AD} \parallel \overline{BC}$

Condition for locus P:

- Equidistant from $\overline{AD}$ and $\overline{BC}$

$\therefore$ Locus P is a line parallel to $\overline{AD}$ and $\overline{BC}$, and is halfway between them.

Condition for locus R:

- RD = RC

$\therefore$ Locus R is a perpendicular bisector of $\overline{CD}$

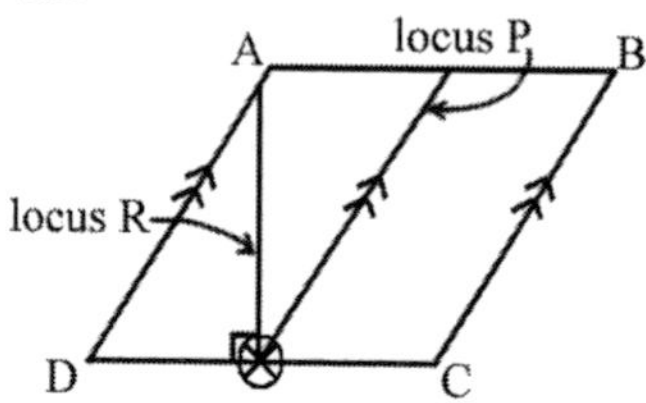

$\therefore$ Locus of points that are equidistant from $\overline{AD}$ and $\overline{BC}$, and perpendicularly bisects $\overline{CD}$ is a single point (⊗) shown in diagram above.

$\therefore$ Mrs. Sergey will find the second pair of key at the midpoint of $\overline{CD}$. //

22. The late Mr. Locus has left his granddaughter, Sophia a will for her to inherit his entire estate. However, before she can execute the will she must find where the will has been hidden. Sophia has engaged your help to locate where the will has been hidden using the following clues:

a) There are 5 garden lights in Mr. Locus' backyard. The garden lights are labeled, P, Q, R, S, and T. PQRST is a regular pentagon (see Figure 8.8).

b) Point A moves such that it is always equidistant from $\overline{PT}$ and $\overline{ST}$, within PQRST.
c) Point B moves within PQRST, such that BP = BR.
d) The intersection of the loci is where the will has been hidden.

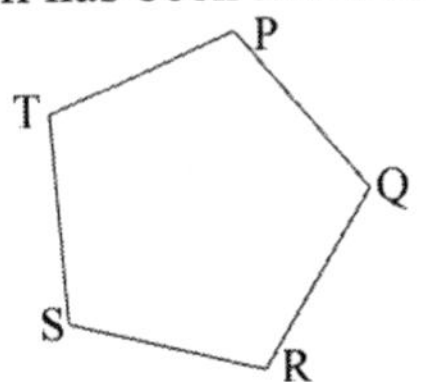

Figure 8. 8

Answer:
Given PQRST = regular pentagon
Condition for locus A:

- Equidistant from $\overline{PT}$ and $\overline{ST}$

$\therefore$ Locus A is an angle bisector of $\angle PTS$. Since PQRST is a regular pentagon, (internal angle $= \frac{(5-2)\times 180^\circ}{5} = 108^\circ$), locus A divides $\angle PTS$ into 2 congruent angles, $\angle PTA \cong \angle ATS = 54^\circ$.
Condition for locus B:

- BP = BR

$\therefore$ Locus B is a perpendicular bisector of $\overline{PR}$.

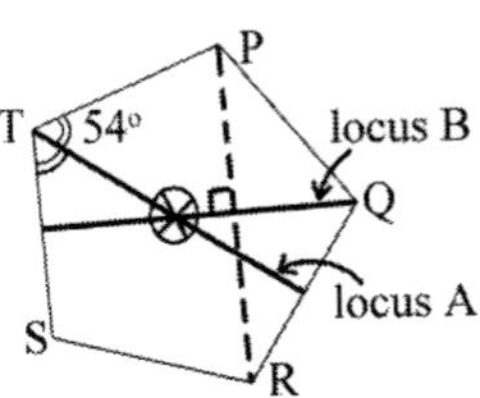

$\therefore$ Locus of points that are equidistant from $\overline{PT}$ and $\overline{ST}$, and BP = BR is a single point (⊗) as shown in the diagram above. Thus Sophia will be able to find the will at ⊗. //

23. Every year Mrs. Applesauce will host an Easter tea party where guests are encouraged to participate in an Easter egg hunt. The highlight of the party is that any guest who finds any of the winning Easter eggs will win cash reward. You have been invited to Mrs. Applesauce Easter tea party and below are the clues to the winning Easter eggs:

a) Points A, B, C and D represent the 4 oak trees in her garden. ABCD is a rectangle (see Figure 8.9).
b) The perpendicular distance of locus H from $\overline{AD}$ is always the same as $\overline{CD}$.
c) Within ABCD, locus K moves such that its distance from A is always AD in length.
d) The intersections of loci H and K are where the winning Easter eggs are located.

Find where the winning Easter eggs are hidden.

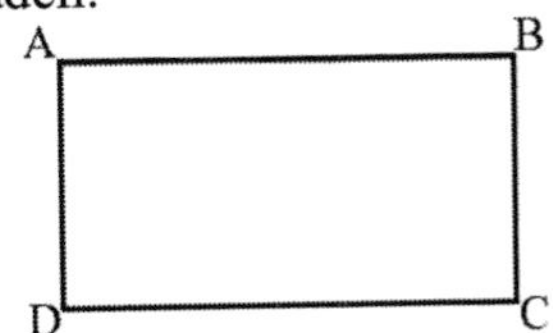

Figure 8. 9

Answer:
Given ABCD = rectangle
Condition for locus H:

- Equidistant from intersecting lines, $\overline{AD}$ and $\overline{CD}$

$\therefore$ Locus H is an angle bisector. Locus H divides $\angle ADC$ into 2 congruent angles, $\angle ADH$ and $\angle CDH$, each 45°.
Condition for locus K:

- AD units from point A

$\therefore$ Locus K is a circle center at A and radius AD length.

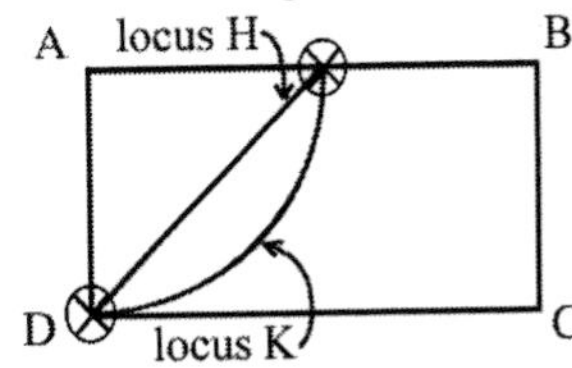

$\therefore$ Locus of points that are equidistant from intersecting lines $\overline{AD}$ and $\overline{CD}$, and always AD units from A are 2 points (⊗) as shown in diagram above. Therefore there are 2 winning Easter eggs and they have been hidden in the areas marked by ⊗. //

24. A major robbery has taken place in your town recently. Mr. Oakleigh a part-time detective who is also a retired math teacher has provided you the following lead

to the stashed loot. Below are Mr. Oakleigh's instructions:

a) A, B, C, and D represent churches with weather vanes. ABCD is a square and AB = 8 miles. Further known P, Q, R and S are midpoints (see Figure 8.10). Point E represents an abandoned flour silo in a deserted field.
b) Locus X moves such that its distance from $\overline{QS}$ is always 2 miles
c) Locus Y moves such that it is always 3 miles from E
d) The loots have been hidden in the intersection points of the loci described above.

As the town's sheriff, your immediate task is to identify where the loot is stashed so that it can be restored to its owner.

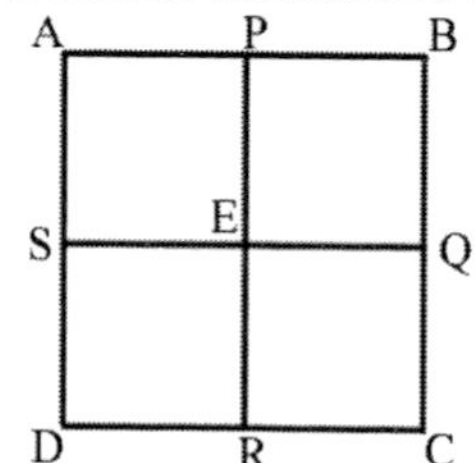

Figure 8. 10

Answer:
Given ABCD = square
P, Q, R and S = midpoints
=> APES ≅ PBQE ≅ EQCR ≅ SERD
Condition for locus X:

- 2 miles from $\overline{QS}$

∴ Locus X is two parallel lines to $\overline{QS}$.
Each line is 2 miles from $\overline{QS}$.
Condition for locus Y:

- 3 miles from E

∴ Locus Y is a circle whose center is E and radius is 3 miles.

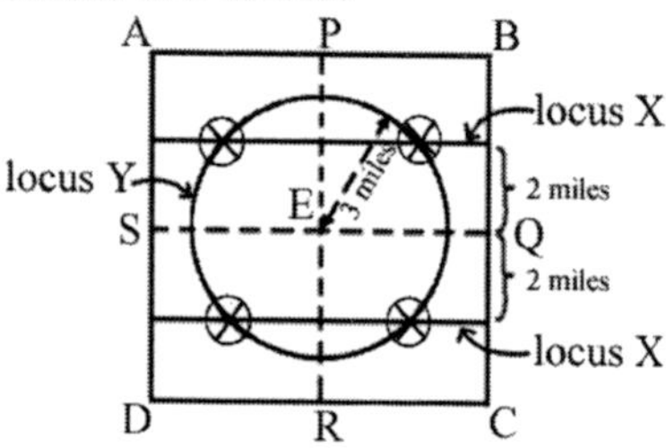

∴ Locus of points that are 2 miles from $\overline{QS}$ and 3 miles from point E are 4 points (⊗) as shown in the diagram above. Hence the town robbery loot has been divided and stashed in 4 different locations as marked by ⊗. //

25. Town A's police station is located at the coordinate (3, 5) and its area of coverage extends for 15 miles. Town B's police station is located at coordinate (8, 9) and its area of coverage extends for 20 miles. In a Cartesian plane, show the area of coverage that overlaps between the two towns' police forces. [Assume 1 unit on the Cartesian plane = 5 miles]

Answer:
Given A = (3, 5), area of coverage = 15
B = (8, 9), area of coverage = 20

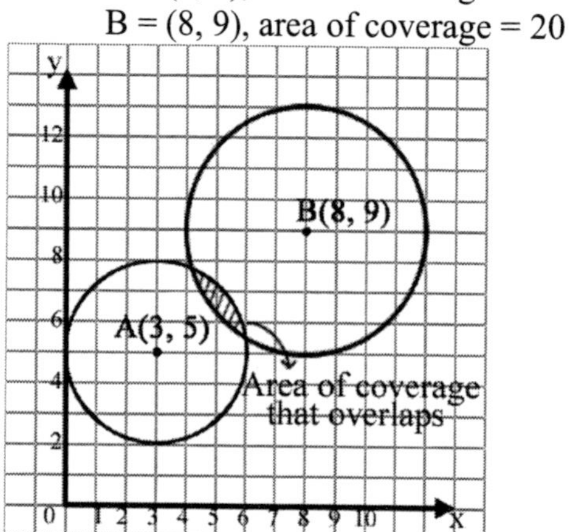

∴ The shaded region in the Cartesian plane above represents the area of coverage that overlaps between the two police stations. //

Chapter 9 Transformations

Transformation – a function that maps an **object** to its **image**

Isometric transformation – object and its image are of equal measure.

3 types of **isometric** transformations:

i. **Translation**

- Object is congruent to its image.
- Represented in the form $\begin{pmatrix} h \\ k \end{pmatrix}$, where all points in the object are moved h units along the x-axis and k units along the y-axis.
- If object is P(x, y) then image P'(x + h, y + k).

ii. **Reflection**

- Object's orientation is reversed in its image over an axis of reflection.

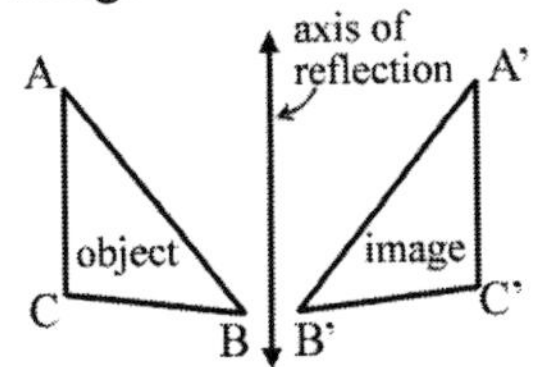

- **Axis/line of reflection** (sometimes called mirror line) is the perpendicular bisector of object to its image. For object located on the axis of reflection, points on the axis of reflection remain unchanged.

iii. **Rotation**

- All points in the object are slide over an angle, following a direction (clockwise or counterclockwise) from the **point of rotation**.
- Object's distance to point of rotation = Image's distance to point of rotation.
- Orientation of object and image are equal.

Non-isometric transformation:

iv. **Dilation**

- All points in the object are moved conforming to the factor of dilation from a fixed point.
- Object and image are similar but not necessary congruent.
- **Scale factor** of dilation, f:

$$= \frac{\text{Length of image's side}}{\text{Length of object's side}} = \frac{B_1C_1}{BC}$$

$$= \frac{\text{Distance of image from P}}{\text{Distance of object from P}} = \frac{PC_1}{PC}$$

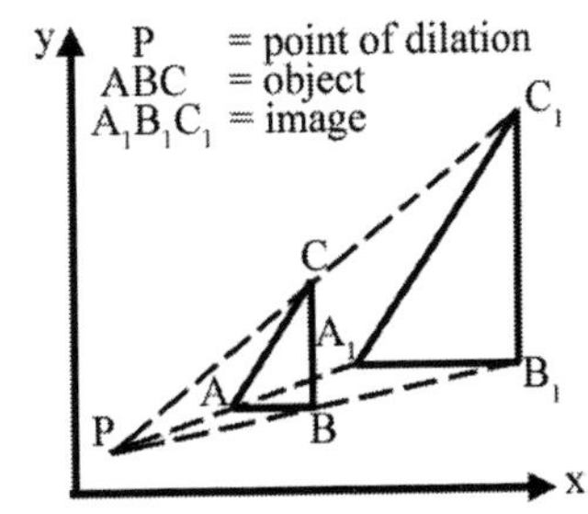

For f > 0:

f > 1 – object enlarged

0 < f < 1 – object dilated

f = 1 – object's size unchanged

Combination of transformations:

BA means transformation ***A*** is *before* transformation ***B***. While ***AB*** means transformation ***B*** is *before* transformation ***A***. However ***BA*** does *not always equal* to ***AB***.

1. In Figure 9.1 object △ABC is transformed to image △A'B'C' through a translation $\begin{pmatrix} h \\ k \end{pmatrix}$.

a) Determine the values of h and k
b) Find the coordinates of image P based on the translation $\begin{pmatrix} h \\ k \end{pmatrix}$

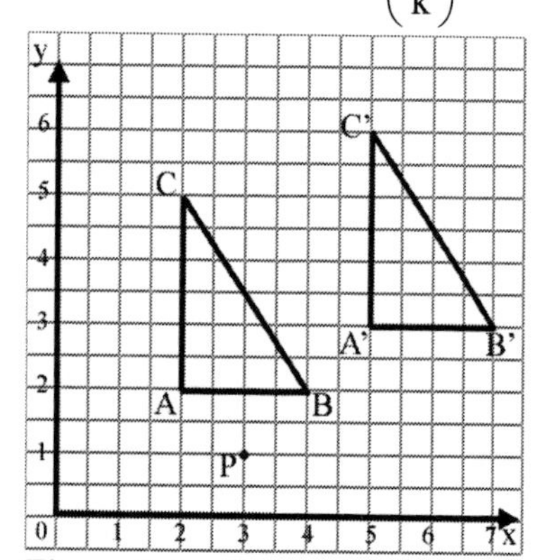

Figure 9. 1

Answers:
a) △ABC is translated to △A'B'C'
Consider point A(2, 2):
[Note: Using point B or C will also produce the same answer]
Object A has been translated to A':
3 units right => h = 3 //
1 unit up => k = 1 //

$\therefore$ Translation = $\begin{pmatrix} 3 \\ 1 \end{pmatrix}$

b) From part a): translation = $\begin{pmatrix} 3 \\ 1 \end{pmatrix}$

Given P = (3, 1)
After transformation:
=> Image, P' = (3 + h, 1 + k)
= (3 + 3, 1 + 1)
= (6, 2) //

2. In Figure 9.2, object PQRS has been transformed to image P'Q'R'S' through the translation $\begin{pmatrix} a \\ b \end{pmatrix}$. Find the values of a and b.

Hence if an object K(3, 4) is transformed under the same translation, what is its image, K'?

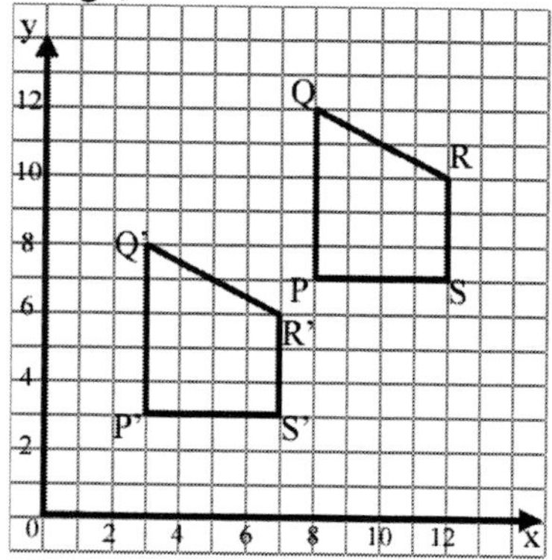

Figure 9. 2

Answers:
PQRS is translated to P'Q'R'S'
Consider point, P(8, 7)
Object P has been translated to P':
5 units left => $a = -5$ //
4 units down => $b = -4$ //

$\therefore$ Translation = $\begin{pmatrix} -5 \\ -4 \end{pmatrix}$

Alternatively:
Object P(8, 7) has been translated to image, P'(3, 3):
Image, $P' = (8 + a, 7 + b)$
$(3, 3) = (8 + a, 7 + b)$
Compare like parts:

x-coordinate:
$3 = 8 + a$
$\therefore a = -5$ //

y-coordinate:
$3 = 7 + b$
$\therefore b = -4$ //

Given K = (3, 4)

Under translation $\begin{pmatrix} -5 \\ -4 \end{pmatrix}$:

Image, $K' = (3 + a, 4 + b)$
$= (3 + (-5), 4 + (-4))$
$= (3 - 5, 4 - 4)$
$= (-2, 0)$ //

3. In a Cartesian plane an object K is translated $\begin{pmatrix} 5 \\ -3 \end{pmatrix}$ to its image, K'(8, 5). Find the coordinates of object, K.

Answer:
Given image, K' = (8, 5)

Also given translation = $\begin{pmatrix} 5 \\ -3 \end{pmatrix}$

Before translation:
Object, $K = (8 - 5, 5 - (-3))$
$= (8 - 5, 5 + 3)$
$= (3, 8)$ //

4. An object T(6, −1) is transformed to its image, T'(3, 4) under a single translation $\begin{pmatrix} x \\ y \end{pmatrix}$. Find the values of x and y.

Answers:
Given object, T = (6, −1)
image, T' = (3, 4)
Translation, T to T':
Image, $T' = (6 + x, -1 + y)$
$(3, 4) = (6 + x, -1 + y)$
Compare like components:
x-coordinate:
$3 = 6 + x$
$\therefore x = -3$ //
y-coordinate:
$4 = -1 + y$
$\therefore y = 5$ //

5. In Figure 9.3, object $\triangle ABC$ is transformed to its image $\triangle A'B'C'$ through a certain reflection. Determine the axis of reflection for this transformation.

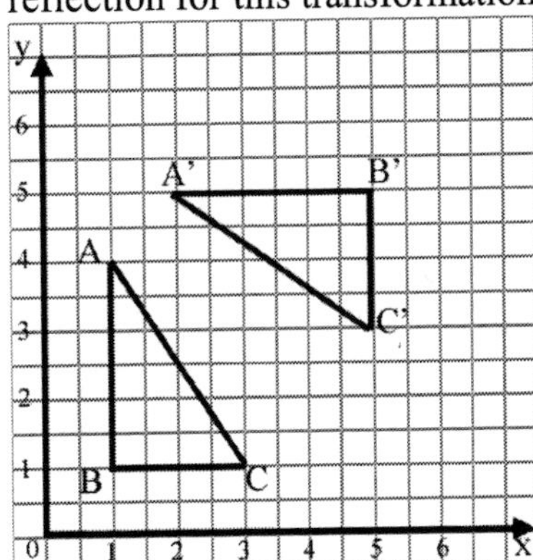

Figure 9. 3

Answer:
Figure 9.3 has been reproduced for labeling:

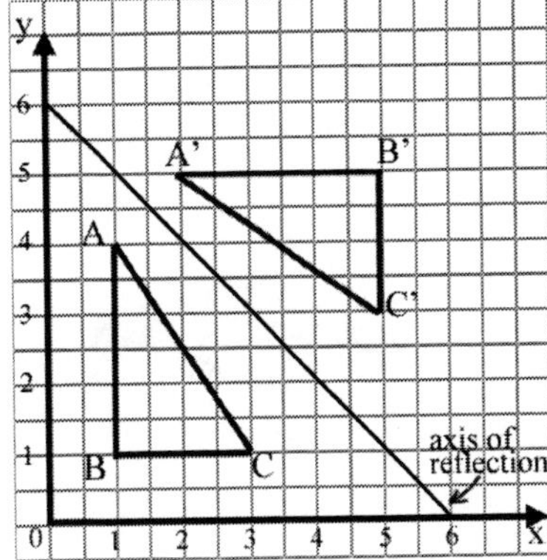

Axis of reflection:

Slope, $m = \dfrac{y_2 - y_1}{x_2 - x_1}$

$$= \frac{6-0}{0-6}$$
$$= -1$$

Using point-slope form, (6, 0) & m = -1:

$y - y_1 = m(x - x_1)$
$y - 0 = -1(x - 6)$
$y = -x + 6$
$x + y = 6$
$\therefore$ Axis of reflection is $x + y = 6$ //

6. In Figure 9.4, object M has been reflected to itself. Determine the axis of reflection for this transformation.

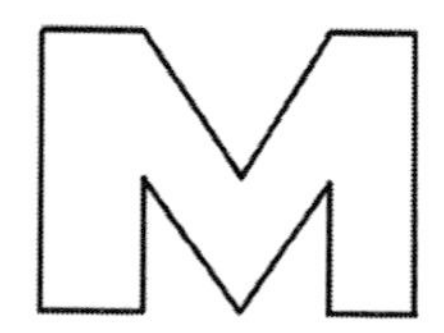

Figure 9. 4

Answer:

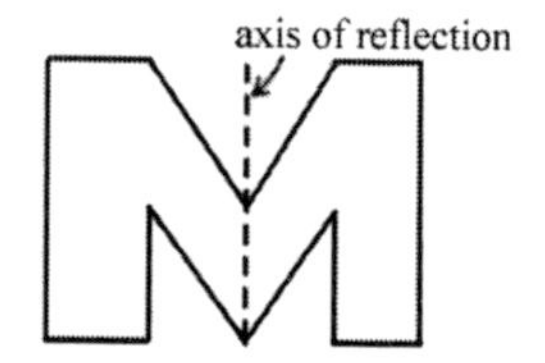

$\therefore$ The axis of reflection is the vertical line of symmetry. //

7. In Figure 9.5, image A' is the reflection of object A through a single transformation. Find the axis of reflection. Hence determine the image of object, P(4, 2) which has been transformed through the same transformation process.

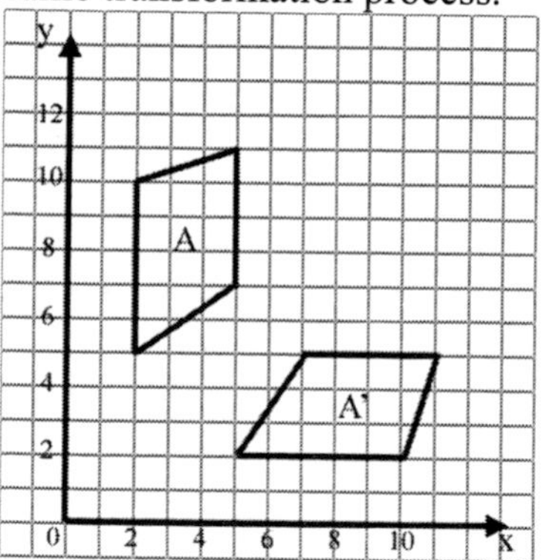

Figure 9. 5

Answers:

Figure 9.5 has been reproduced for labeling

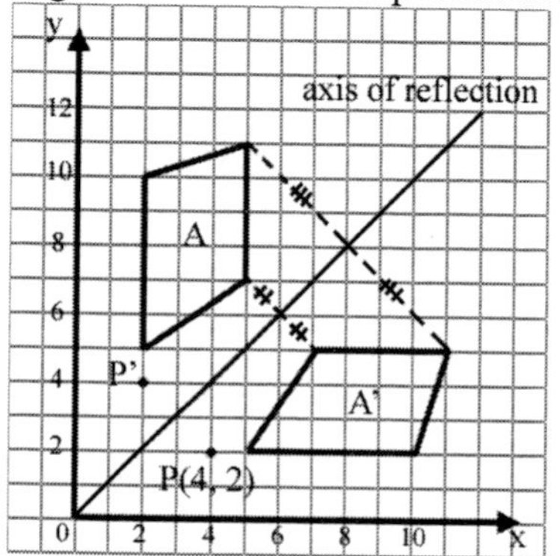

Axis of reflection:

$$\text{Slope, } m = \frac{y_2 - y_1}{x_2 - x_1}$$
$$= \frac{8-0}{8-0}$$
$$= 1$$

Using point slope form: (0, 0) and m = 1

$y - y_1 = m(x - x_1)$
$y - 0 = 1(x - 0)$
$y = x$
$\therefore$ Axis of reflection is $y = x$. //

Given object P = (4, 2)
Let P' = image of P
Under the same transformation: reflection on $y = x$
Image, P' = (2, 4) //
[Coordinates P' has been obtained by plotting the mirror image of P over $y = x$. Notice that the distance from P to $y = x$ is the same as the distance from P' to $y = x$.]

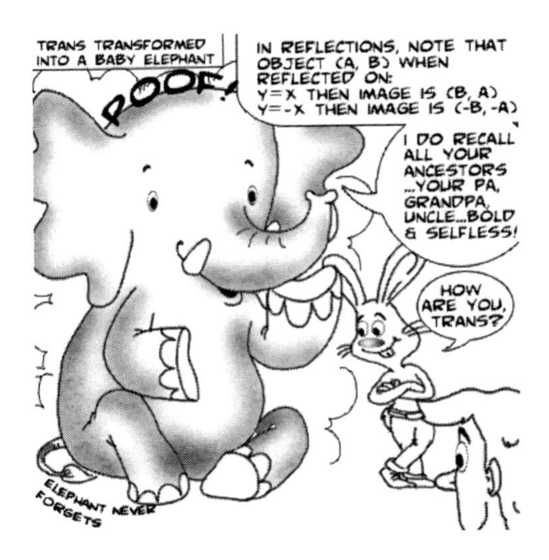

8. What is the image of object A(5, 6) if it is reflected on x = 3?

Answer:
Given A = (5, 6)
Let A' = image of A
Also given axis of reflection: x = 3

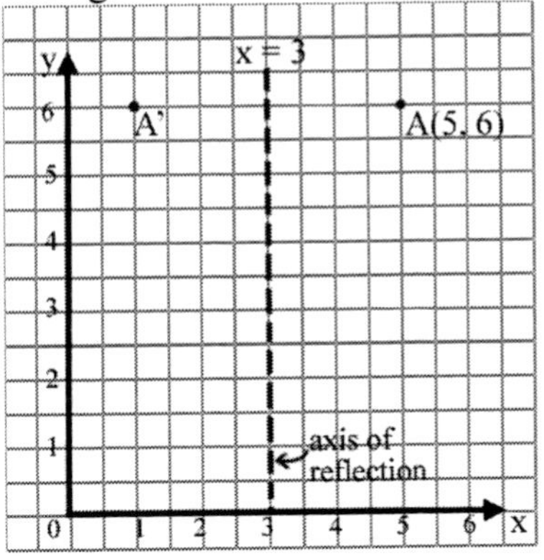

∴ Image, A' is (1, 6). //

9. Determine the object of image, R'(2, 6) that has been reflected on y = x + 1.

Answer:
Given R'(2, 6)
Let R = object of image R'
Also given axis of reflection: y = x + 1
[Note: Plot R'(2, 6) and draw y = x + 1 on the Cartesian plane. Next, find the reflection of image, R'. R can be located as if R' has been flipped over the line of reflection y = x + 1.]

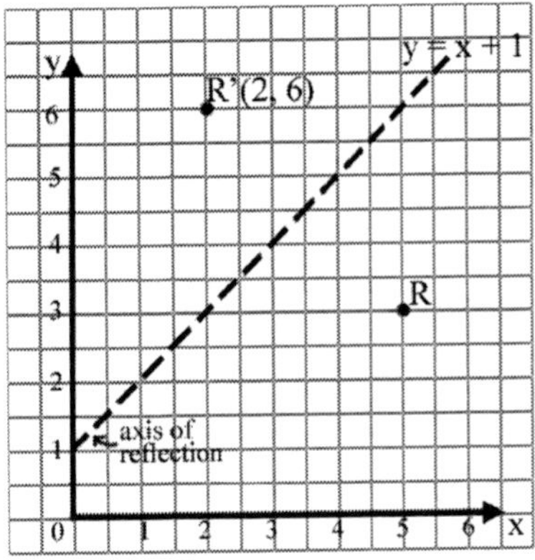

∴ Object, R is (5, 3) //

10. Smith was intrigued when his nephew told him that the word "CHECK" after a transformation still reads "CHECK". Describe the transformation to Smith.

CHECK

Figure 9. 6

Answer:
The word "CHECK" remains unchanged after a transformation because it has been reflected over its horizontal line of symmetry.

Notice that the horizontal line of symmetry divides the word "CHECK" into 2 mirror-image halves. Hence the horizontal line of symmetry in this situation is also the axis of reflection. //

11. In Figure 9.7, object A has been transformed to its image A' through a clockwise rotation at point C(7, 5). Determine the angle of rotation. If point K(1, 1) is transformed under the same rotation at point C, find image K'.

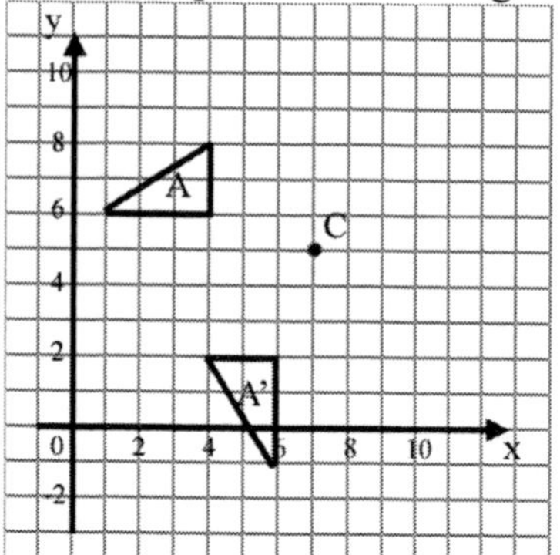

Figure 9. 7

Answers:
Given A transformed to A'
Also given C = point of rotation
Given direction of rotation: clockwise
=> Angle of rotation: 270° //

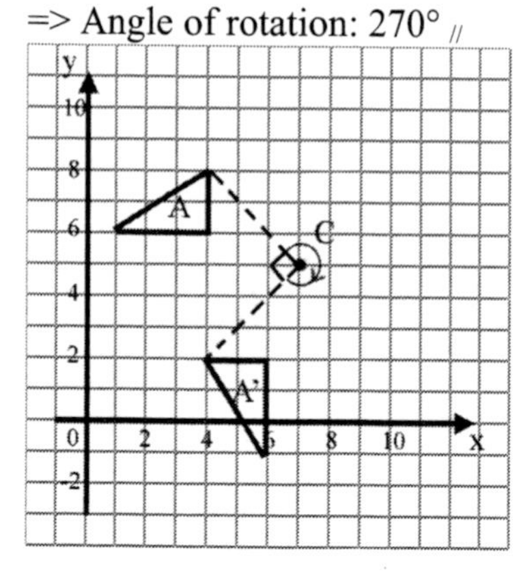

Given K = (1, 1)
After clockwise rotation 270° at point C, image, K' is (11, −1). //

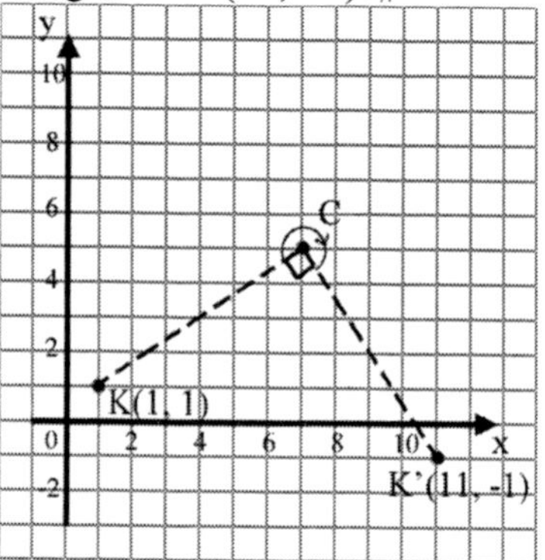

12. In Figure 9.8, object A is rotated 90° clockwise at point P(3, 4). Find the image, A' on the Cartesian plane.

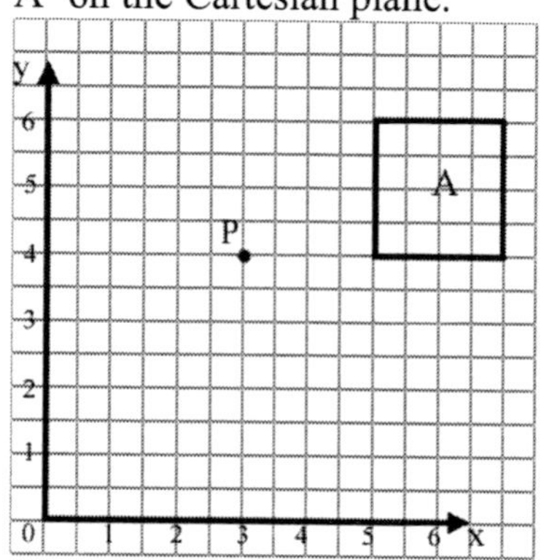

Figure 9. 8

Answer:
Given P(3, 4) = center of rotation
Transformation: clockwise rotation 90° at P
Thus, diagram below illustrates the location of image, A' after clockwise rotation 90° from P.

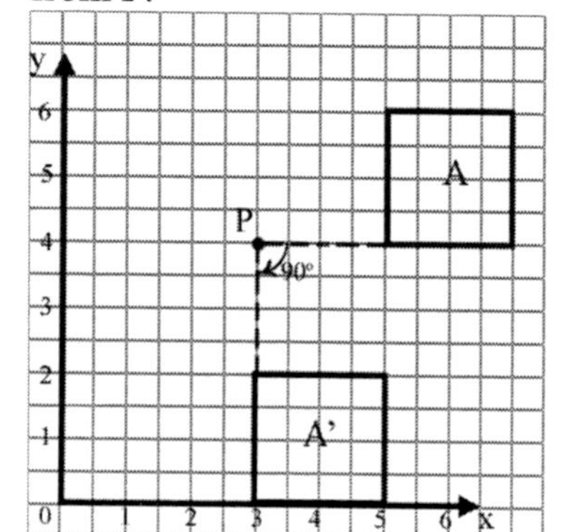

13. In Figure 9.9, object A has been transformed to its image A' in a single rotation. Find the point of rotation, angle of rotation and direction of rotation.

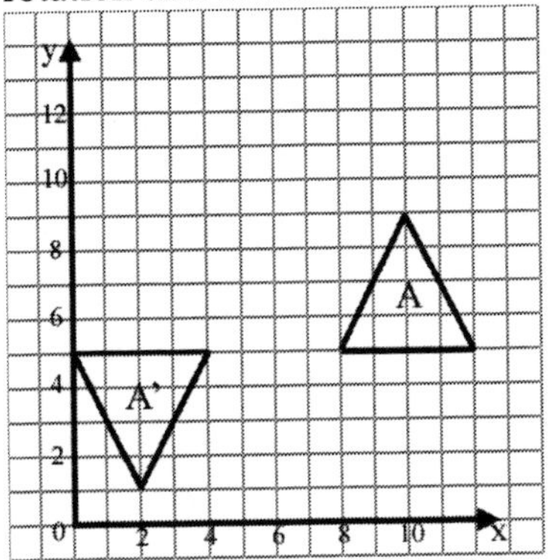

Figure 9. 9

Answers:
Given A = object
A' = image
Also given, transformation => 1 rotation
Let P = point of rotation
Below, Figure 9.9 has been reproduced for labeling:

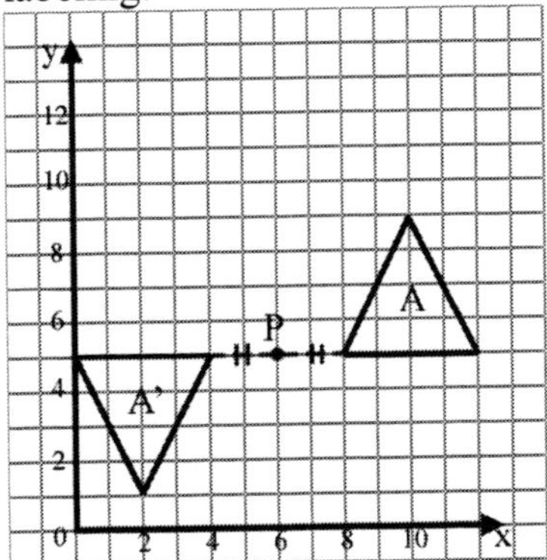

Based on the diagram above;
=> Angle of rotation = 180°//
=> Direction of rotation = clockwise or counterclockwise //
P = midpoint of (8, 5) and (4, 5)

$$= \left(\frac{x_1 + x_2}{2}, \frac{y_1 + y_2}{2}\right)$$

$$= \left(\frac{8+4}{2}, \frac{5+5}{2}\right)$$

$$= \left(\frac{12}{2}, \frac{10}{2}\right)$$

$$= (6, 5)$$

∴ Point of rotation is (6, 5). //

14. In Figure 9.10, object ABCD has been transformed to its image A'B'C'D' through dilation.

a) Determine the center of dilation
b) Find the scale factor of the dilation

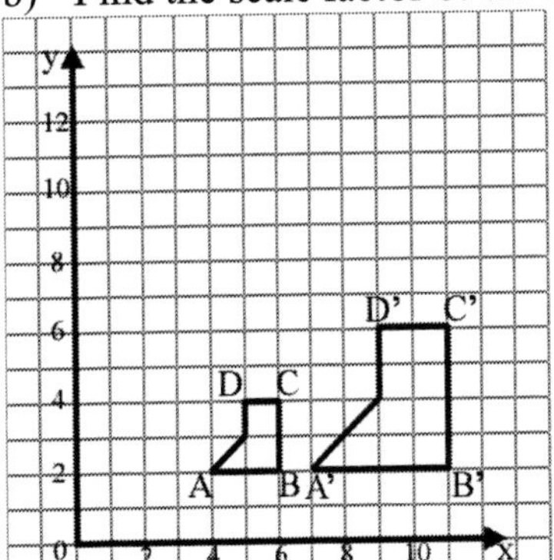

Figure 9. 10

Answers:
Given ABCD = object
A'B'C'D' = image
Also given, transformation=> dilation
a) Let P = center of dilation

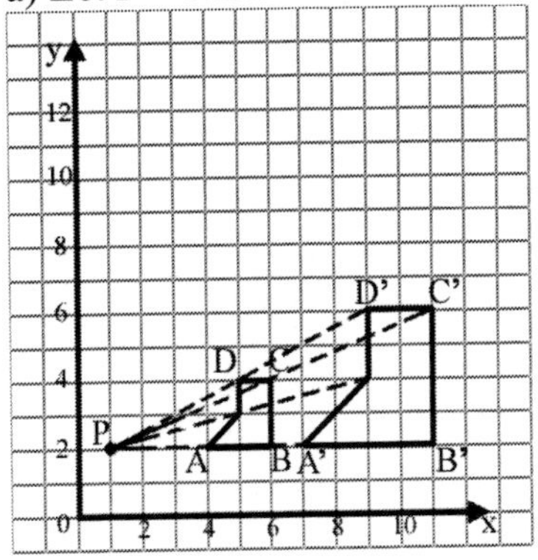

All points of object ABCD and image A'B'C'D' concur at the center of dilation, P (see the reproduced Figure 9.10 on page 151).

$\therefore$ Center of dilation, P = (1, 2) //

b) Let f = scale factor of dilation

$$f = \frac{\text{Length of image's side}}{\text{Length of object's side}}$$

$$= \frac{A'B'}{AB}$$

$$= \frac{4}{2}$$

$$= 2$$

$\therefore$ Scale factor of dilation is 2. //

Alternatively:

$$f = \frac{\text{Distance from image to P}}{\text{Distance from object to P}}$$

$$= \frac{PA'}{PA}$$

$$= \frac{6}{3}$$

$$= 2 \ //$$

15. In Figure 9.11, $\triangle$ABC has been transformed to $\triangle$A'B'C' under a single transformation, T. Find the:
a) Center of dilation
b) Scale factor of dilation
c) Area of $\triangle$A'B'C' if it is known that the area of $\triangle$ABC is 4 units^2.

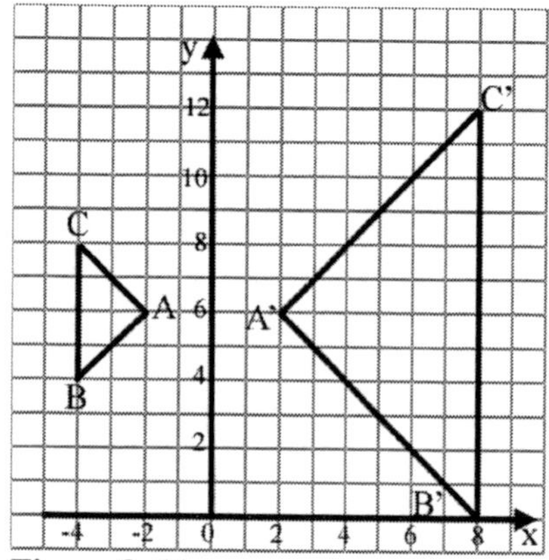

Figure 9. 11

Answers:
Given $\triangle$ABC = object
Also given $\triangle$A'B'C' = image

a)Let P = center of dilation

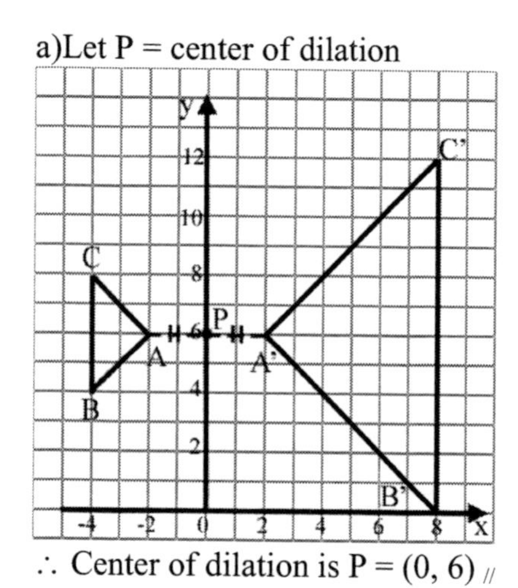

$\therefore$ Center of dilation is P = (0, 6) //

b) Let f = scale factor of dilation

$$f = -\frac{\text{Length of } \Delta A'B'C'}{\text{Length of } \Delta ABC}$$

$$= -\frac{B'C'}{BC}$$ ⇦ Negative as image and object are in opposite directions

$$= -\frac{12}{4}$$

$$= -3 \ //$$

c) Given area of $\triangle$ABC = 4
Thus, area of image, $\triangle$A'B'C':
= $f^2 \times$ area of $\triangle$ABC
= $(-3)^2 \times 4$
= 36 units^2 //

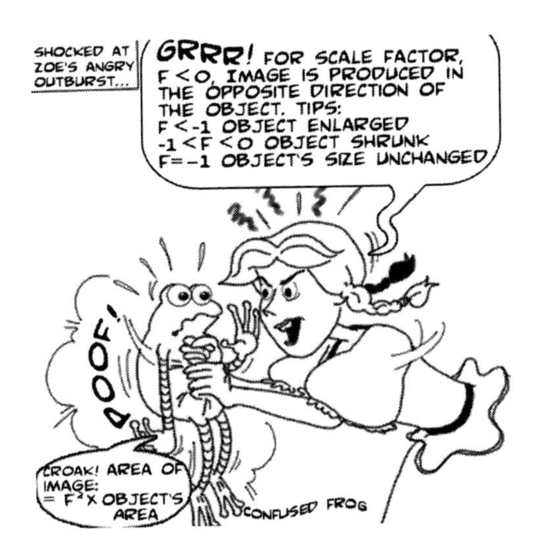

16. In Figure 9.12, △ABC has been dilated to its image △AB'C' from center of dilation A. If AB = 2 in. and AB' = 8 in., find area of △AB'C' given area of △ABC is 3 in^2.

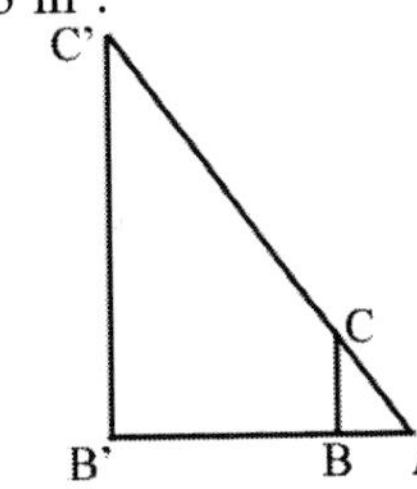

Figure 9. 12

Answer:
Given object = △ABC
Image = △AB'C'
Also given, AB = 2
AB' = 8
Let f = scale factor of dilation
Thus,

$$f = \frac{AB'}{AB}$$

$$= \frac{8}{2}$$

$$= 4$$

Since it is given area of △ABC = 3
Area of △AB'C':
= f^2 × area of △ABC
= $(4)^2 \times 3$
= 16×3
= 48 in^2
∴ Area of image △AB'C' is 48 in^2. //

17. In a combination of transformations, MN, object A(1, 4) has been transformed to image B(5, 1). It is further known that transformations, M and N are translations. If point C(3, 2) has been transformed under the transformation NM, find its image, C'.

Answer:
Given object, A = (1, 4)
Image, B = (5, 1)
Object A has been transformed to its image B through the combinations of translations MN.
Let L = combination of translations, MN
Thus, translation L:

$$= \begin{pmatrix} 5-1 \\ 1-4 \end{pmatrix}$$

$$= \begin{pmatrix} 4 \\ -3 \end{pmatrix}$$

Also given C = (3, 2)
Since translations NM = translations MN
=> translations NM = translation L
Thus, after transformation NM:

$$C' = \begin{pmatrix} 3+4 \\ 2-3 \end{pmatrix}$$

$$= \begin{pmatrix} 7 \\ -1 \end{pmatrix}$$

= (7, −1)
∴ Image, C' is (7, −1). //

Alternatively:
In Cartesian plane:

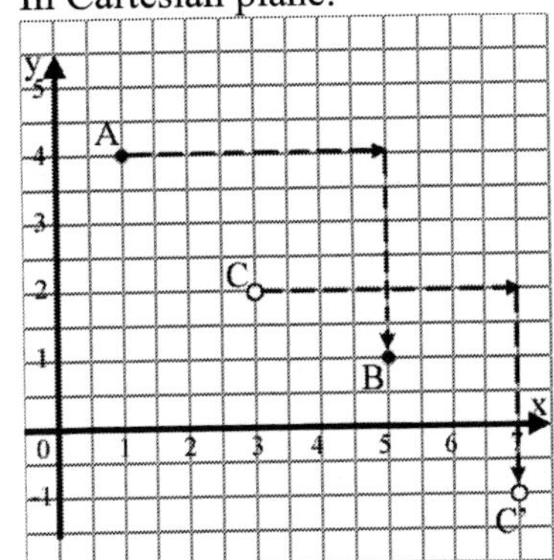

Object A has been transformed to its image B through combinations of transformations, MN: (see diagram above)
=> 4 units right
=> 3 units down

$MN = \begin{pmatrix} 4 \\ -3 \end{pmatrix}$

Since M and N are translations, then MN =

$NM = \begin{pmatrix} 4 \\ -3 \end{pmatrix}$

From diagram above, image C' = (7, −1) //

18. Given transformations, H and K are translations, where $H = \begin{pmatrix} 3 \\ 2 \end{pmatrix}$ and $K = \begin{pmatrix} -2 \\ 3 \end{pmatrix}$. Determine the final images of point P(2, 1) after the combination of transformations HK and KH.

Answers:

Given $H = \begin{pmatrix} 3 \\ 2 \end{pmatrix}$

$K = \begin{pmatrix} -2 \\ 3 \end{pmatrix}$

Also given point P = (2, 1)
Let P' = after first transformation
P'' = final image of point P
After transformation K:

$$P' = \begin{pmatrix} 2+(-2) \\ 1+3 \end{pmatrix} = \begin{pmatrix} 0 \\ 4 \end{pmatrix}$$

After transformation HK:

$$P'' = \begin{pmatrix} 0+3 \\ 4+2 \end{pmatrix} = \begin{pmatrix} 3 \\ 6 \end{pmatrix}$$

∴ After transformation HK, image P'' is (3, 6). //

After transformation H:

$$P' = \begin{pmatrix} 2+3 \\ 1+2 \end{pmatrix} = \begin{pmatrix} 5 \\ 3 \end{pmatrix}$$

After transformation KH:

$$P'' = \begin{pmatrix} 5+(-2) \\ 3+3 \end{pmatrix} = \begin{pmatrix} 3 \\ 6 \end{pmatrix}$$

∴ After transformation KH, image P'' is (3, 6). //

19. In Figure 9.13, object A is transformed to its image A' under the combination of transformations NM. It is further known that M and N are translations and translation M is $\begin{pmatrix} -3 \\ -2 \end{pmatrix}$. What is translation N? If point R(2, 3) undergoes the same combination of transformations NM, find its image R'.

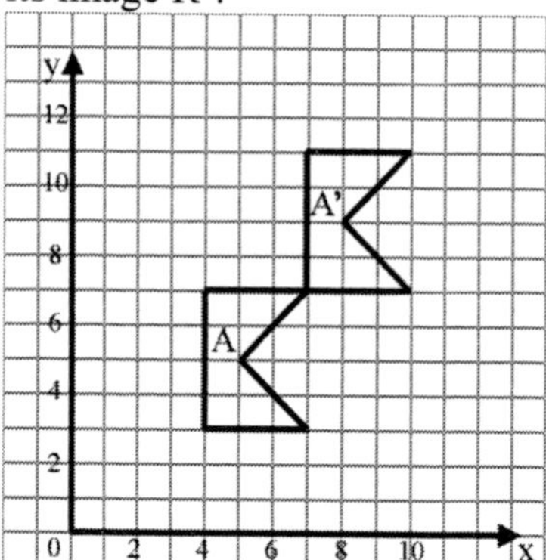

Figure 9. 13

Answers:
Given M and N = translations

$M = \begin{pmatrix} -3 \\ -2 \end{pmatrix}$

Also given under NM object A has been transformed to image A'
From Figure 9.13, combination of transformation NM:
=> 3 units right
=> 4 units up

$NM = \begin{pmatrix} 3 \\ 4 \end{pmatrix}$ …*

Thus, translation N: (i.e. NM – M)

$$=\begin{pmatrix}3-(-3)\\4-(-2)\end{pmatrix}$$

$$=\begin{pmatrix}6\\6\end{pmatrix}_{//}$$

Alternatively:

Let A_1 = Object, A after transformation M

Diagram below has been reproduced from Figure 9.13 for illustration:

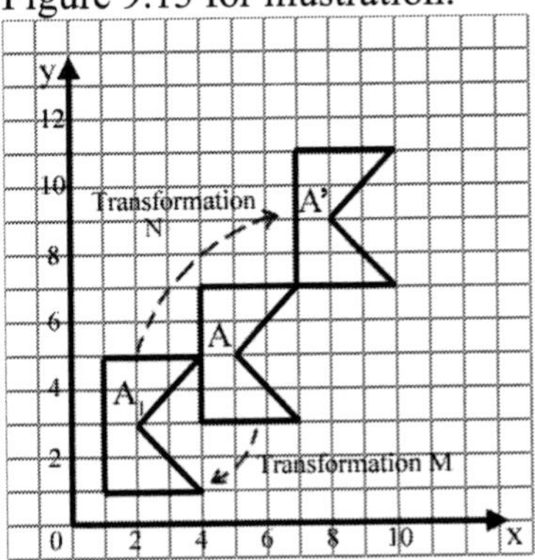

From diagram above, A_1 has been translated to A' through:

=> 6 units right

=> 6 units up

$$\therefore N=\begin{pmatrix}6\\6\end{pmatrix}_{//}$$

Also given R= (2, 3)

We have found that $NM=\begin{pmatrix}3\\4\end{pmatrix}$ ⇦ refer *

After transformation NM:

$$R'=\begin{pmatrix}2+3\\3+4\end{pmatrix}=\begin{pmatrix}5\\7\end{pmatrix}$$

$\therefore$ R' = (5, 7) //

Answers:

Given P = (3, 6)

Translation, $T=\begin{pmatrix}-1\\5\end{pmatrix}$ and $U=\begin{pmatrix}2\\3\end{pmatrix}$

Thus, after transformation U:

$$P_1=\begin{pmatrix}3+2\\6+3\end{pmatrix}=\begin{pmatrix}5\\9\end{pmatrix}$$

After combination of transformations TU:

$$P_1=\begin{pmatrix}5+(-1)\\9+5\end{pmatrix}=\begin{pmatrix}4\\14\end{pmatrix}$$

$\therefore$ P_1 is (4, 14) //

For P_1 after the transformation T:

$$P_2=\begin{pmatrix}4+(-1)\\14+5\end{pmatrix}=\begin{pmatrix}3\\19\end{pmatrix}$$

After combination of transformations UT:

$$P_2=\begin{pmatrix}3+2\\19+3\end{pmatrix}=\begin{pmatrix}5\\22\end{pmatrix}$$

$\therefore$ P_2 is (5, 22) //

20. P(3, 6) is the center of a circle. Under a combination of transformations, TU, where T is $\begin{pmatrix}-1\\5\end{pmatrix}$ and U is $\begin{pmatrix}2\\3\end{pmatrix}$, P has been transformed to P_1. Find the coordinate of the new circle's center, P_1. In a further experiment, P_1 is transformed to P_2 through a combination of transformations UT. What is the location of the new circle's center P_2?

21. Point A(3, 5) is transformed to its image B(6, 6) under the combination of transformations TU. It is given that transformations, T and U are translations. If U represents translation $\begin{pmatrix}-2\\-3\end{pmatrix}$, what is translation T?

Answer:

Let C = image A after translation U

Given A = (3, 5)

Given translation $U = \begin{pmatrix} -2 \\ -3 \end{pmatrix}$

Thus, A after translation U:

$$C = \begin{pmatrix} 3+(-2) \\ 5+(-3) \end{pmatrix} = \begin{pmatrix} 1 \\ 2 \end{pmatrix}$$

Let translation $T = \begin{pmatrix} x \\ y \end{pmatrix}$

After translation TU:

$$B = \begin{pmatrix} 1+x \\ 2+y \end{pmatrix} \quad \ldots *$$

Substitute B = (6, 6) into *:
(6, 6) = (1 + x, 2 + y)
Comparing like components:
x-component:
6 = 1 + x
$\therefore$ x = 6 – 1 = 5
y-component:
6 = 2 + y
$\therefore$ y = 6 – 2 = 4

$\therefore T = \begin{pmatrix} 5 \\ 4 \end{pmatrix}$ //

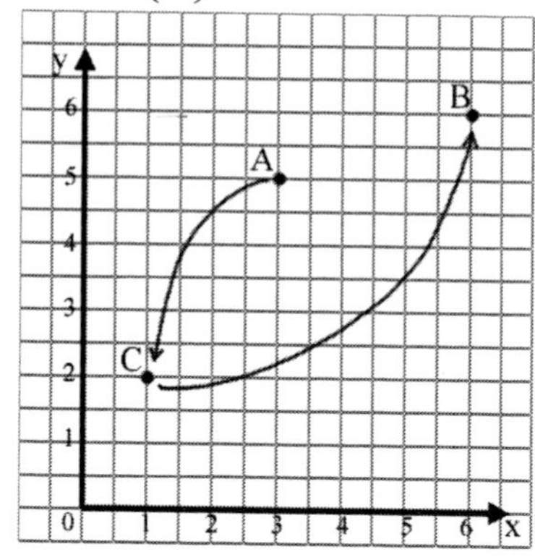

22. Image B'(3, 4) has been obtained from the combination of transformations, TU. If T represents translation $\begin{pmatrix} 2 \\ 2 \end{pmatrix}$ while U is translation $\begin{pmatrix} -2 \\ -3 \end{pmatrix}$, find the coordinate of object B.

Answer:
Given image B' = (3, 4)

Also given translations, $T = \begin{pmatrix} 2 \\ 2 \end{pmatrix}$

$U = \begin{pmatrix} -2 \\ -3 \end{pmatrix}$

After translation TU, B' = (3, 4)
Before translation T (i.e. TU – T):

$$B' = \begin{pmatrix} 3-2 \\ 4-2 \end{pmatrix} = \begin{pmatrix} 1 \\ 2 \end{pmatrix}$$

Before translation U:

$$B = \begin{pmatrix} 1-(-2) \\ 2-(-3) \end{pmatrix} = \begin{pmatrix} 3 \\ 5 \end{pmatrix}$$

$\therefore$ Object B is (3, 5) //

23. Point H has been transformed to its image K(6, 4) under a combination of transformations BA. It is further known that A represents translation $\begin{pmatrix} -4 \\ 2 \end{pmatrix}$, while B represents reflection on the straight line, y = x. Find the coordinate of H.

Answer:
Given image, K = (6, 4)

Also given, A = translation $\begin{pmatrix} -4 \\ 2 \end{pmatrix}$

B = reflection on line, y = x
Let J = after transformation A on object H
Thus,
Before transformation B (i.e. BA – B):
J = (4, 6) ⇦ (a, b) reflected on y = x is (b, a)
Before translation A:

$$H = \begin{pmatrix} 4-(-4) \\ 6-2 \end{pmatrix} = \begin{pmatrix} 8 \\ 4 \end{pmatrix}$$

$\therefore$ Object H is (8, 4). //

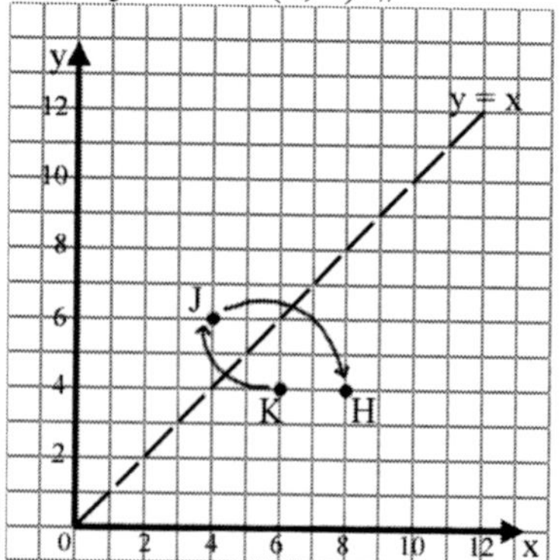

24. In Figure 9.14, PQRU is an isosceles trapezoid. $\overline{PT}$ and $\overline{QS}$ are perpendicular lines to $\overline{PQ}$ and $\overline{RU}$. If it is known that $\triangle QRS$ is the image of object $\triangle PUT$, detail the two transformations that have moved $\triangle PUT$ to $\triangle QRS$.

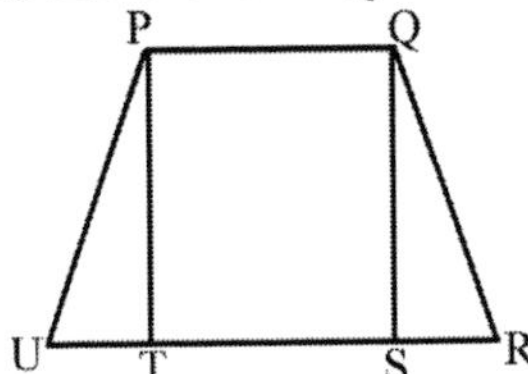

Figure 9. 14

Answers:
Given $\triangle PUT$ = object
$\triangle QRS$ = image
Also given, $\triangle PUT$ has been transformed to $\triangle QRS$ by 2 transformations.
Let A and B = the 2 transformations

i. A is translation $\begin{pmatrix} TS \\ 0 \end{pmatrix}$, that is $\triangle PUT$ has been translated horizontally by length TS.

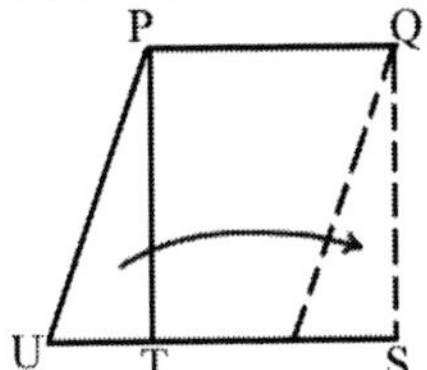

ii. B is reflection on $\overline{QS}$, that is $\triangle PUT$ is then reflected on the straight line, $\overline{QS}$.

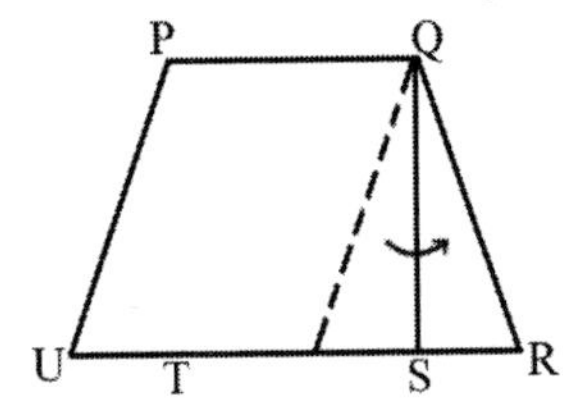

$\therefore$ The 2 transformations are translation $\begin{pmatrix} TS \\ 0 \end{pmatrix}$ and reflection on the straight line, $\overline{QS}$. BA will result in $\triangle PUT$ being transformed to its image, $\triangle QRS$. //

Note: if B is reflection on $\overline{PT}$, then the combination of transformations is AB.

25. In Figure 9.15, H is the image of K after translation, E followed by a transformation, F.

a) Determine the displacement vector of the translation, E.
b) Describe transformation F.

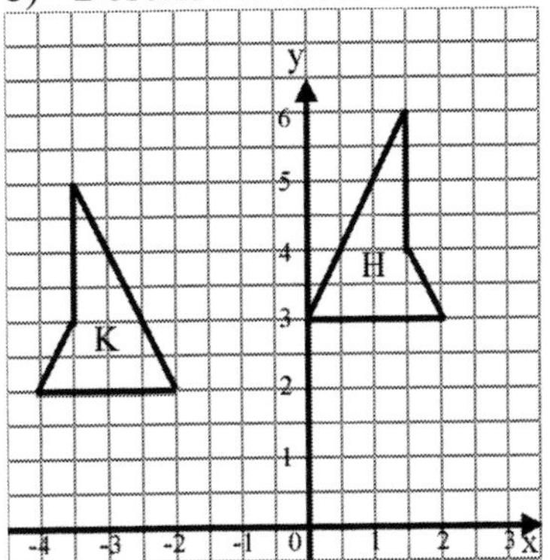

Figure 9. 15

Answers:
Given H = image
K = object
a) Let J = image of K after translation E

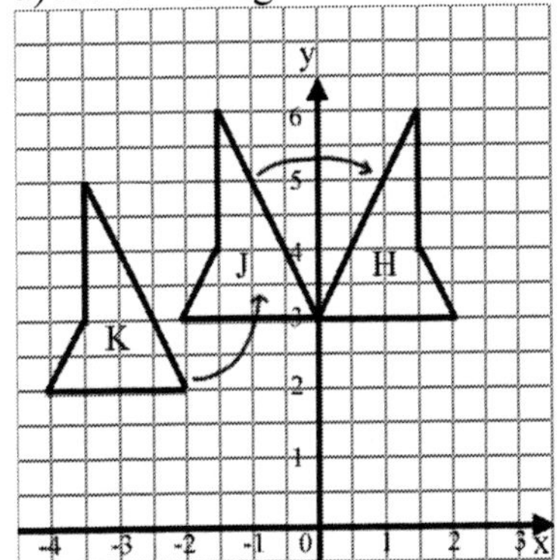

Thus K to J
=> 2 units right
=> 1 unit up

$\therefore$ Translation, E = $\begin{pmatrix} 2 \\ 1 \end{pmatrix}$ //

b) For J to be transformed into image H, transformation F must be a reflection on the y axis (where x = 0). //

26. In Figure 9.16, RWVU is a rhombus that has been segregated into 16 smaller congruent rhombuses. A is the shaded rhombus. If A is displaced by a combination of transformations, MN, where is its image B? N represents

translation $\overline{PR}$ and M represents reflection on $\overline{UW}$.

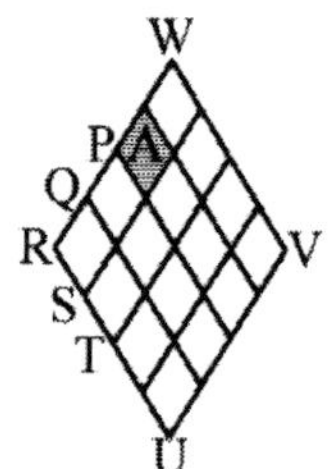

Figure 9. 16

Answer:
Given A = object
B = image
Also given the 16 small rhombuses are ≅
Transformation N = translation $\overline{PR}$
M= reflection $\overline{UW}$
Thus, after combination of transformations MN, image B is located on:

27. Figure 9.17, depicts object A in a Cartesian plane. In the same Cartesian plane sketch the following transformations:

a) Object A is rotated 90° clockwise at origin to image A_1
b) Object A_1 is rotated 90° clockwise at origin to image A_2

Hence, describe a single transformation that combines transformation in part a) and b).

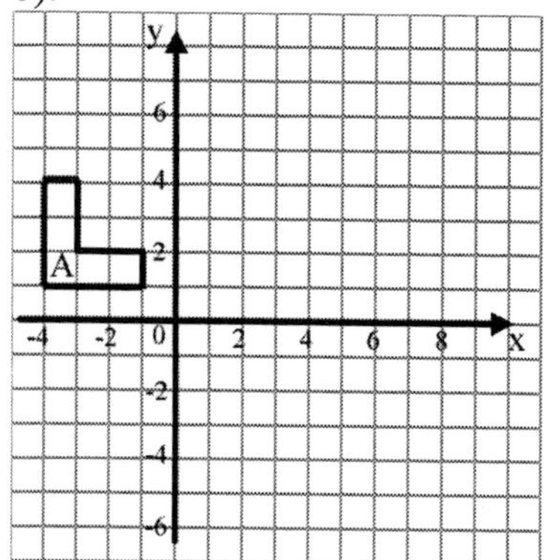

Figure 9. 17

Answers:
Transformations:

a) Object A rotated 90° clockwise at origin to image A_1
b) Object A_1 rotated 90° clockwise at origin to image A_2

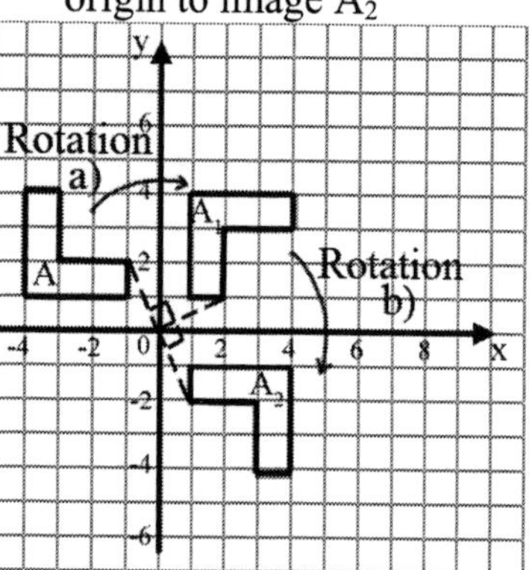

∴ A single transformation that combines transformations a) and b) is object A is rotated 180° clockwise/counterclockwise at the origin to image A_2. //

28. Point P(6, 5) is located on the vertex of a triangle. If the triangle is transformed to its image through a combination of transformations;

A represents translation $\begin{pmatrix} 3 \\ -2 \end{pmatrix}$

B represents rotation 90° clockwise from the origin

C represents rotation 90° counterclockwise from origin

D represents rotation 180° from origin

Find the coordinate P' in the following transformations:

a) BA
b) CA
c) BC
d) DA

Answers:
Given object P = (6, 5)
P' = image
Let P_1 = image after first transformation
Also given:

$$A = \text{translation} \begin{pmatrix} 3 \\ -2 \end{pmatrix}$$

B = clockwise rotation 90° from origin
C = counterclockwise rotation 90° from origin
D = 180° rotation from origin

a) After translation A:

$$P_1 = \begin{pmatrix} 6+3 \\ 5-2 \end{pmatrix} = \begin{pmatrix} 9 \\ 3 \end{pmatrix}$$

$P_1 = (9, 3)$ …*
After transformation BA:
P' = (3, −9) //

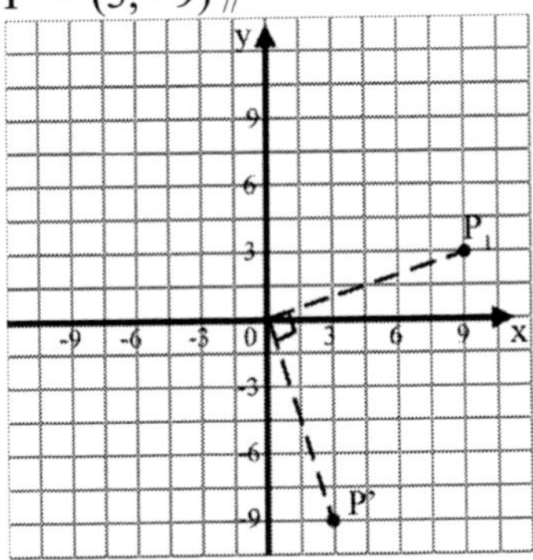

b) After translation A: [refer to *]
$P_1 = (9, 3)$
After transformation CA:
P' = (-3, 9) //

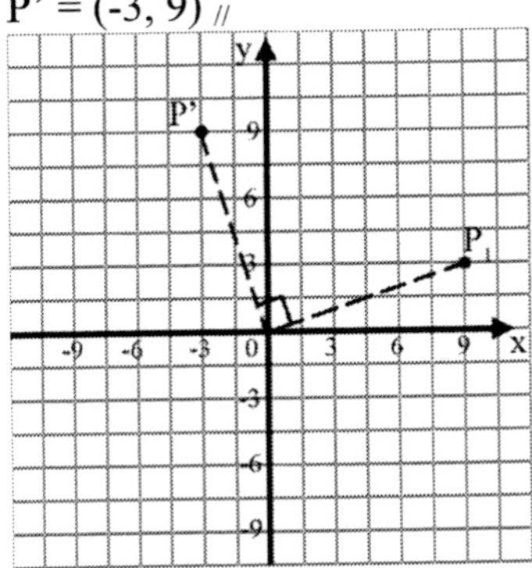

c) After transformation C:
$P_1 = (-5, 6)$
After transformation BC:
P' = (6, 5) //

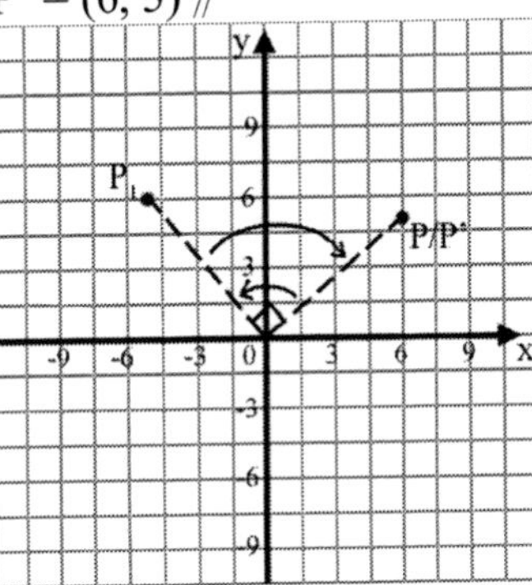

d) After translation A: [refer to *]
$P_1 = (9, 3)$
After transformation DA:
P' = (−9, −3) //

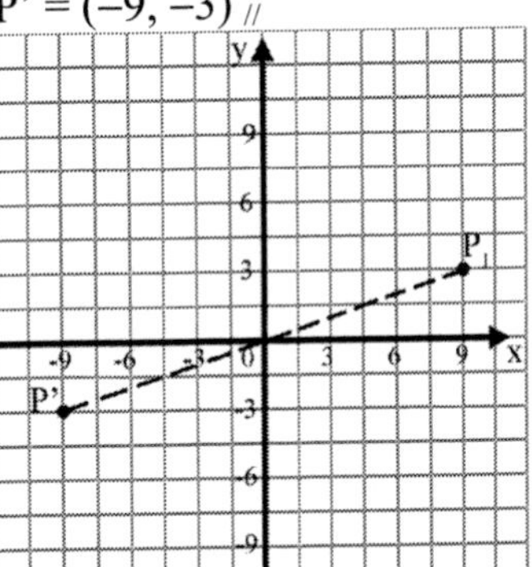

29. Figure 9.18, represents a regular octagon. Describe a single transformation that transforms △OEF to:

a) Image △OAB

b) Image $\triangle ODE$

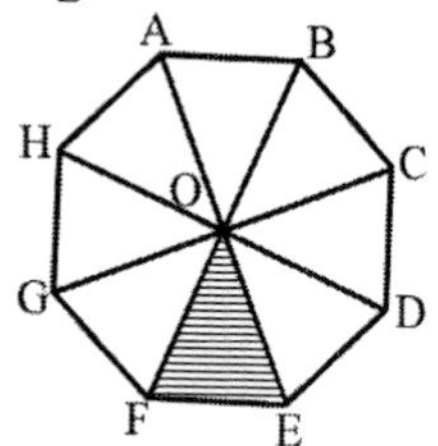

Figure 9. 18

Answers:

a) A single transformation that transforms object $\triangle OEF$ to image $\triangle OAB$ is a rotation 180° from center point, O. //

b) A single transformation that transforms object $\triangle OEF$ to image $\triangle ODE$ is rotation 45° (i.e. 360° ÷ 8 = 45°) counterclockwise at center point, O. //

Alternatively:

A single transformation that transforms object $\triangle OEF$ to image $\triangle ODE$ is reflection on straight line, $\overline{AE}$. //

30. In Figure 9.19, C is the image of B after a combination of transformations MN. If N represents counterclockwise rotation 90° from the origin, hence describe transformation M. Subsequently, find a single transformation that will map object B to its image C.

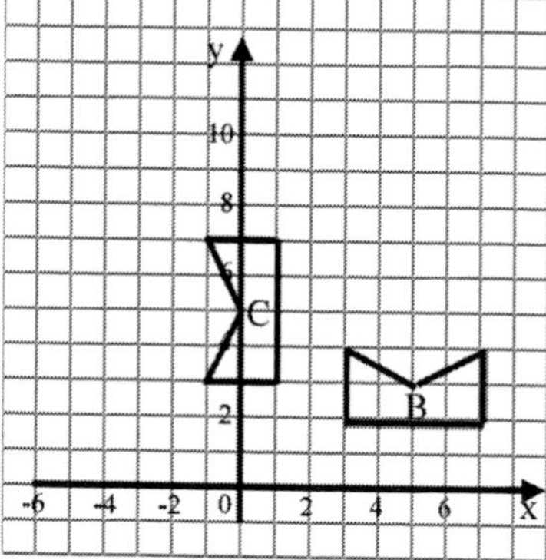

Figure 9. 19

Answers:

Given B = object

C = image

Transformation:

N = counterclockwise rotation 90° from origin

Let B' = image after transformation N:

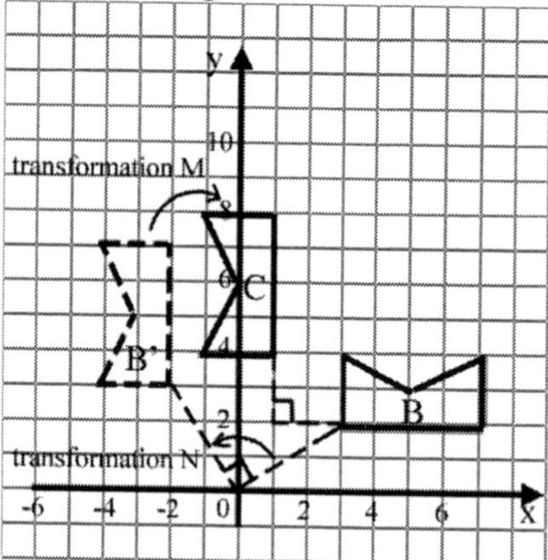

From diagram above, transformation M transforms B' to C. Thus, transformation M represents translation $\begin{pmatrix} 3 \\ 1 \end{pmatrix}$. //

Let L = a single transformation that is equivalent to MN

From the Cartesian plane above, transformation L is a rotation 90° counterclockwise at rotation point (1, 2). //

31. K(4, 7) is located at the vertex of a triangle. It is further known that transformations, A represents reflection on the y-axis and B represents rotation 180° from origin. If the triangle is transformed to its image by a combination of transformations, AB, find the new location of image, K'.

Answer:

Given object, K = (4, 7)

K' = image

Transformations:

A = reflection on y-axis

B = rotation 180° from origin

Let K_1 = image after transformation B

After transformation B:

$K_1 = (-4, -7)$

After transformation AB:

K' = (4, −7)

∴ Combination of transformations, AB maps K to its image K'(4, −7). //

32. In Figure 9.20, PQRT is a rhombus. $\overline{PR}$ is a diagonal whose length equals the side of the rhombus. Further given, $\triangle PQR$ has been transformed to $\triangle STU$ through a combination of transformations; detail the 2 transformations that were involved.

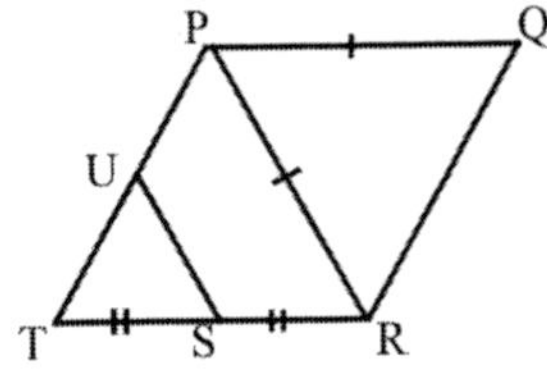

Figure 9. 20

Answers:
Given PQRT = rhombus
=> PQ = QR = RT = PT
Also given PR = PQ
$\triangle PQR$ = object
$\triangle STU$ = image
Since it is known that 2 transformations have transformed $\triangle PQR$ to image $\triangle STU$:
Transformation 1:
Rotation 60° clockwise at P (or rotation 60° counterclockwise at R) //
Transformation 2:
Dilation at point T with scale factor of dilation $= \frac{1}{2}$ //

Rotate 60° clockwise at P, Q, 60°, Dilate, T, S, R

Alternatively:
Transformation 1:
Reflection at straight line, $\overline{PR}$ //
Transformation 2:
Dilation at point T with scale factor of dilation $= \frac{1}{2}$ //

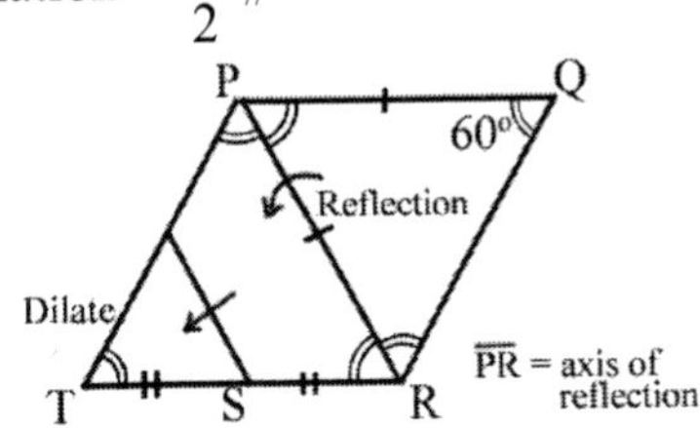

33. In Figure 9.21, quadrilateral PQRS has been transformed to P'QR'S' through a combination of transformations BA. If A represents reflection, determine:
a) Mirror line
b) Transformation B
Hence find a single transformation that maps PQRS to P'QR'S'.

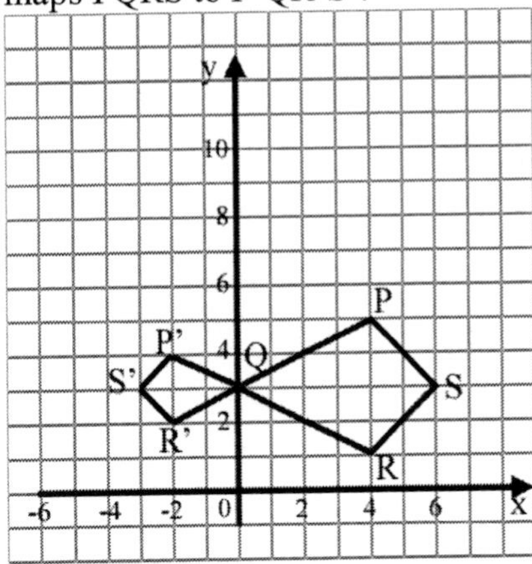

Figure 9. 21

Answers:
Given PQRS = object
P'QR'S' = image
Also given transformation, A = reflection
a)
Mirror line is y-axis (i.e. x = 0) //
b)
B represents dilation. The center of dilation is Q(0, 3) and the scale factor of dilation is $\frac{1}{2}\left(\text{scale factor} = \frac{QS'}{QS} = \frac{3}{6}\right)$. //

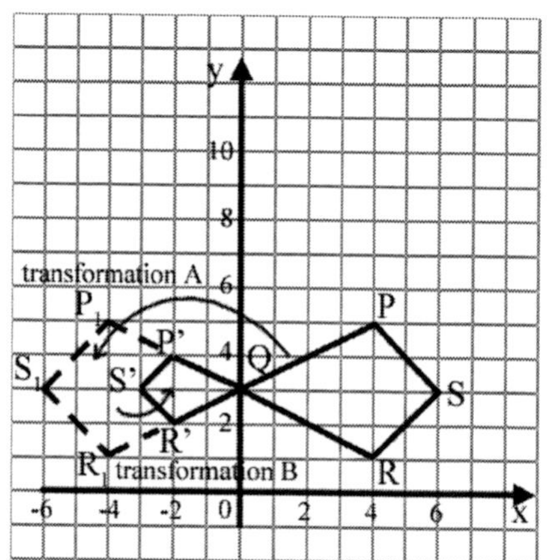

A single transformation that equals the combine transformation BA is dilation at Q(0, 3) and scale factor of dilation is $-\frac{1}{2}\left(\text{Scale factor} = -\frac{QS'}{QS} = -\frac{3}{6}\right)$ //

34. In Figure 9.22, △ABC has been dilated to its image △AB'C' through a combination of transformations T^2 at center of dilation, A. Find the scale factor of dilation for transformation, T.

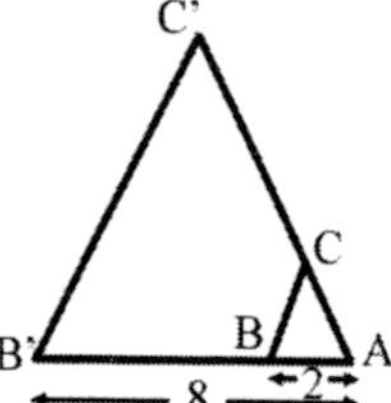

Figure 9. 22

Answer:

Given $\triangle ABC \xrightarrow{T^2} \triangle AB'C'$

Let f = scale factor of dilation T

Thus, scale factor:

$$f^2 = \frac{AB'}{AB}$$

$$= \frac{8}{2} = 4$$

$$f = \sqrt{4} = 2 \;//$$

35. In Figure 9.23, ABCD and PQRC are similar rectangles. It is given that △ABC has been transformed to △CPQ through a combination of transformations, KL. If L represents rotation, find:

a) Angle of rotation

b) Center of rotation

Hence detail transformation K.

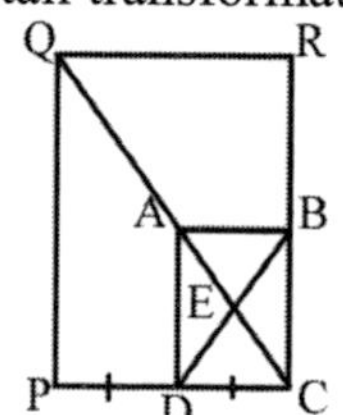

Figure 9. 23

Answers:

Given △ABC = object

△CPQ = image

ABCD and PQRC = similar rectangles

Also given transformation, L = rotation

a) Angle of rotation = 180° //

b) Center of rotation = E //

Hence, transformation K represents dilation. C is the center of dilation and scale factor of dilation is $2\left(\frac{CP}{CD} = \frac{2CD}{CD}\right)$ //

Alternatively:

b) Center of rotation = A //

Hence, transformation K represents dilation. Q is the center of dilation and scale factor of dilation is $2\left(\frac{QC}{QA} = \frac{2QA}{QA}\right)$ //

Chapter 10
Surface Areas & Volumes

Surface area – the sum of area of each of its faces in a geometric solid

Volume – space occupied by a geometric solid

Lateral surface – the side of a geometric solid that is not its base

Lateral surface area – area of lateral surfaces of a solid

Sphere:

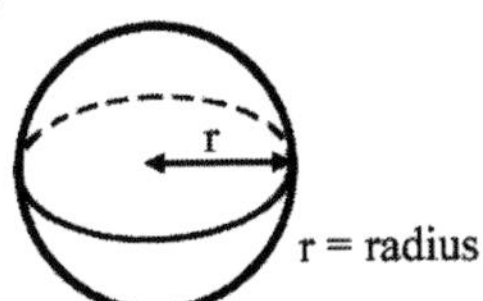

Surface area = $4\pi r^2$

Volume = $\frac{4}{3}\pi r^3$

Circular cone:

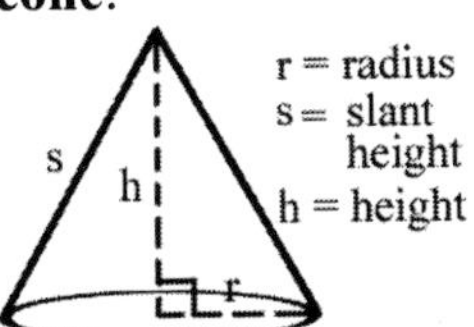

Surface area = $\pi r(s + r)$

Volume = $\frac{1}{3}\pi r^2 h$

Pyramid:

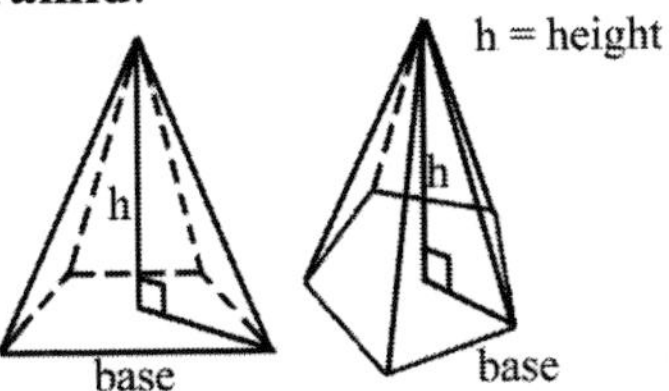

Surface area = base + lateral surface area

Volume = $\frac{1}{3} \times \text{base} \times \text{h}$

Cylinder:

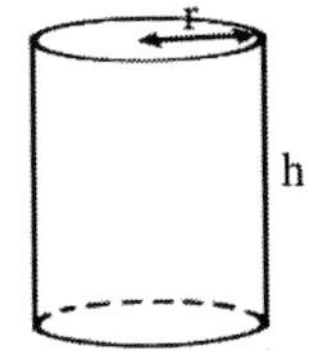

Surface area = $2\pi r(h + r)$

Volume = $\pi r^2 h$

Rectangular block:

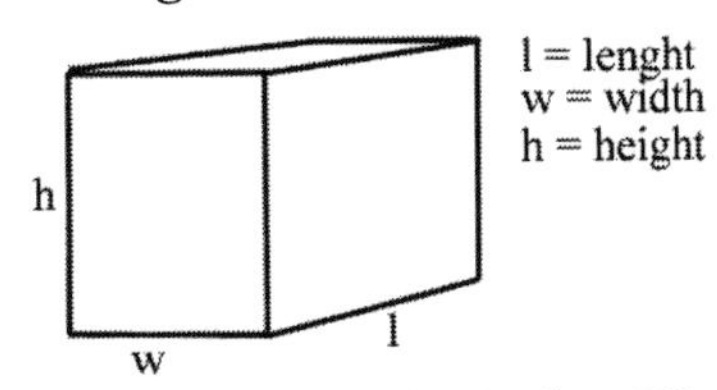

Surface area = 2wl + 2wh + 2lh

Volume = wlh

Cube:

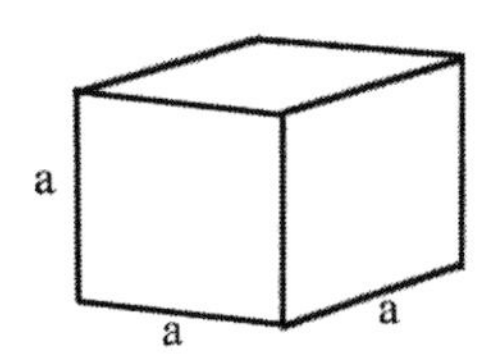

Surface area = $6a^2$

Volume = a^3

Prism: (Two congruent and parallel bases and lateral surfaces are parallelograms)

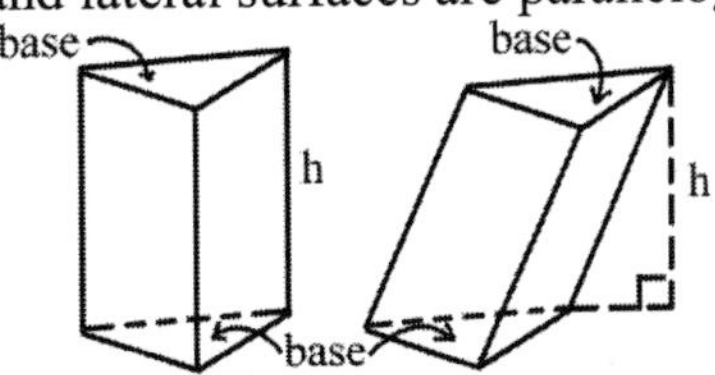

Surface area = 2 × base + sum of parallelogram faces

1. Figure 10.1 shows a right prism. What is the area of its lateral surface?

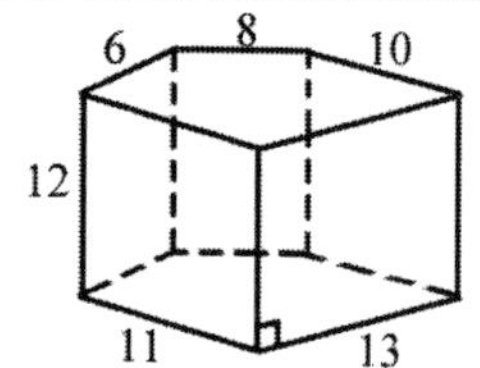

Figure 10. 1

Answer:
For right prism:
Lateral area = perimeter of base × height
$= (6 + 8 + 10 + 13 + 11) \times 12$
$= 48 \times 12$
$= 576 \text{ units}^2$ //

2. What is the surface area of a closed hemisphere with diameter 10 units? (Let $\pi = 3.142$)

Answer:
Given diameter = 10
=> radius, r = diameter ÷ 2
$= 10 \div 2 = 5$
Since hemisphere is closed:

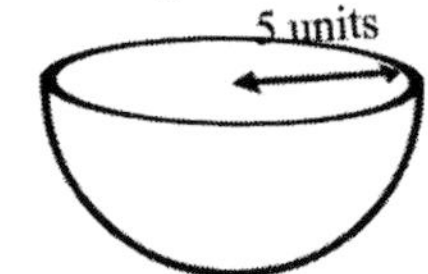

Surface area of closed hemisphere:
$= \text{base} + \frac{1}{2}(\text{surface area of sphere})$
$= \pi r^2 + \frac{1}{2}\left(4\pi r^2\right)$
$= \pi(5)^2 + 2\pi(5)^2$
$= 3.142 \times 25 + 2 \times 3.142 \times 25$
$= 78.55 + 157.1$
$= 235.65 \text{ units}^2$ //

3. What is the surface area of an ice cream cone sleeve in Figure 10.2?

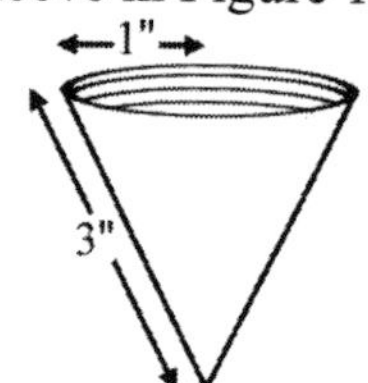

Figure 10. 2

Answer:
Given ice cream cone sleeve = hollow cone
Also given radius, r = 1
Slant height, s = 3
Surface area of ice cream cone sleeve:
$= \pi rs$
$= \frac{22}{7} \times 1 \times 3$
$= 9.429 \text{ in}^2$ //

4. Arthur needs to build a right circular cone for his science project using foil for both the lateral surface area and base. If the specifications for the right circular cone are; radius 7 inches, and height 24 inches, what is the surface area of the foil he will need? ($\pi = \frac{22}{7}$)

Answer:
Given radius, r = 7
Height, h = 24

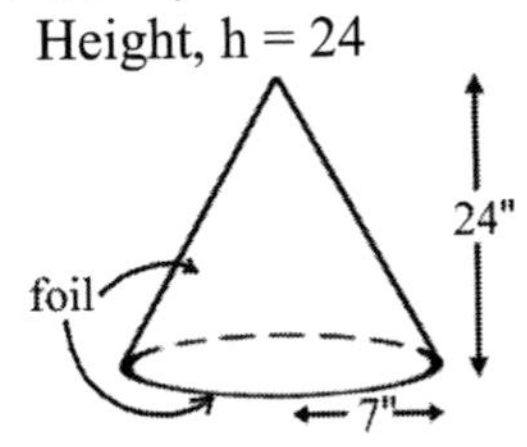

Let s = slant height
Using Pythagorean Theorem:

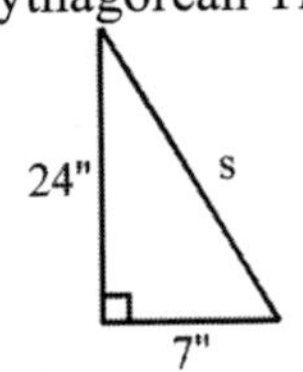

$s^2 = r^2 + h^2$
$= 7^2 + 24^2$
$= 49 + 576$
$= 625$
$s = \sqrt{625} = 25$ in
Lateral surface area of right circular cone:
$= \pi rs$
$= \frac{22}{7} \times 7 \times 25$
$= 550$ in^2
Base:
$= \pi r^2$
$= \frac{22}{7} \times 7^2$
$= 154$ in^2
Total foil area needed:
= lateral area of circular cone + base
= 550 + 154
$= 704$ in^2 //

5. Find the surface area of a traffic cone with a square base as shown in Figure 10.3. (Assume that the traffic cone is a hollow cone)

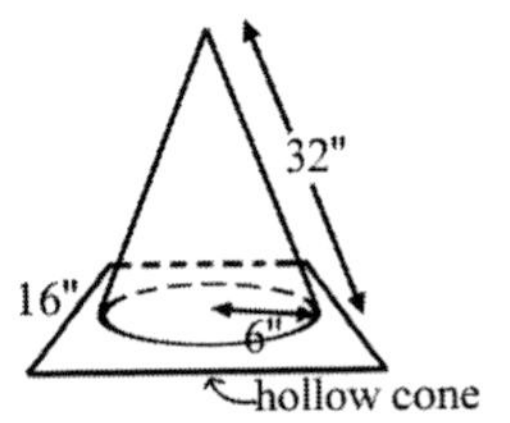

Figure 10. 3

Answer:
Given traffic cone => hollow cone
=> square base
Also given radius, r = 6
Slant height, s = 32
Length of square base, a = 16
Surface area:
= surface area of cone + square base – cone base
$= \pi rs + a^2 - \pi r^2$
$= \frac{22}{7} \times 6 \times 32 + 16^2 - \frac{22}{7} \times 6^2$
$= 603.4286 + 256 - 113.1429$
$= 746.2857$ in^2 //

6. Figure 10.4 shows a rectangular block. Find the surface area of the rectangular block. Hence what is the difference between its surface area and lateral surface area?

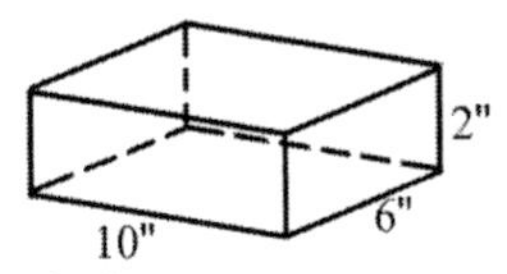

Figure 10. 4

Answers:
Given rectangular block:
Also given width, w = 6
Length, l = 10
Height, h = 2
Thus, surface area:
= 2wl + 2wh + 2lh
= 2×6×10 + 2×6×2 + 2×10×2
= 120 + 24 + 40
= 184 in^2 //

Lateral surface area:
= 2wh + 2lh
= 2×6×2 + 2×10×2
= 24 + 40
= 64 in^2

Difference between surface area & lateral surface area:
= surface area – lateral surface area
= 184 – 64
= 120 in^2 //

Alternatively:
Difference between surface area & lateral surface area:
= 2 × base
= 2 × wl
= 2×6×10
= 120 in^2 //

7. Figure 10.5 shows a child's toy barn. What is the surface area of the barn?

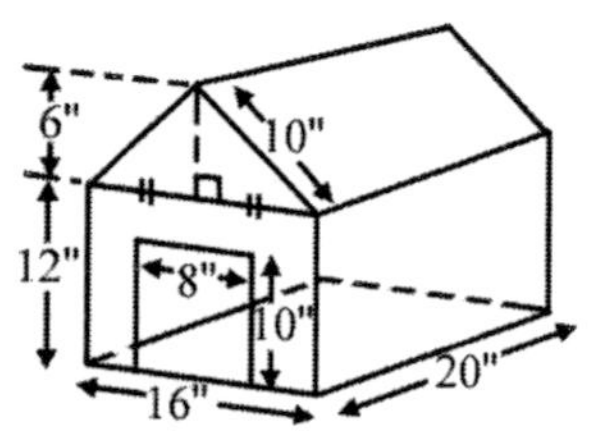

Figure 10. 5

Answer:
Surface area:
$= 16\times20 + 2(12\times20) + 2(12\times16) - 8\times10 + 2(10\times20) + 2(\frac{1}{2}\times6\times16)$
= 320 + 480 + 384 – 80 + 400 + 96
= 1600 in^2
∴ Surface area of the barn is 1600 in^2 //

8. Find the surface area of a ping pong ball (also called a table tennis ball) whose diameter is 4 mm.

Answer:
Given diameter = 4
=> radius, r = diameter ÷ 2
= 4 ÷ 2 = 2
Since ping pong ball = sphere
Surface area of ping pong ball:
$= 4\pi r^2$
$= 4\times\frac{22}{7}\times2^2$
= 50.2857 mm^2
∴ Surface area of ping pong ball is 50.2857 mm^2 //

9. Figure 10.6 shows an opened can. What is the surface area of the hollow can?

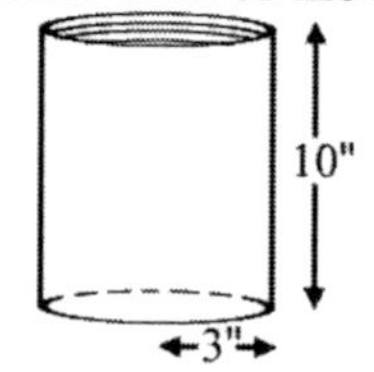

Figure 10. 6

Answer:
Given opened can = cylinder with one base
Radius, r = 3
Height, h = 10

Thus,
Surface area of opened can:
$= 2\pi rh + \pi r^2$
$= 2\times\frac{22}{7}\times 3\times 10+\frac{22}{7}\times 3^2$
$= 188.5714 + 28.2857$
$= 216.8571\ in^2$ //

10. Find the volume of the rectangular block in Figure 10.7.

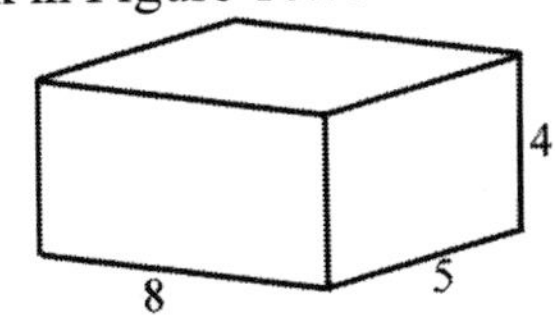

Figure 10. 7

Answer:
Given rectangular block
Thus, volume of rectangular block:
= width × length × height
= 5 × 8 × 4
$= 160\ units^3$ //

11. Find the volume of the rectangular chest shown in Figure 10.8.

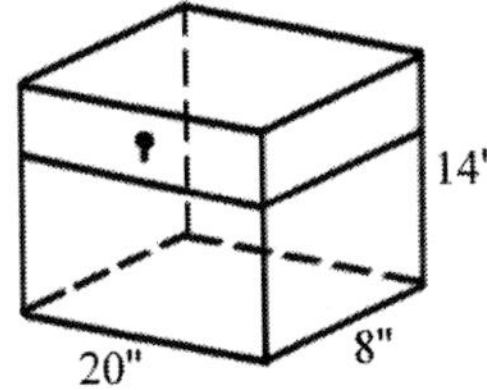

Figure 10. 8

Answer:
Given chest = rectangular block
Volume of chest:
= width × length × height
= 8 × 20 × 14
$= 2240\ in^3$ //

12. An artist models a right pyramid whose base is a square, as shown in Figure 10.9. If it is further known that the side of the square base is 12 inches and the slant height is 10 inches, what is the lateral surface area and height? Hence find the volume of the model.

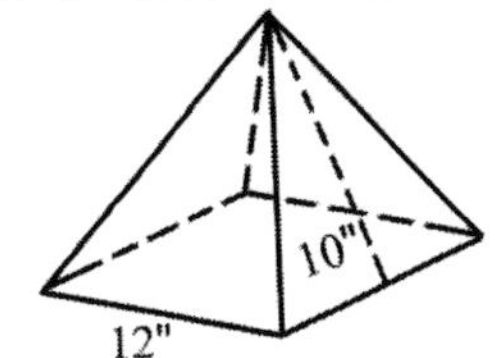

Figure 10. 9

Answers:
Given right pyramid whose base is square
=> All lateral faces are congruent
=> All sides of the base, a = 12
Also given slant height, s = 10
Lateral surface area:
= 4 × lateral face
$= 4 \times (\frac{1}{2} \times a \times s)$
$= 4 \times (\frac{1}{2} \times 12 \times 10)$
$= 240\ in^2$ //

Let h = height of pyramid

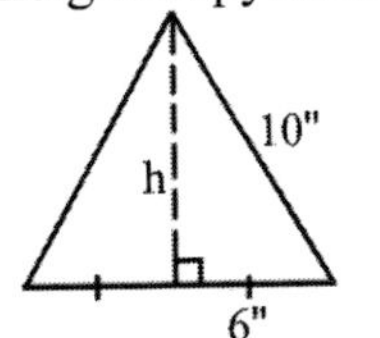

Using Pythagorean Theorem:
$h^2 = 10^2 - 6^2$
$= 100 - 36$
$= 64$
$h = \sqrt{64} = 8$ in
∴ Height of pyramid is 8 inches. //

Volume of pyramid:

$= \frac{1}{3} \times h \times \text{base}$

$= \frac{1}{3} \times h \times a^2$

$= \frac{1}{3} \times 8 \times 12^2$

$= 384 \text{ in}^3$

$\therefore$ Volume of pyramid model is 384 in^3. //

13. A concrete fixture has a shape of a right pyramid. If it is further known that its base is a square whose length is 9″ and height is 15″, find the volume of the pyramid fixture.

Answer:
Given right pyramid whose base is square
Also given side of base, a = 9
Height, h = 15
Volume of pyramid:

$= \frac{1}{3} \times h \times a^2$

$= \frac{1}{3} \times 15 \times 9^2$

$= 405 \text{ in}^3$ //

14. A miniature globe whose diameter is 14″ is made of fiber glass. Determine the volume of the globe.

Answer:
Given globe = sphere
Also given, diameter = 14
=> radius, r = diameter ÷ 2
= 14 ÷ 2 = 7
Thus, volume of globe:

$= \frac{4}{3} \pi r^3$

$= \frac{4}{3} \times \frac{22}{7} \times 7^3$

$= 1437.3333 \text{ in}^3$ //

15. Find the volume of a gym ball (also called an exercise ball) whose diameter is 12 inches. (Answer should be rounded to the nearest inch.)

Answer:
Given gym ball = sphere
Also given, diameter = 12
=> radius, r = diameter ÷ 2
= 12 ÷ 2 = 6
Volume of gym ball:

$= \frac{4}{3} \pi r^3$

$= \frac{4}{3} \times \frac{22}{7} \times 6^3$

$= 905.1429 \text{ in}^3$

$\approx 905 \text{ in}^3$

$\therefore$ Volume of the gym ball is 905 in^3. //

16. Find the volume of a hemisphere whose radius is 8″. (Let $\pi = 3.142$)

Answer:
Given radius, r = 8
Volume of hemisphere:

$= \frac{1}{2} \times \frac{4}{3} \pi r^3$

$= \frac{2}{3} \times 3.142 \times 8^3$

$= 1072.4693 \text{ in}^3$ //

17. The volume of an oblique cone is 48π units^3 and its height is 4 units. What is the radius of the cone?

Answer:
Given volume of cone, $V = 48\pi$
Height, h = 4
Let r = radius
Thus,

Volume of cone $= \frac{1}{3} \pi r^2 h$

$48\pi = \frac{1}{3} \pi r^2 4$

$48\cancel{\pi} = \frac{1}{3} \cancel{\pi} r^2 4$

$r^2 = 36$

$r = \sqrt{36}$

r = 6 units

$\therefore$ Radius of the cone is 6 units. //

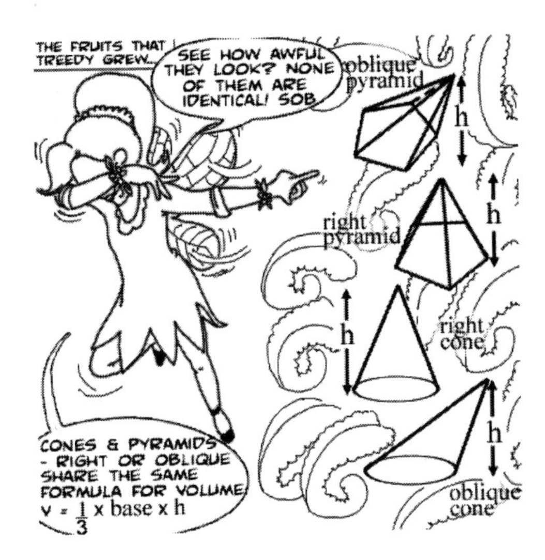

18. Figure 10.10 shows a clown's stool which is the shape of a frustum (i.e. a truncated cone). If it is known that the diameters of the upper and lower bases are 16 in and 20 in and the altitude is 21 in, what is the volume of the stool?

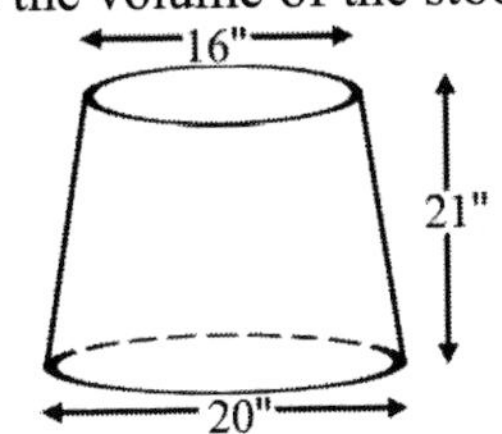

Figure 10. 10

Answer:
Given clown's stool = frustum
Also given,
Radius of upper base, $r_U = 16 \div 2 = 8$
Radius of lower base, $r_L = 20 \div 2 = 10$
Altitude, $h = 21$
Volume of clown's stool:

$= \frac{\pi}{3} h \left(r_U^2 + r_L^2 + r_U r_L \right)$

$= \frac{1}{3} \times \frac{22}{7} \times 21 \left(8^2 + 10^2 + 8 \times 10 \right)$

$= 22(64 + 100 + 80)$

$= 22 \times 244$

$= 5368 \text{ in}^3$

$\therefore$ Volume of clown's stool is 5368 in^3. //

19. Figure 10.11 shows a paper cup. What is the volume of the cup? If the thickness of the paper used to make the cup is negligible, what is the surface area of the paper needed to make the cup? ($\pi = 3.142$)

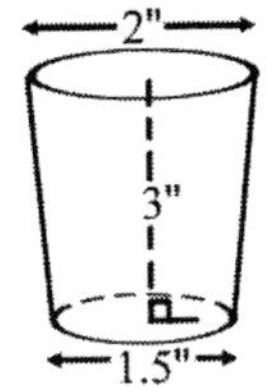

Figure 10. 11

Answers:
Given paper cup = frustum
Height, $h = 3$
Radius of upper base, $r_U = 2 \div 2 = 1$
Radius of lower base, $r_L = 1.5 \div 2 = 0.75$
Volume of cup:

$= \frac{1}{3} \pi h \left(r_U^2 + r_L^2 + r_U r_L \right)$

$= \frac{1}{3} \times 3.142 \times 3 \times \left(1^2 + 0.75^2 + 1 \times 0.75 \right)$

$= 3.142 \times (1 + 0.5625 + 0.75)$

$= 3.142 \times 2.3125$

$= 7.2659 \text{ in}^3$

$\therefore$ Volume of paper cup is 7.2659 in^3. //

Let s = slant height
Figure 10.11 has been reproduced for labeling and illustration:

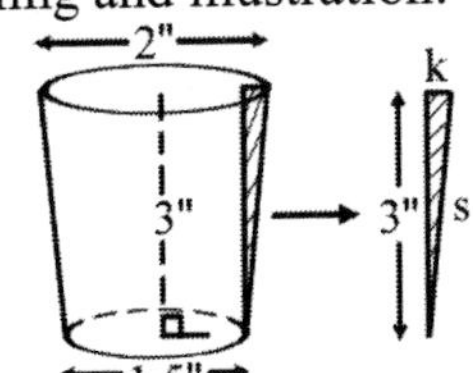

$k = r_U - r_L$
$= 1 - 0.75 = 0.25$
Using Pythagorean Theorem:
$s^2 = 0.25^2 + 3^2$
$= 0.0625 + 9$
$= 9.0625$
$s = \sqrt{9.0625} = 3.0104$ in
Thus,
Surface area of paper cup:
= surface area of frustum (with one base)
$= \pi s(r_U + r_L) + \pi (r_L^2)$
$= 3.142 \times 3.0104 \times (1 + 0.75)$
$+ 3.142 \times (0.75^2)$
$= 16.5527 + 1.7674$

$= 18.32 \text{ in}^2$

$\therefore$ Surface area of the paper cup equals to 18.32 in^2.//

20. Find the volume of the Swiss gold bar as shown in Figure 10.12.

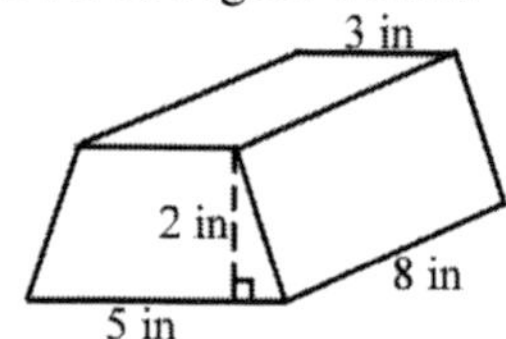

Figure 10. 12

Answer:
Given Swiss gold bar = prism
Given height of base, $h_b = 2$
Upper base, a = 3
Lower base, b = 5
Height of bar, h = 8
Area of base:
= Area of trapezoid

$= \frac{1}{2} h_b (a + b)$

$= \frac{1}{2} \times 2 \times (3 + 5)$

$= 8 \text{ in}^2$

Thus,
Volume of Swiss gold bar:
= volume of prism
= height × area of base
= 8 × 8
$= 64 \text{ in}^3$ //

21. Figure 10.13 shows a water tank. Find the maximum volume of water the tank holds before it overflows. (Assume the thickness of the material used to construct the tank is negligible)

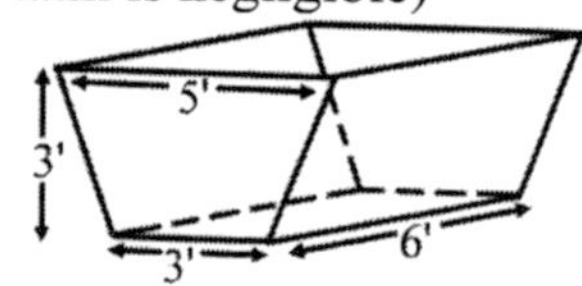

Figure 10. 13

Answer:
Given water tank = prism
Height, h = 6
Upper base, a = 5
Lower base, b = 3
Height of base, $h_b = 3$
Area of trapezoid face:

$= \frac{1}{2} h_b (a + b)$

$= \frac{1}{2} \times 3 \times (5 + 3)$

$= 12 \text{ feet}^2$

Volume of water tank:
= volume of prism
= h × area of trapezoid face
= 6 × 12
$= 72 \text{ feet}^3$ //

22. A ceramic vase has an external diameter of 5 in and external height of 14.5 in (see Figure 10.14). If the walls and base of the vase is 0.5 in thick, determine the total cubic inches of ceramic that was used to make the vase.

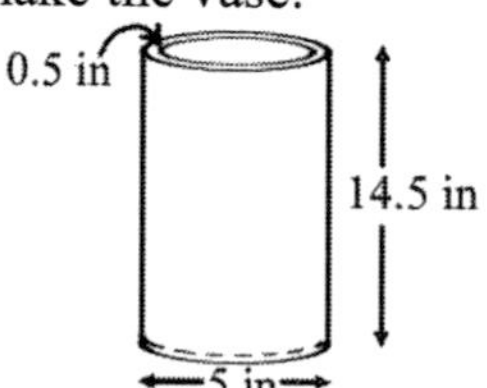

Figure 10. 14

Answer:
Given thickness of walls, h = 0.5
Given external radius, $r_e = 5 \div 2 = 2.5$
Internal radius, $r_i = r_e - h$
$= 2.5 - 0.5$
$= 2$

Thus,
Base of vase:
Volume, V_B = area of base × h
$= \pi \times r_e^2 \times 0.5$
$= \frac{22}{7} \times 2.5^2 \times 0.5$
$= 9.8214 \text{ in}^3$
Walls of vase:
Volume, $V_W = (\pi r_e^2 - \pi r_i^2) \times 14$
$= (\pi 2.5^2 - \pi 2^2) \times 14$
$= (6.25\pi - 4\pi) \times 14$
$= 2.25 \times \frac{22}{7} \times 14$
$= 99 \text{ in}^3$
Volume of vase:
$= V_B + V_W$
$= 9.8214 + 99$
$= 108.8214 \text{ in}^3$ //

23. Figure 10.15 shows an unthreaded square nut. What is the volume of the nut?

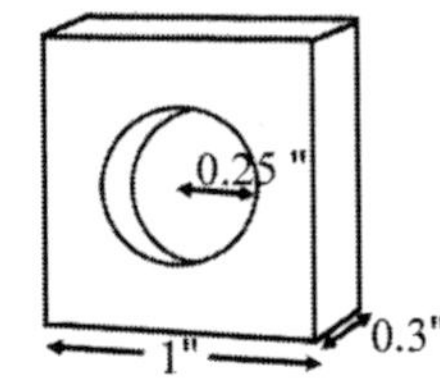

Figure 10. 15

Answer:
Given radius, r = 0.25
Height, h = 0.3
Since Figure 10.15 is a square nut:
=> Side of nut, a = 1
Thus,
Volume of the square nut:
$= a^2h - \pi r^2 h$
$= 1^2 \times 0.3 - \frac{22}{7} \times 0.25^2 \times 0.3$
$= 0.3 - 0.0589$
$= 0.2411 \text{ in}^3$ //

24. Figure 10.16 shows a circular cylinder plastic pipe. If the pipe is 140 inches in length, 2 inches thick and has a diameter of 20 inches, what is the volume of the plastic that was used to construct the pipe?

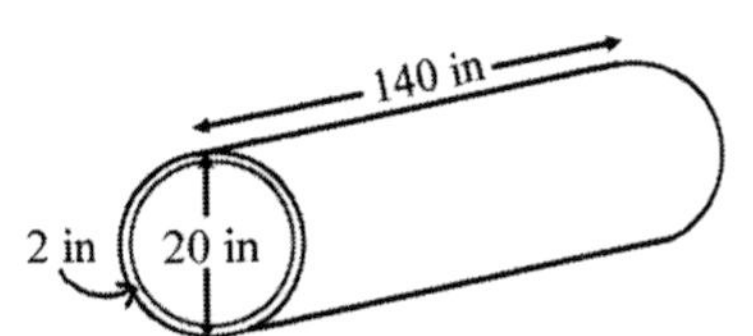

Figure 10. 16

Answer:
Given diameter, d = 20
=> Radius, $r_1 = d \div 2$
$= 20 \div 2 = 10$
Also given length, h = 140
Thickness, w = 2
=> Radius of internal circle, $r_2 = r_1 - 2$
$= 10 - 2 = 8$
Volume of plastic used:
$= \pi r_1^2 h - \pi r_2^2 h$
$= \frac{22}{7} \times 10^2 \times 140 - \frac{22}{7} \times 8^2 \times 140$
$= 44000 - 28160$
$= 15840 \text{ in}^3$ //

25. Find the volume of the antique basket shown in Figure 10.17.

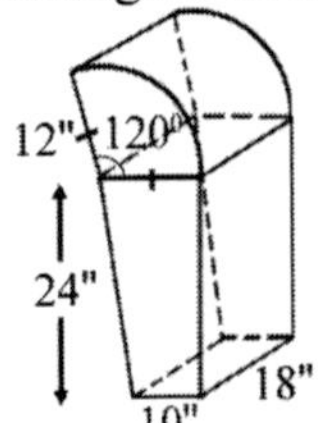

Figure 10. 17

Answer:
Given radius, r = 12

Height, $h = 18$
Lower base of trapezoid, $a = 10$
Upper base of trapezoid, $b = 12$
Height of trapezoid, $h_b = 24$

Volume of the antique basket:

$= \text{Volume of section cylinder} + \text{Volume of prism}$

$$= \frac{120^\circ}{360^\circ}\pi r^2 h + \frac{1}{2}h_b h(a+b)$$

$$= \frac{120^\circ}{360^\circ}\times\frac{22}{7}\times 12^2\times 18 + \frac{1}{2}\times 24\times 18\,(10+12)$$

$= 2715.4286 + 4752$

$= 7467.4286 \text{ in}^3$ //

26. Figure 10.18 shows two containers, A whose lower base is a square and B which is a circular cylinder. If container A is filled to the brim with liquid and the liquid content is subsequently poured into container B, find the height of the liquid content in container B.

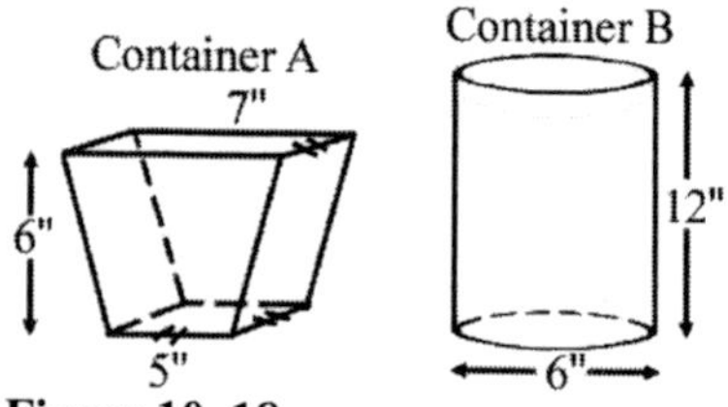

Figure 10. 18

Answer:

Given container A = prism
Lower base of trapezoid, $a = 5$
Upper base of trapezoid, $b = 7$
Height, $h_A = 6$
Thickness of container A, $w = 5$

Volume of container A:

$$V_A = \frac{1}{2}h_A w(a+b)$$

$$= \frac{1}{2}\times 6\times 5\times(5+7)$$

$= 180 \text{ in}^3$

Given diameter of container B's base = 6

=> Radius, $r = \text{diameter} \div 2$
$= 6 \div 2$
$= 3$

Also given height of container B, $h_B = 12$

Let h = height of liquid (from container A) in container B

Thus, when the liquid from container A is poured into container B:

$\pi r^2 h$ = volume of container A

$$\frac{22}{7}\times 3^2\times h = 180$$

$$h = 180\times\frac{1}{3^2}\times\frac{7}{22}$$

$\therefore h = 6.3636$ in //

27. A toy manufacturer specializes in play food for kids, has included miniature plastic 'snow cone' as part of its production. Figure 10.19 shows the plastic snow cone. What is the volume of a plastic cone?

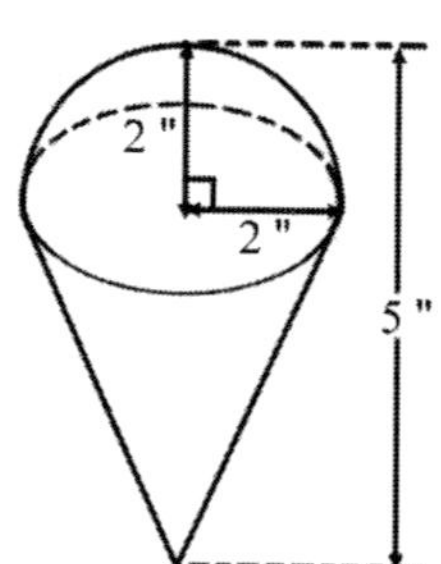

Figure 10. 19

Answer:

Given hemisphere and circular cone

Radius of hemisphere, $r = 2$

Radius of base of cone, $r = 2$

Height of cone, $h = 5 - r$
$= 5 - 2$
$= 3$

Volume of hemisphere, V_h:

$= \frac{1}{2} \times \text{sphere}$

$= \frac{1}{2} \times \frac{4}{3} \times \pi r^3$

$= \frac{1}{2} \times \frac{4}{3} \times \frac{22}{7} \times 2^3$

$= 16.7619 \text{ in}^3$

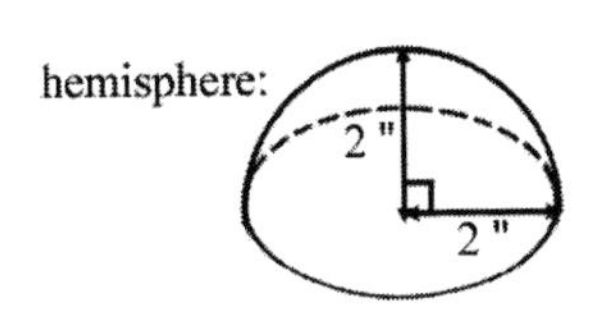

Volume of cone, V_c:

$= \frac{1}{3} \times \text{base} \times \text{height}$

$= \frac{1}{3} \times \pi r^2 \times h$

$= \frac{1}{3} \times \frac{22}{7} \times 2^2 \times 3$

$= 12.5714 \text{ in}^3$

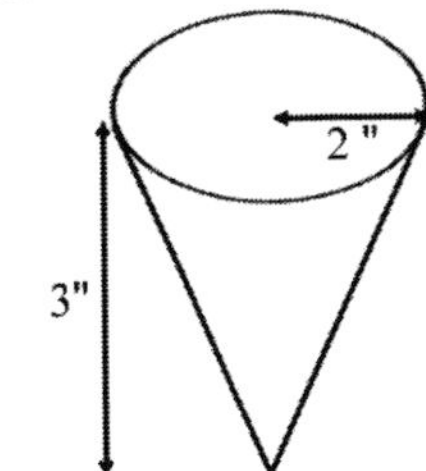

Volume of snow cone:

$= V_h + V_c$

$= 16.7619 + 12.5714$

$= 29.3333 \text{ in}^3$ //

28. Figure 10.20 represents a child's foam slide that can be found in a play gym. If section A is half a cylinder, find the volume of the slide.

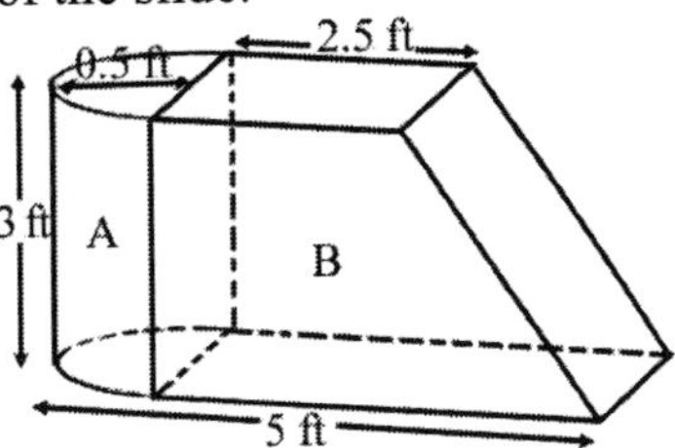

Figure 10. 20

Answer:

Given section A = half cyclic cylinder

Radius of cyclic cylinder, r = 0.5

Height of cyclic cylinder, h = 3

Volume of section A, V_A:

$= \frac{1}{2} \times \text{Volume of cyclic cylinder}$

$= \frac{1}{2} \times \pi r^2 h$

$= \frac{1}{2} \times \frac{22}{7} \times 0.5^2 \times 3$

$= 1.1786 \text{ ft}^3$

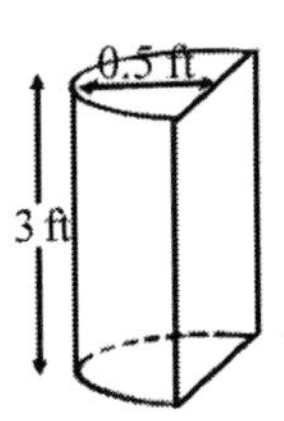

Given section B = prism

Volume of section B, V_B:

= area of trapezoid base × height

$= \frac{1}{2} \times 3 \times (2.5 + 4.5) \times (1)$

$= \frac{1}{2} \times 3 \times 7 \times 1$

$= 10.5 \text{ ft}^3$

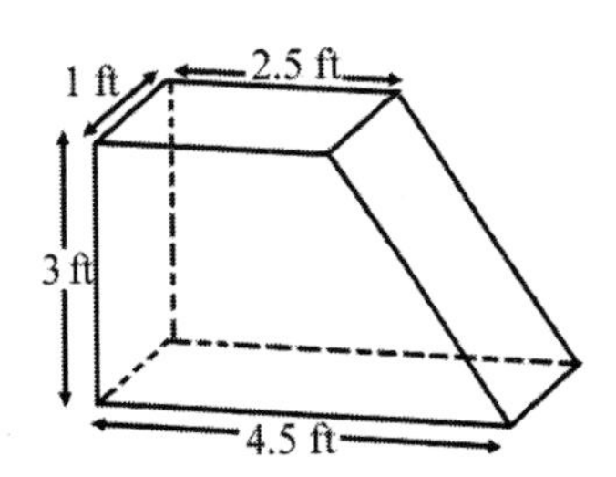

Thus,
Volume of child's foam slide:
$= V_A + V_B$
$= 1.1786 + 10.5$
$= 11.6786 \text{ ft}^3$ //

29. Figure 10.21 shows an unfinished wood craft in the shape of a cylinder. A hemisphere whose radius is 5 inches has been removed from the top of the wood craft. Find the volume of the wood craft. ($\pi = \frac{22}{7}$)

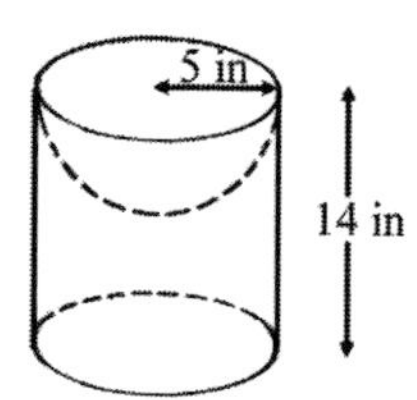

Figure 10. 21

Answer:
Given cylinder:
Radius, r = 5
Height, h = 14
Volume of cylinder, V_C:
$= \pi r^2 h$
$= \frac{22}{7} \times 5^2 \times 14$
$= 1100 \text{ in}^3$

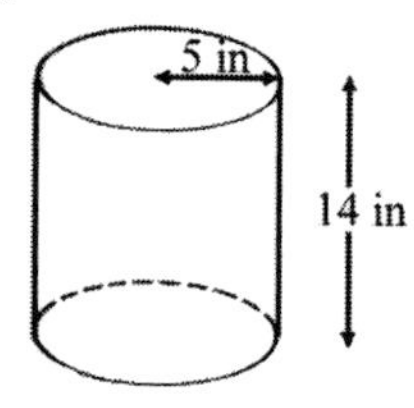

Given hemisphere:
Radius, r = 5
Volume of hemisphere that was removed, V_H:
$= \frac{1}{2} \times \text{volume of sphere}$
$= \frac{1}{2} \times \frac{4}{3} \pi r^3$
$= \frac{1}{2} \times \frac{4}{3} \times \frac{22}{7} \times 5^3$
$= 261.9048 \text{ in}^3$

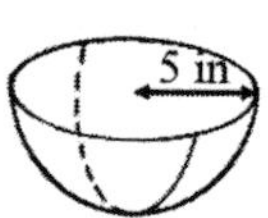

Thus, volume of unfinished wood craft:
$= V_C - V_H$
$= 1100 - 261.9048$
$= 838.0952 \text{ in}^3$

$\therefore$ Volume of unfinished wood craft is 838.0952 in^3. //

Chapter 11 Planes

3-dimension Cartesian space – denotes by the x-, y- and z-axis that are mutually perpendicular. Points on the 3-dimension Cartesian plane are expressed as **ordered triple** (x_1, y_1, z_1)

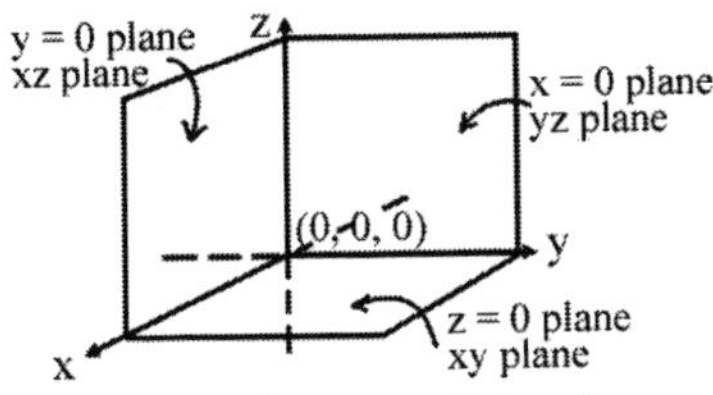

Distance of two points, (x_1, y_1, z_1) and (x_2, y_2, z_2):

$$= \sqrt{(x_2 - x_1)^2 + (y_2 - y_1)^2 + (z_2 - z_1)^2}$$

Midpoint of (x_1, y_1, z_1) and (x_2, y_2, z_2):

$$= \left(\frac{x_1 + x_2}{2}, \frac{y_1 + y_2}{2}, \frac{z_1 + z_2}{2}\right)$$

Line in space – line extending indefinitely

Equation of line with **direction vector** [A, B, C] and passes through point (x_1, y_1, z_1)

Cartesian equation or symmetric form:

$$\frac{(x - x_1)}{A} = \frac{(y - y_1)}{B} = \frac{(z - z_1)}{C}$$

where $A \neq 0, B \neq 0, C \neq 0$

Parametric form:

$x = x_1 + At$

$y = y_1 + Bt$

$z = z_1 + Ct$

where t is called parameter and t = real value

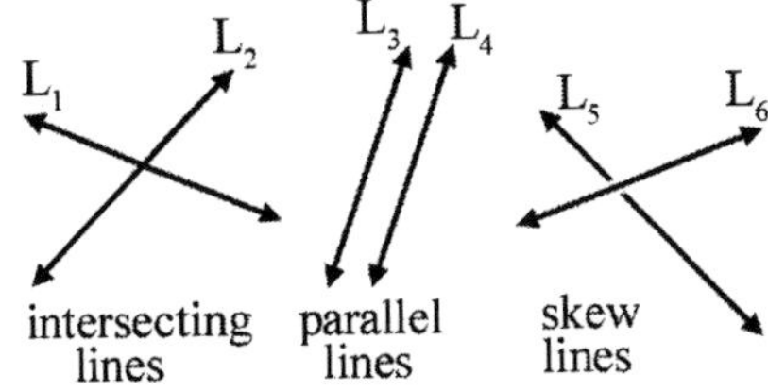

Skew lines – lines that do not intersect and are not parallel. Thus skew lines are on different planes

Collinear – points that lie on the same straight line

Planes – flat surface extending indefinitely in all directions. A plane is thus determined by points or lines that lie on the plane

Equation of plane with **normal vector**, $\underset{\sim}{n}$ [A, B, C] and containing point (x_1, y_1, z_1):

General form:

$A(x - x_1) + B(y - y_1) + C(z - z_1) = 0$

or

$Ax + By + Cz = Ax_1 + By_1 + Cz_1 = D$

where D = constant

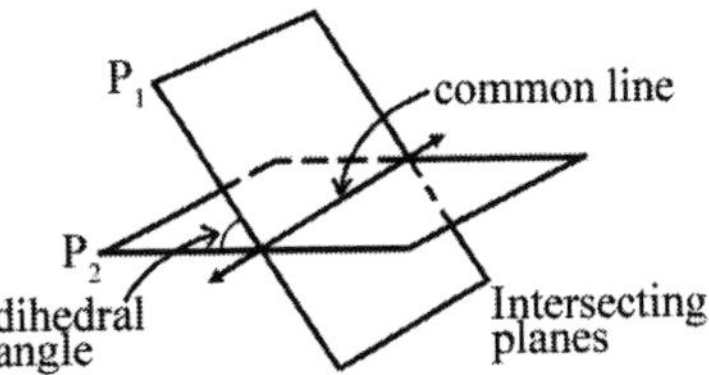

Dihedral angle – angle between two intersecting planes

Coplanar – lines, points that are located on the same plane

Distance from a point (x_1, y_1, z_1) to a plane with normal vector [A, B, C]:

$$= \frac{|Ax_1 + By_1 + Cz_1 + D|}{\sqrt{A^2 + B^2 + C^2}}$$

Equation of **sphere** in space, with center (x_1, y_1, z_1) and radius r:

$(x - x_1)^2 + (y - y_1)^2 + (z - z_1)^2 = r^2$

1. Plot point A whose ordered triple is (2, 3, 5) on a three dimensional Cartesian space.

Answer:
Given A = (2, 3, 5)

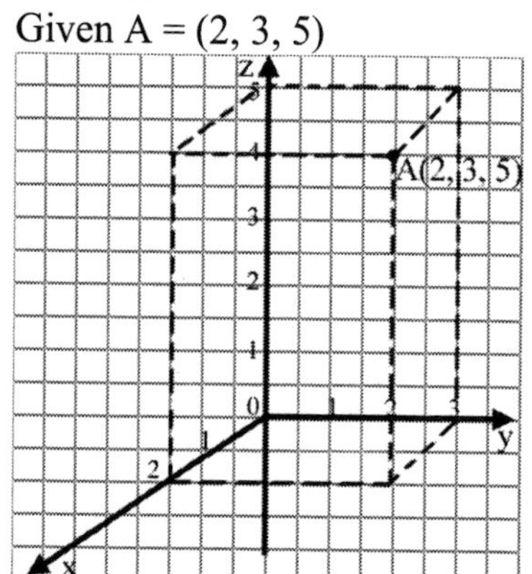

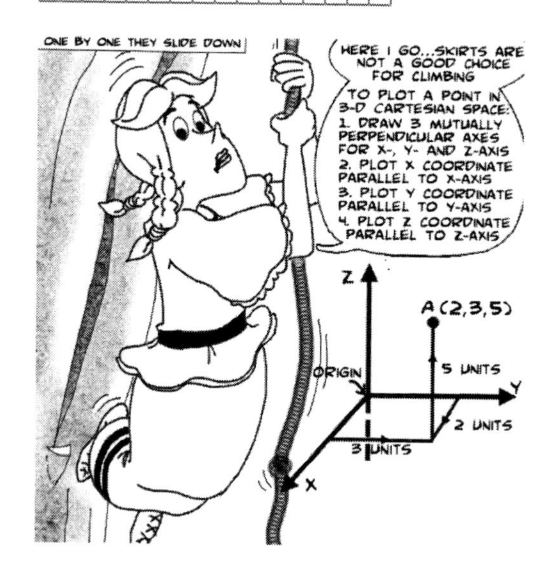

2. Find the distance between P(2, 3, 4) and Q(5, 6, 7). Hence what is the midpoint of the line segment $\overline{PQ}$?

Answers:
Given P = (2, 3, 4)
Q = (5, 6, 7)
Thus,

$$PQ = \sqrt{(x_2 - x_1)^2 + (y_2 - y_1)^2 + (z_2 - z_1)^2}$$
$$= \sqrt{(5-2)^2 + (6-3)^2 + (7-4)^2}$$
$$= \sqrt{3^2 + 3^2 + 3^2}$$
$$= \sqrt{27} \quad \Leftarrow \text{Note: } \sqrt{27} = \sqrt{9 \times 3}$$
$$= 3\sqrt{3} \text{ units } //$$

Midpoint of $\overline{PQ}$:

$$= \left(\frac{x_1 + x_2}{2}, \frac{y_1 + y_2}{2}, \frac{z_1 + z_2}{2}\right)$$
$$= \left(\frac{2+5}{2}, \frac{3+6}{2}, \frac{4+7}{2}\right)$$
$$= \left(\frac{7}{2}, \frac{9}{2}, \frac{11}{2}\right) //$$

3. Find the midpoint P of the line segment $\overline{AB}$, if A(2, 2, 3) and B(4, 6, 9).

Answer:
Given A = (2, 2, 3)
B = (4, 6, 9)
Midpoint $\overline{AB}$ = P

$$P = \left(\frac{x_1 + x_2}{2}, \frac{y_1 + y_2}{2}, \frac{z_1 + z_2}{2}\right)$$
$$= \left(\frac{2+4}{2}, \frac{2+6}{2}, \frac{3+9}{2}\right)$$
$$= \left(\frac{6}{2}, \frac{8}{2}, \frac{12}{2}\right)$$
$$= (3, 4, 6) //$$

4. Find the distance of point P(5, 3, 2) from the origin.

Answer:
Given P = (5, 3, 2)
O = origin = (0, 0, 0)
Thus,

$$PO = \sqrt{(x_2 - x_1)^2 + (y_2 - y_1)^2 + (z_2 - z_1)^2}$$
$$= \sqrt{(5-0)^2 + (3-0)^2 + (2-0)^2}$$
$$= \sqrt{25+9+4}$$
$$= \sqrt{38} \text{ units} \ //$$

5. If P(3, 6, 5) and Q(2, 1, 8) are two points in space, find:
a) Distance from P to Q
b) Midpoint of $\overline{PQ}$

Answers:
Given P = (3, 6, 5)
Q = (2, 1, 8)
a) Distance from P to Q = PQ

$$PQ = \sqrt{(x_2 - x_1)^2 + (y_2 - y_1)^2 + (z_2 - z_1)^2}$$
$$= \sqrt{(3-2)^2 + (6-1)^2 + (5-8)^2}$$
$$= \sqrt{1^2 + 5^2 + (-3)^2}$$
$$= \sqrt{1+25+9}$$
$$= \sqrt{35} \text{ units} \ //$$

b) Midpoint of $\overline{PQ}$:

$$= \left(\frac{x_1 + x_2}{2}, \frac{y_1 + y_2}{2}, \frac{z_1 + z_2}{2}\right)$$
$$= \left(\frac{3+2}{2}, \frac{6+1}{2}, \frac{5+8}{2}\right)$$
$$= \left(\frac{5}{2}, \frac{7}{2}, \frac{13}{2}\right) //$$

6. A line in space is represented by the Cartesian equation $x - 1 = \frac{y+3}{-2} = \frac{z-2}{5}$.
Find the point on the line with z-coordinate = 7.

Answer:
Given line:

$$x - 1 = \frac{y+3}{-2} = \frac{z-2}{5}$$

When z = 7
Thus,

$$x - 1 = \frac{y+3}{-2} = \frac{7-2}{5} = 1$$

=> x − 1 = 1
∴ x = 2

$$=> \frac{y+3}{-2} = 1$$

y + 3 = −2
∴ y = −5
∴ Point with z = 7 is (2, −5, 7). //

7. Given the Cartesian equation of a line in space is $\frac{x+3}{2} = \frac{1-y}{5} = \frac{z-7}{-3}$. What is the coordinate of the point on the line with y-coordinate = 6?

Answer:
Given line: $\frac{x+3}{2} = \frac{1-y}{5} = \frac{z-7}{-3}$

When y = 6 => $\frac{1-y}{5} = \frac{1-6}{5} = -1$

$$\frac{x+3}{2} = -1 = \frac{z-7}{-3}$$

$$=> \frac{x+3}{2} = -1$$

x + 3 = −2
∴ x = −5

$$=> \frac{z-7}{-3} = -1$$

z − 7 = 3
∴ z = 10
∴ The point is (−5, 6, 10). //

8. P(−1, 3, 8) and Q(3, 2, 5) are two points in space. Determine the equation of the line that passes through P and Q.

Answer:
Given P = (−1, 3, 8), Q = (3, 2, 5)

Direction vector of line:

$\overrightarrow{PQ} = [x_Q - x_P, y_Q - y_P, z_Q - z_P]$

$= [3 - (-1), 2 - 3, 5 - 8]$

$= [4, -1, -3]$

Using point P(−1, 3, 8), and A = 4, B = −1, C = −3:

Equation of line (symmetric form):

$$\frac{x+1}{4} = \frac{y-3}{-1} = \frac{z-8}{-3} \;//$$

Alternatively:

Equation of line (parametric form):

$x = -1 + 4t$

$y = 3 - t$

$z = 8 - 3t$ // where t = real value

9. Find the distance of P(2, 1, 2) to the origin, O and illustrate it on a coordinate plane.

Answers:

Given P = (2, 1, 2)

O = (0, 0, 0)

Distance PO:

$= \sqrt{(x_2 - x_1)^2 + (y_2 - y_1)^2 + (z_2 - z_1)^2}$

$= \sqrt{(2-0)^2 + (1-0)^2 + (2-0)^2}$

$= \sqrt{2^2 + 1^2 + 2^2}$

$= \sqrt{4+1+4}$

$= \sqrt{9}$

= 3 units //

On the coordinate plane:

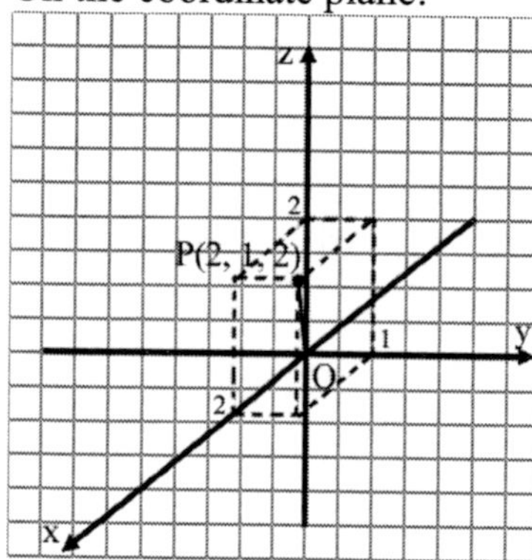

10. Find the distance of P(0, −3, 0) from the origin O and illustrate it on a coordinate plane.

Answers:

Given P = (0, −3, 0)

O = (0, 0, 0)

Distance PO:

$= \sqrt{(x_2 - x_1)^2 + (y_2 - y_1)^2 + (z_2 - z_1)^2}$

$= \sqrt{(0-0)^2 + (-3-0)^2 + (0-0)^2}$

$= \sqrt{(-3)^2}$

$= \sqrt{9}$

= 3 units //

On the coordinate plane:

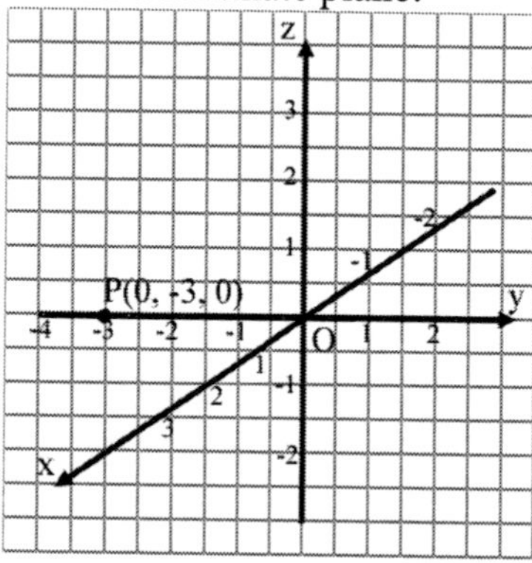

11. Find the distance of P(0, 1, 3) from Q(1, 2, 0) and illustrate it on the coordinate plane.

Answers:

Given P = (0, 1, 3)

Q = (1, 2, 0)

Distance PQ:

$= \sqrt{(x_2 - x_1)^2 + (y_2 - y_1)^2 + (z_2 - z_1)^2}$

$= \sqrt{(1-0)^2 + (2-1)^2 + (0-3)^2}$

$= \sqrt{1^2 + 1^2 + (-3)^2}$

$= \sqrt{11}$ units //

On the coordinate plane:

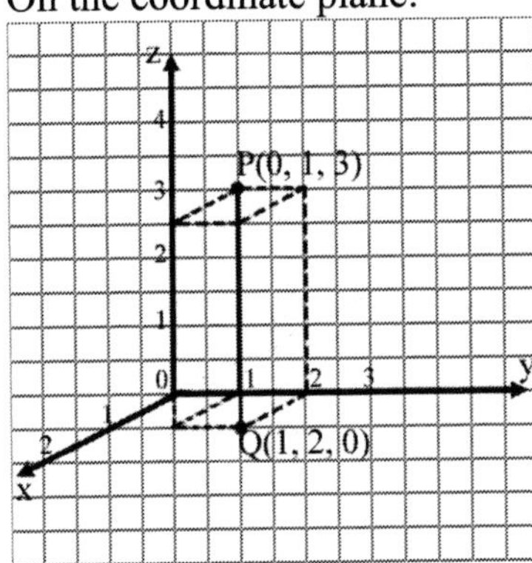

12. Find the equation of the plane with normal vector [2, 1, 5] and contains the point (–3, 2, 1).

Answer:
Given normal vector = [2, 1, 5]
Point on the plane = (–3, 2, 1)
Equation of the plane:

$ax + by + cz = ax_1 + by_1 + cz_1$

$2x + y + 5z = 2(-3) + 1(2) + 5(1)$

$= -6 + 2 + 5$

$2x + y + 5z = 1$ //

Alternatively:

Equation of plane:

$a(x - x_1) + b(y - y_1) + c(z - z_1) = 0$

$2(x - (-3)) + 1(y - 2) + 5(z - 1) = 0$

$2(x + 3) + y - 2 + 5z - 5 = 0$

$2x + 6 + y + 5z - 7 = 0$

$2x + y + 5z - 1 = 0$

$\therefore 2x + y + 5z = 1$ //

13. If a plane contains a point (6, 5, 4) and has a normal vector [2, –3, 5], find the equation of the plane.

Answer:
Given normal vector = [2, –3, 5]
Point on the plane = (6, 5, 4)
Equation of plane:

$ax + by + cz = ax_1 + by_1 + cz_1$

$2x - 3y + 5z = 2(6) - 3(5) + 5(4)$

$= 12 - 15 + 20$

$2x - 3y + 5z = 17$ //

Alternatively:

Equation of plane:

$a(x - x_1) + b(y - y_1) + c(z - z_1) = 0$

$2(x - 6) - 3(y - 5) + 5(z - 4) = 0$

$2x - 12 - 3y + 15 + 5z - 20 = 0$

$2x - 3y + 5z - 17 = 0$

$\therefore 2x - 3y + 5z = 17$ //

14. Draw a graph of the plane, P_1 represented by the equation:

$2x + 2y + z = 4$

Answer:
Given equation of plane: $2x + 2y + z = 4$
Find x-axis intercept:
For $y = 0, z = 0$:

$2x + 2(0) + 0 = 4$

$x = 2$

$\therefore P_1$ cuts x-axis at (2, 0, 0)

Find y-axis intercept:
For $x = 0, z = 0$:

$2(0) + 2y + 0 = 4$

$y = 2$

$\therefore P_1$ cuts y-axis at (0, 2, 0)

Find z-axis intercept:
For $x = 0, y = 0$:

$2(0) + 2(0) + z = 4$
$z = 4$
$\therefore$ P_1 cuts z-axis at (0, 0, 4)
Thus P_1 in 3-dimension Cartesian space is:

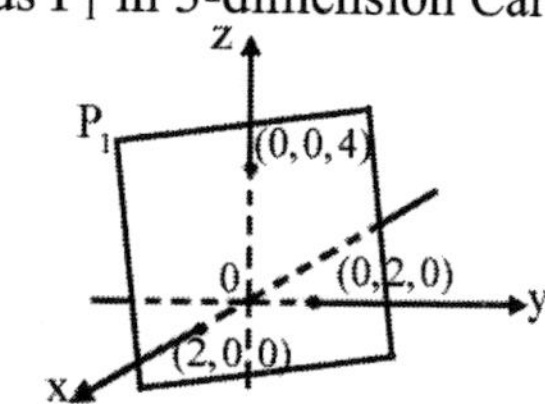

15. Given plane, P_1 has the equation $3x + 5y + 2z - 16 = 0$. Find the z-intercept of P_1.

Answer:
Given equation of plane, P_1:
$3x + 5y + 2z - 16 = 0$
In general equation form:
$ax + by + cz + d = 0$
Thus, z-intercept:

$= \frac{-d}{c}$

$= \frac{-(-16)}{2}$

$= \frac{16}{2}$

$= 8$

$\therefore$ P_1 cuts the z-axis at (0, 0, 8). //

16. The following are the equations of two planes:
P_1: $a_1x + b_1y + c_1z = d_1$
P_2: $a_2x + b_2y + c_2z = d_2$
When solving simultaneously, determine the number of possible solutions.

Answer:
Given equations of 2 planes:
The possible outcomes are:

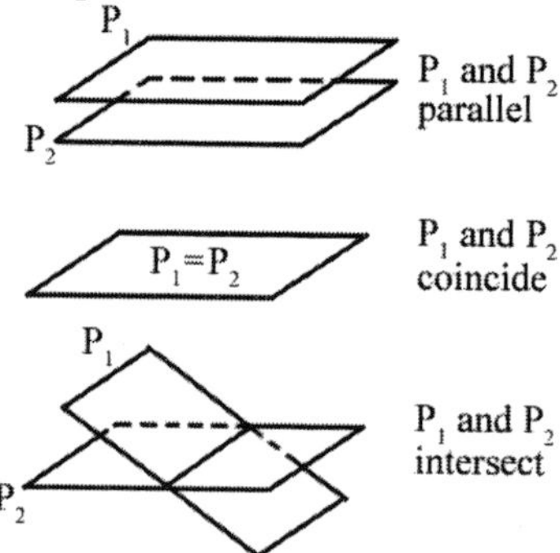

$\therefore$ There are 3 possible solutions. //

17. Figure 11.1 depicts 2 planes: $2x + y + 2z = 15$ and $3x + 2y + z = 16$. Line K is common to both planes. If point (2, 3, 4) lies on line K, find the equation for line K.

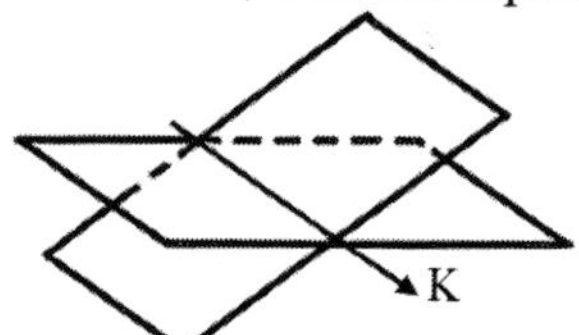

Figure 11. 1

Answer:
Given, $2x + y + 2z = 15$
=> Normal vector, $\underset{\sim}{n}_1 = [2, 1, 2]$
$3x + 2y + z = 16$
=> Normal vector, $\underset{\sim}{n}_2 = [3, 2, 1]$
Direction vector of line K = $\underset{\sim}{n}_1 \times \underset{\sim}{n}_2$
$[a_1, a_2, a_3] \times [b_1, b_2, b_3]$
$= [a_2b_3 - a_3b_2, a_3b_1 - a_1b_3, a_1b_2 - a_2b_1]$
$= [1(1) - 2(2), 2(3) - 2(1), 2(2) - 1(3)]$
$= [1 - 4, 6 - 2, 4 - 3]$
$= [-3, 4, 1]$

$\therefore$ Equation of $\overleftrightarrow{K}$: $\frac{x-2}{-3} = \frac{y-3}{4} = \frac{z-4}{1}$ //

18. P(3, 1, 4), Q(2, 0, 1) and R(6, 3, 6) are three points in space and coplanar. Find the equation of the plane that contains P, Q, and R.

Answer:
Given P = (3, 1, 4)
Q = (2, 0, 1)

$R = (6, 3, 6)$
Direction vector of:
$\overrightarrow{PQ} = [x_2 - x_1, y_2 - y_1, z_2 - z_1]$
$= [2 - 3, 0 - 1, 1 - 4]$
$= [-1, -1, -3]$
$\overrightarrow{PR} = [6 - 3, 3 - 1, 6 - 4]$
$= [3, 2, 2]$
Since P, Q, R are coplanar,
Let plane P_1 = contains P, Q, and R
$\underset{\sim}{n}$ = normal vector of plane P_1

$\underset{\sim}{n}$ is perpendicular to $\overrightarrow{PQ}$ and $\overrightarrow{PR}$
$\underset{\sim}{n} = \overrightarrow{PQ} \times \overrightarrow{PR}$
$= [a_1, a_2, a_3] \times [b_1, b_2, b_3]$
$= [a_2b_3 - a_3b_2, a_3b_1 - a_1b_3, a_1b_2 - a_2b_1]$
$= [-1(2) - (-3)(2), (-3)(3) - (-1)(2),$
$(-1)(2) - (-1)(3)]$
$= [-2 + 6, -9 + 2, -2 + 3]$
$= [4, -7, 1]$
Hence equation of plane, P_1, whose normal vector is $[4, -7, 1]$ and a point on the plane, $P(3, 1, 4)$: ⇦ Use either P, Q or R
$ax + by + cz = ax_1 + by_1 + cz_1$
$4x - 7y + 1z = 4(3) + (-7)(1) + (1)(4)$
$4x - 7y + z = 12 - 7 + 4$
$4x - 7y + z = 9$
∴ Equation of plane P_1 is $4x - 7y + z = 9$. //

19. Given $\overleftrightarrow{A}$: $x + 2 = \dfrac{y - 3}{-2} = \dfrac{z + 1}{5}$ and $\overleftrightarrow{B}: \dfrac{x - 1}{2} = \dfrac{y - 8}{-4} = \dfrac{z}{10}$ are two lines in space. Determine if lines, $\overleftrightarrow{A}$ and $\overleftrightarrow{B}$ are coplanar.

Answer:
Given $\overleftrightarrow{A}: x + 2 = \dfrac{y - 3}{-2} = \dfrac{z + 1}{5}$
Let $\underset{\sim}{a}$ = direction vector of $\overleftrightarrow{A}$
=> $\underset{\sim}{a} = [1, -2, 5]$ ...(1)
Also given, $\overleftrightarrow{B}: \dfrac{x - 1}{2} = \dfrac{y - 8}{-4} = \dfrac{z}{10}$
Let $\underset{\sim}{b}$ = direction vector of $\overleftrightarrow{B}$
=> $\underset{\sim}{b} = [2, -4, 10]$
$= 2[1, -2, 5]$ ...(2)
Substitute (1) into (2):
$\underset{\sim}{b} = 2\underset{\sim}{a}$

Note that vector $\underset{\sim}{a}$ is in the same direction as vector $\underset{\sim}{b}$ but vector $\underset{\sim}{b}$ has twice the magnitude of vector $\underset{\sim}{a}$.
Since vector $\underset{\sim}{a}$ and vector $\underset{\sim}{b}$ are in the same direction $[1, -2, 5]$, hence lines, $\overleftrightarrow{A}$ and $\overleftrightarrow{B}$ must be parallel to each other.
∴ As parallel lines are coplanar, thus we have shown that $\overleftrightarrow{A}$ and $\overleftrightarrow{B}$ are coplanar lines. //

20. Lines, $\overleftrightarrow{P}$ and $\overleftrightarrow{Q}$ are two lines in space.
The parametric equation of $\overleftrightarrow{P}$ is:
$x = 4 + 2t$
$y = 1 + 3t$
$z = 2 + t$
And the symmetric equation of $\overleftrightarrow{Q}$ is:
$$\frac{x + 3}{2} = \frac{y + 4}{2} = z + 1$$
Determine if $\overleftrightarrow{P}$ and $\overleftrightarrow{Q}$ are coplanar.

Answer:
Let $\underset{\sim}{p}$ = direction vector of $\overleftrightarrow{P}$

Given $\overleftrightarrow{P}$:
$x = 4 + 2t$
$y = 1 + 3t$
$z = 2 + t$ ...(1)
=> vector $\underset{\sim}{p} = [2, 3, 1]$

Let $\underset{\sim}{q}$ = direction vector of $\overrightarrow{Q}$

Given $\overrightarrow{Q}: \frac{x+3}{2} = \frac{y+4}{2} = z+1$ ...(2)

=> vector $\underset{\sim}{q}$ = [2, 2, 1]

Since $\underset{\sim}{p} \neq \underset{\sim}{q}$, $\overrightarrow{P}$ and $\overrightarrow{Q}$ are not parallel lines

To test if $\overrightarrow{P}$ and $\overrightarrow{Q}$ intersect:

Substitute (1) into (2):

$$\frac{4+2t+3}{2} = \frac{1+3t+4}{2} = 2+t+1$$

Solving x and y components:

$$\frac{4+2t+3}{2} = \frac{1+3t+4}{2}$$

$4 + 2t + 3 = 1 + 3t + 4$

$\therefore t = 2$

Solving for y and z components:

$$\frac{1+3t+4}{2} = 2+t+1$$

$1+3t + 4 = 2(2 + t + 1)$

$5 + 3t = 4 + 2t + 2$

$\therefore t = 1$

Thus t is not the common solution. The lines do not intersect. Intersecting or parallel lines are coplanar. $\overrightarrow{P}$ and $\overrightarrow{Q}$ are skew lines as they do not intersect and are not parallel. Hence we have shown that $\overrightarrow{P}$ and $\overrightarrow{Q}$ are not coplanar. //

21. Show that line $\overrightarrow{P}: \frac{x-2}{3} = y-5 = \frac{z-1}{4}$ and line $\overrightarrow{Q}: \frac{x-2}{-1} = \frac{y-1}{1} = \frac{z-3}{-2}$ are two coplanar lines. Hence find the equation of the plane.

Answers:

Let $\underset{\sim}{p}$ = direction vector of $\overrightarrow{P}$

Given $\overrightarrow{P}$:

$$\frac{x-2}{3} = y-5 = \frac{z-1}{4} \quad ...(1)$$

=> $\underset{\sim}{p}$ = [3, 1, 4]

Let $\underset{\sim}{q}$ = direction vector of $\overrightarrow{Q}$

Given $\overrightarrow{Q}$:

$$\frac{x-2}{-1} = \frac{y-1}{1} = \frac{z-3}{-2} \quad ...(2)$$

=> $\underset{\sim}{q}$ = [−1, 1, −2]

Since $\underset{\sim}{p} \neq \underset{\sim}{q}$, $\overrightarrow{P}$ and $\overrightarrow{Q}$ are not parallel lines.

To test if $\overrightarrow{P}$ and $\overrightarrow{Q}$ intersect:

For $\overrightarrow{Q}$, change symmetric equation to parametric equation:

$x = 2 - t$

$y = 1 + t$

$z = 3 - 2t$ ...(3)

Substitute (3) into (1):

$$\frac{2-t-2}{3} = 1+t-5 = \frac{3-2t-1}{4}$$

Solving x and y components:

$$\frac{2-t-2}{3} = 1+t-5$$

$2 - t - 2 = 3(1 + t - 5)$

$-t = 3 + 3t - 15$

$-4t = -12$

$\therefore t = 3$

Solving y and z components:

$$1+t-5 = \frac{3-2t-1}{4}$$

$4(1 + t - 5) = 3 - 2t - 1$

$4 + 4t - 20 = 2 - 2t$

$6t = 18$

$\therefore t = 3$

Intersecting point of $\overrightarrow{P}$ and $\overrightarrow{Q}$:

Substitute t = 3 into (3):

$x = 2 - 3 = -1$

$y = 1 + 3 = 4$

$z = 3 - 2(3) = 3 - 6 = -3$

$\therefore$ Intersecting point is (−1, 4, −3)

$\therefore$ t is the common solution. $\overrightarrow{P}$ and $\overrightarrow{Q}$ intersect at t = 3 at point (−1, 4, −3). Intersecting lines are coplanar hence we have shown that $\overrightarrow{P}$ and $\overrightarrow{Q}$ are coplanar. //

From above the direction vectors are:

$\underset{\sim}{p}$ = [3, 1, 4]

$\underset{\sim}{q}$ = [−1, 1, −2]

Let $\underset{\sim}{n}$ = normal vector of the plane that contains $\overrightarrow{P}$ and $\overrightarrow{Q}$

$\underset{\sim}{n} = \underset{\sim}{p} \times \underset{\sim}{q}$

$= [a_1, a_2, a_3] \times [b_1, b_2, b_3]$
$= [a_2b_3 - a_3b_2, a_3b_1 - a_1b_3, a_1b_2 - a_2b_1]$
$= [1(-2) - 4(1), 4(-1) - 3(-2),$
$3(1) - 1(-1)]$
$= [-2 - 4, -4 + 6, 3 + 1]$
$= [-6, 2, 4]$

Equation of plane with normal vector, $\underset{\sim}{n}$ $[-6, 2, 4]$ and point $(-1, 4, -3)$:

$a(x - x_1) + b(y - y_1) + c(z - z_1) = 0$
$-6(x - (-1)) + 2(y - 4) + 4(z - (-3)) = 0$
$-6x - 6 + 2y - 8 + 4z + 12 = 0$
$-6x + 2y + 4z = 2$ //

22. Point P has coordinates (5, 6, 7). Find the ordered triple of the projection of P on:
a) XY plane
b) XZ plane
c) YZ plane

Answers:
Given P = (5, 6, 7)
a) On the XY plane, consider the coordinates x and y.
Notice that on the XY plane, z = 0

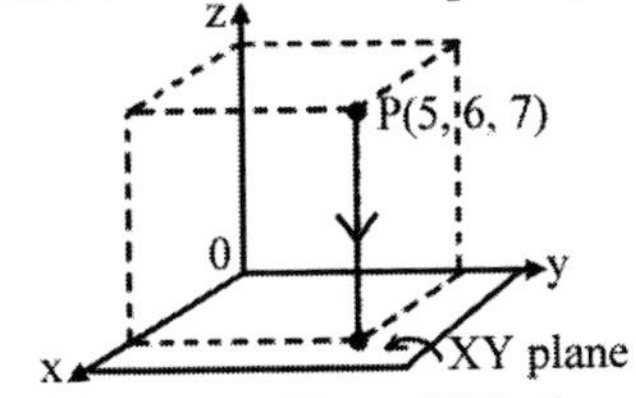

∴ Projection of P on XY plane is (5, 6, 0) //

b) On the XZ plane, consider the coordinates x and z.
Notice that on the XZ plane, y = 0

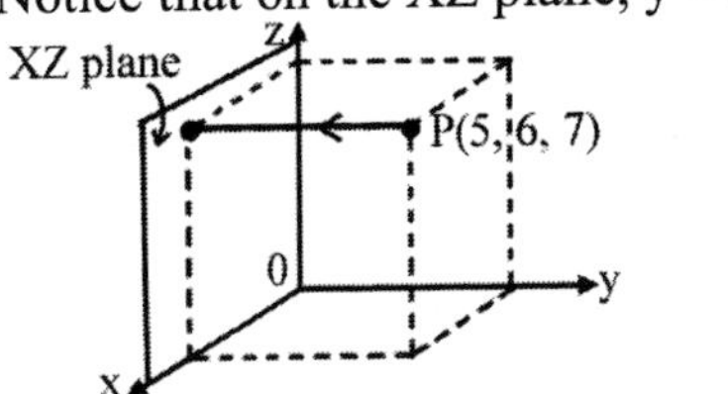

∴ Projection of P on XZ plane is (5, 0, 7) //

c) On the YZ plane consider the coordinates y and z.
Notice that on the YZ plane, x = 0

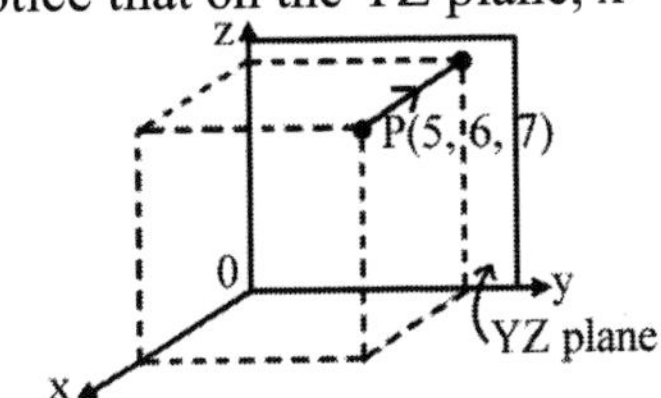

∴ Projection of P on YZ plane is (0, 6, 7) //

23. Point P has coordinate (−2, 5, 6). What is the shortest distance from P to:
a) XY plane
b) XZ plane
c) YZ plane

Answers:
Given P = (−2, 5, 6)
a) Shortest distance from P to XY plane, consider P's z-coordinate:

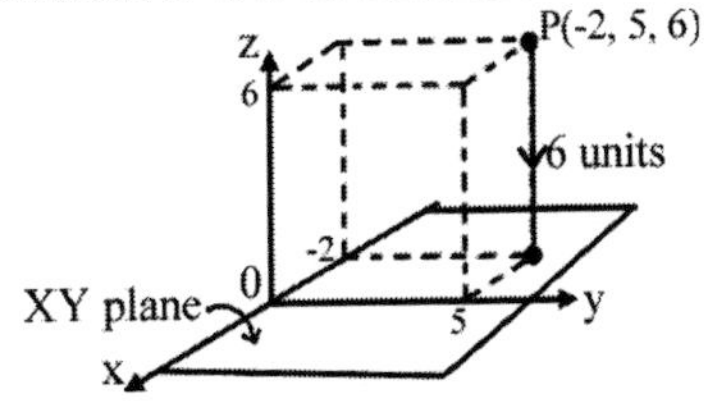

∴ P is 6 units above XY plane. //

b) Shortest distance from P to XZ plane, consider P's y-coordinate:

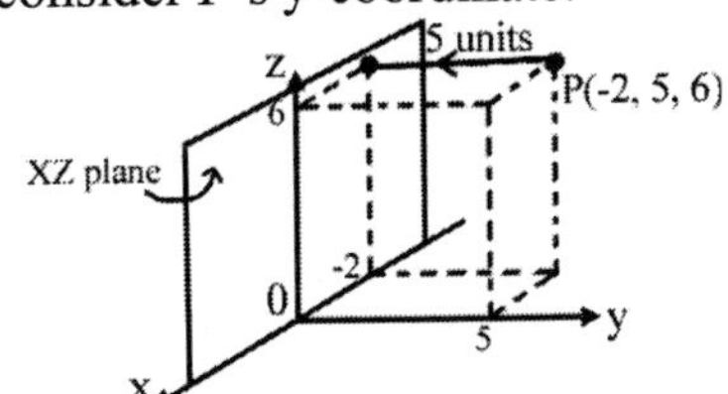

∴ P is 5 units from the XZ plane. //

c) Shortest distance from P to YZ plane, consider P's x coordinate.

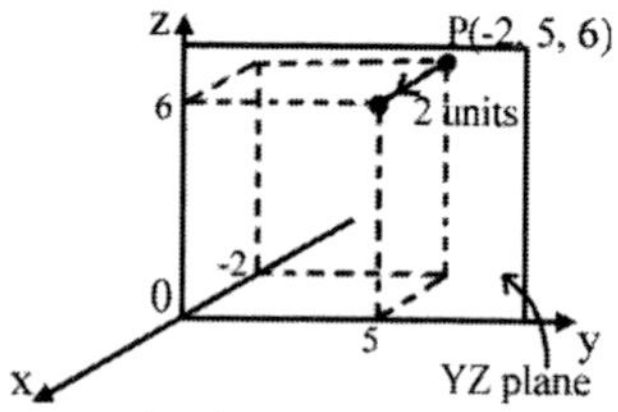

$\therefore$ P is 2 units from YZ plane. //

24. Line M is projected onto plane R. Determine the image of the projection if:

a) $\overleftrightarrow{M}$ is perpendicular to plane R

b) $\overleftrightarrow{M}$ is not perpendicular to plane R

Answers:

Given $\overleftrightarrow{M}$ = line segment

R = plane

a) Given $\overleftrightarrow{M}$ is perpendicular to plane R

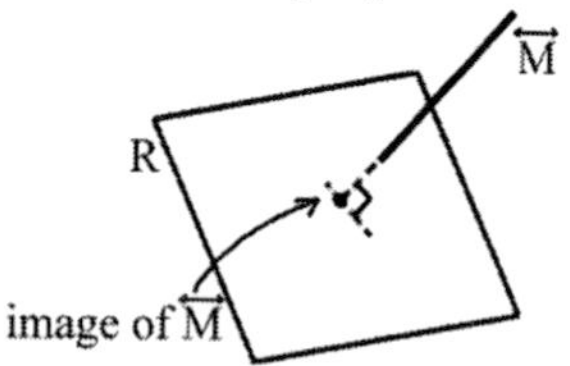

$\therefore$ The projected image is a point on plane R. //

b) Given $\overleftrightarrow{M}$ is not perpendicular to plane R

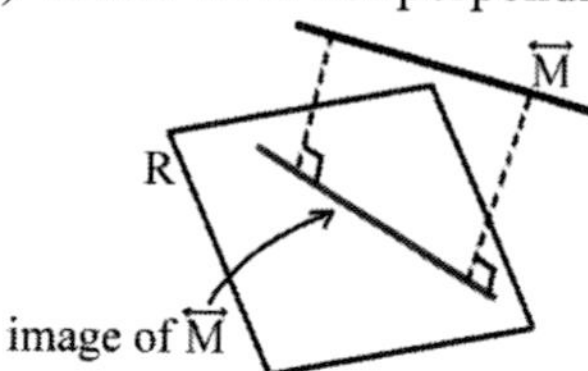

$\therefore$ The projected image is a line on plane R.//

25. Find shortest distance of T(3, 1, 6) to:

a) Origin, O

b) Y-axis

c) XZ plane

Answers:

Given T = (3, 1, 6)

a) Origin, O = (0, 0, 0)

Distance of TO:

$= \sqrt{(x_2 - x_1)^2 + (y_2 - y_1)^2 + (z_2 - z_1)^2}$

$= \sqrt{(0-3)^2 + (0-1)^2 + (0-6)^2}$

$= \sqrt{(-3)^2 + (-1)^2 + (-6)^2}$

$= \sqrt{9+1+36}$

$= \sqrt{46}$

= 6.7823 units //

b) Distance of T to y-axis

Let K = the point at y-axis

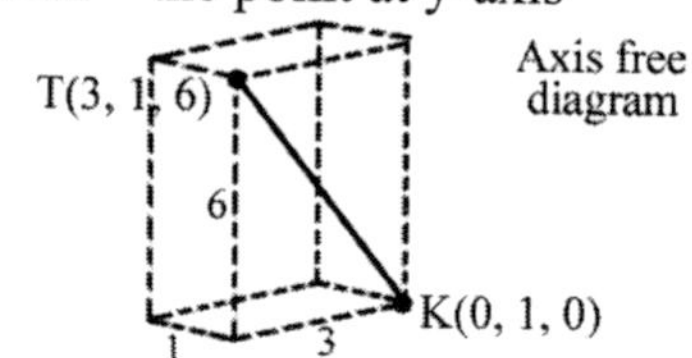

$TK = \sqrt{(x_2 - x_1)^2 + (y_2 - y_1)^2 + (z_2 - z_1)^2}$

$= \sqrt{(0-3)^2 + (1-1)^2 + (0-6)^2}$

$= \sqrt{(-3)^2 + 0^2 + (-6)^2}$

$= \sqrt{9+36}$

$= \sqrt{45}$

= 6.7082 units //

Alternatively:

Using Pythagorean Theorem:

$TK^2 = 6^2 + 3^2$

$= 36 + 9$

$= 45$

$TK = \sqrt{45}$ = 6.7082 units //

c) Distance from T to XZ plane

Let N = a point on XZ which is perpendicular to T (shortest distance)

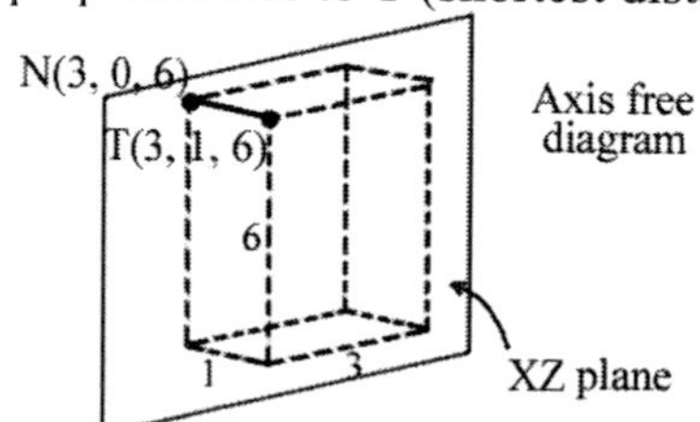

$TN = \sqrt{(x_2 - x_1)^2 + (y_2 - y_1)^2 + (z_2 - z_1)^2}$

$= \sqrt{(3-3)^2 + (0-1)^2 + (6-6)^2}$

$= \sqrt{1^2}$

= 1 unit //

26. Determine the shortest distance of Q(5, –3, –6) from the:

a) YZ plane
b) XY plane

Answers:
Given Q = (5, –3, –6)
a) Distance from Q to YZ plane:

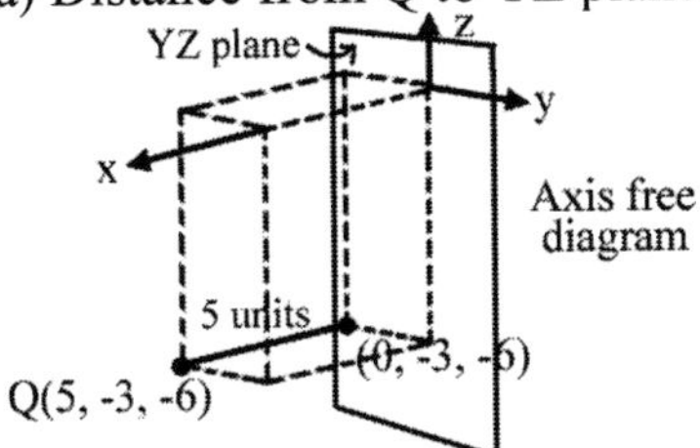

Q to YZ plane:

$= \sqrt{(x_2 - x_1)^2 + (y_2 - y_1)^2 + (z_2 - z_1)^2}$

$= \sqrt{(5-0)^2 + (-3+3)^2 + (-6+6)^2}$

$= \sqrt{5^2}$

= 5 units //

b) Distance from Q to XY plane:

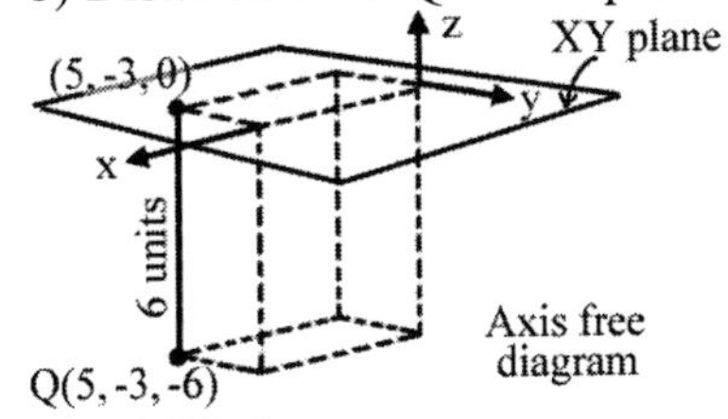

Q to XY plane:

$= \sqrt{(x_2 - x_1)^2 + (y_2 - y_1)^2 + (z_2 - z_1)^2}$

$= \sqrt{(5-5)^2 + (-3+3)^2 + (0+6)^2}$

$= \sqrt{6^2}$

= 6 units //

27. Consider the point A(2, –2, 5). Draw a diagram to locate the position of A in space. Find the shortest distance from A to the x-axis and XZ plane.

Answers:
Given A = (2, –2, 5)
Position of A in a 3-dimensional Cartesian plane:

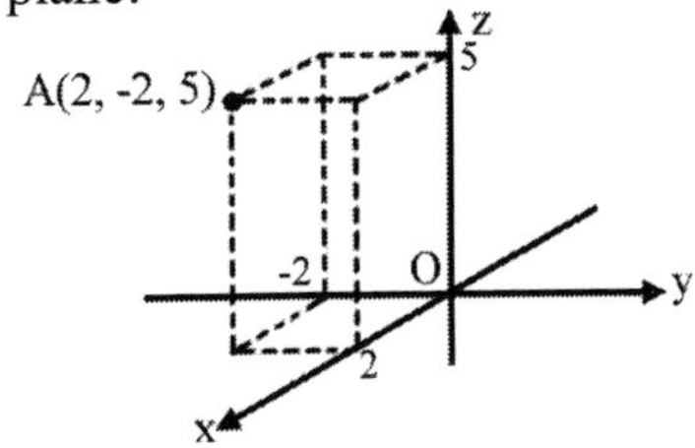

Distance from A to x-axis:

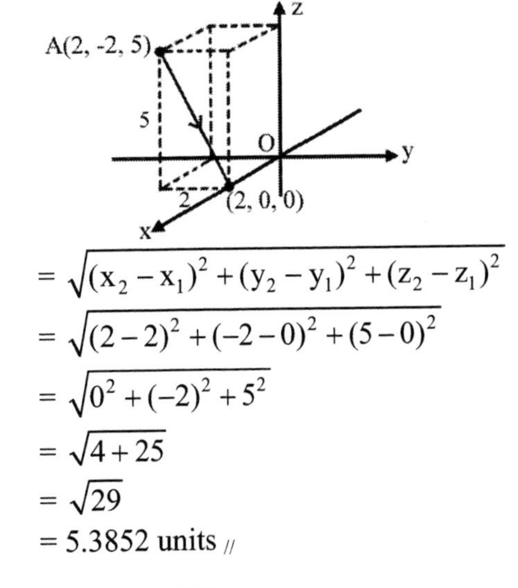

$= \sqrt{(x_2 - x_1)^2 + (y_2 - y_1)^2 + (z_2 - z_1)^2}$

$= \sqrt{(2-2)^2 + (-2-0)^2 + (5-0)^2}$

$= \sqrt{0^2 + (-2)^2 + 5^2}$

$= \sqrt{4 + 25}$

$= \sqrt{29}$

= 5.3852 units //

Alternatively:
Using Pythagorean Theorem:
Let d = Distance from A to x-axis

$d^2 = 5^2 + 2^2$
$= 25 + 4$
$= 29$

$d = \sqrt{29}$
= 5.3852 units //

Distance from A to XZ plane:

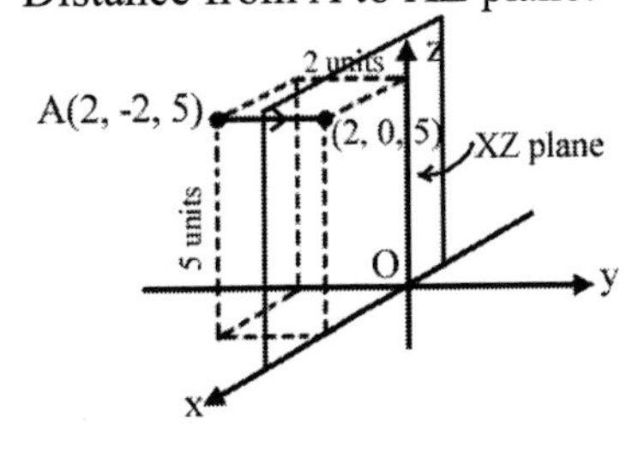

Consider the y-coordinate:
∴ Distance from A to XZ plane is 2 units. //

Alternatively:
Distance from A to XZ plane:
$= \sqrt{(x_2 - x_1)^2 + (y_2 - y_1)^2 + (z_2 - z_1)^2}$
$= \sqrt{(2-2)^2 + (-2-0)^2 + (5-5)^2}$
$= \sqrt{(-2)^2} = \sqrt{4} = 2$ units //

28. Show that A(3, 6, 1), B(2, 8, 3) and C(5, 7, –1) is an isosceles triangle.

Answer:
Given A = (3, 6, 1)
B = (2, 8, 3)
C = (5, 7, –1)
Length of AB:
$AB = \sqrt{(x_2 - x_1)^2 + (y_2 - y_1)^2 + (z_2 - z_1)^2}$
$= \sqrt{(2-3)^2 + (8-6)^2 + (3-1)^2}$
$= \sqrt{(-1)^2 + 2^2 + 2^2}$
$= \sqrt{1+4+4} = \sqrt{9}$
= 3 units
$BC = \sqrt{(5-2)^2 + (7-8)^2 + (-1-3)^2}$
$= \sqrt{3^2 + (-1)^2 + (-4)^2}$
$= \sqrt{9+1+16}$
$= \sqrt{26}$ units

$AC = \sqrt{(5-3)^2 + (7-6)^2 + (-1-1)^2}$
$= \sqrt{2^2 + 1^2 + (-2)^2}$
$= \sqrt{4+1+4} = \sqrt{9}$
= 3 units
∴ Since AB = AC = 3 units, we have shown that △ABC is an isosceles triangle.//

29. $\overline{PQ}$ is the diameter of a sphere O, where P(3, 5, 6) and Q(–1, 3, 2). Find the center point O and the radius of the sphere.

Answers:
Given P = (3, 5, 6) and Q = (–1, 3, 2)
Also given $\overline{PQ}$ = diameter of sphere O
=> Midpoint of $\overline{PQ}$ = center point, O
Center point, O:
$= \left(\frac{x_1 + x_2}{2}, \frac{y_1 + y_2}{2}, \frac{z_1 + z_2}{2}\right)$
$= \left(\frac{3+(-1)}{2}, \frac{5+3}{2}, \frac{6+2}{2}\right)$
$= \left(\frac{2}{2}, \frac{8}{2}, \frac{8}{2}\right)$
= (1, 4, 4)
∴ Center point, O is (1, 4, 4). //

Length of diameter, PQ = distance P to Q
$PQ = \sqrt{(x_2 - x_1)^2 + (y_2 - y_1)^2 + (z_2 - z_1)^2}$
$= \sqrt{(3-(-1))^2 + (5-3)^2 + (6-2)^2}$
$= \sqrt{4^2 + 2^2 + 4^2}$
$= \sqrt{36} = 6$ units
Radius = diameter ÷ 2 = PQ ÷ 2
= 6 ÷ 2
= 3 units
∴ Radius of sphere O is 3 units. //

30. P and Q are located on the z-axis. If P and Q are 8 units from point R where R(6, 3, 1), what is the ordered triple of P and Q?

Answers:
Given P and Q on z-axis
Let P = (0, 0, z_p) and Q = (0, 0, z_q)

Also given R = (6, 3, 1)

PR = QR = 8

Thus, PR:

$$\sqrt{(x_2 - x_1)^2 + (y_2 - y_1)^2 + (z_2 - z_1)^2} = 8$$

$$\sqrt{(6-0)^2 + (3-0)^2 + (1-z_p)^2} = 8$$

$6^2 + 3^2 + (1 - z_p)^2 = 8^2$

$36 + 9 + 1 - 2z_p + z_p^2 = 64$

$z_p^2 - 2z_p - 18 = 0$

Using quadratic formula to solve z_p:
{where $z_p^2 - 2z_p - 18 = 0$ in the form $ax^2 + bx + c = 0$, thus a = 1, b = −2, c = −18}

$$z_p = \frac{-b \pm \sqrt{b^2 - 4ac}}{2a}$$

$$= \frac{-(-2) \pm \sqrt{(-2)^2 - 4(1)(-18)}}{2(1)}$$

$$= \frac{2 \pm \sqrt{4 + 72}}{2}$$

$$= \frac{2 \pm \sqrt{76}}{2}$$

$$= \frac{2 \pm 8.7178}{2}$$

= 5.3589 or −3.3589

∴ P(0, 0, 5.3589) and Q(0, 0, −3.3589) //

31. Find the distance of plane P whose equation is x − 5y + 3z = −8 from the origin.

Answer:

Given P: x − 5y + 3z = −8

Rearrange to general form:

=> ax + by + cz + d = 0

x − 5y + 3z + 8 = 0

=> a = 1, b = −5, c = 3, d = 8

Also given origin, O = (0, 0, 0)

Distance from plane P to O:

$$= \frac{|ax_1 + by_1 + cz_1 + d|}{\sqrt{a^2 + b^2 + c^2}}$$

$$= \frac{|1(0) + (-5)(0) + (3)(0) + 8|}{\sqrt{1^2 + (-5)^2 + 3^2}}$$

$$= \frac{|8|}{\sqrt{1 + 25 + 9}}$$

$$= \frac{8}{\sqrt{35}}$$

= 1.3522 units //

32. Point K(1, 3, −5) floats above plane P whose equation is, 2x + y − 2z = 5. What is the distance between K and plane P?

Answer:

Given plane, P: 2x + y − 2z = 5

Rearrange to general form:

=> ax + by + cz + d = 0

2x + y − 2z − 5 = 0

=> a = 2, b = 1, c = −2, d = −5

Also given K = (1, 3, −5)

Distance from K to plane P:

$$= \frac{|ax_1 + by_1 + cz_1 + d|}{\sqrt{a^2 + b^2 + c^2}}$$

$$= \frac{|2(1) + 1(3) + (-2)(-5) + (-5)|}{\sqrt{2^2 + 1^2 + (-2)^2}}$$

$$= \frac{|2 + 3 + 10 - 5|}{\sqrt{4 + 1 + 4}}$$

$$= \frac{|10|}{\sqrt{9}}$$

$$= \frac{10}{3} \text{ units}$$ //

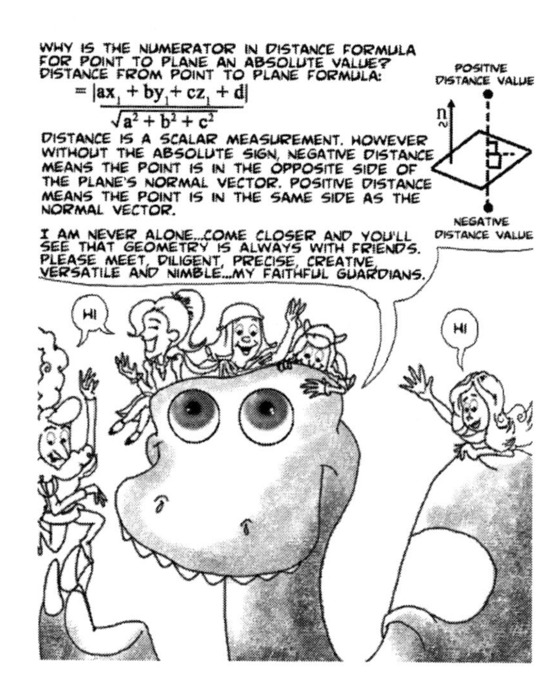

33. The symmetric equation of a line is 2x = 3y + 1 = $\frac{6-z}{2}$. If P is a point on the line where the line meets the XY plane, determine the coordinate for P.

Answer:

Given 2x = 3y + 1 = $\frac{6-z}{2}$

At the XY plane, z = 0 => $\frac{6-0}{2} = 3$

2x = 3y + 1 = 3

For x coordinate:

=> 2x = 3

$\therefore x = \frac{3}{2}$

For y coordinate:

=> 3y + 1 = 3

3y = 3 – 1 = 2

$\therefore y = \frac{2}{3}$

$\therefore P = \left(\frac{3}{2}, \frac{2}{3}, 0\right)$ //

$2x = 3y + 1 = \frac{6-z}{2}$

$P(\frac{3}{2}, \frac{2}{3}, 0)$

XY plane

Axis free diagram

34. Find the coordinate of the point T where the line with symmetric equation, $x + 1 = \frac{3y-2}{4} = z - 1$ meets the YZ plane.

Answer:

Given $x + 1 = \frac{3y-2}{4} = z - 1$ ...(1)

At the YZ plane, x = 0 => 0 + 1 = 1

Substitute x = 0 into (1):

$1 = \frac{3y-2}{4} = z - 1$

Thus,

For y coordinate:

$\frac{3y-2}{4} = 1$

3y – 2 = 4

3y = 4 + 2 = 6

$\therefore$ y = 2

For z coordinate:

z – 1 = 1

$\therefore$ z = 1 + 1= 2

$\therefore$ T(0, 2, 2) //

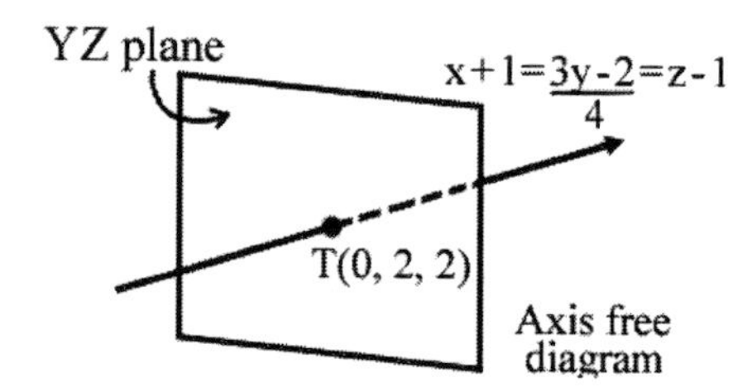

35. Line M meets the XZ plane at point R. If the symmetric equation of $\overleftrightarrow{M}$ is $\frac{3x-1}{5}$ = y + 4 = 2z – 2, find the ordered triple of R.

Answer:

Given $\frac{3x-1}{5} = y + 4 = 2z - 2$ ...(*)

At XZ plane, y = 0 => 0 + 4 = 4

Substitute into (*):

=> $\frac{3x-1}{5} = 4 = 2z - 2$

Thus,

For x coordinate:

$\frac{3x-1}{5} = 4$

3x – 1 = 20

$3x = 20 + 1 = 21$
$\therefore x = 7$
For z coordinate:
$=> 2z - 2 = 4$
$2z = 4 + 2 = 6$
$\therefore z = 3$
$\therefore R(7, 0, 3)$ //

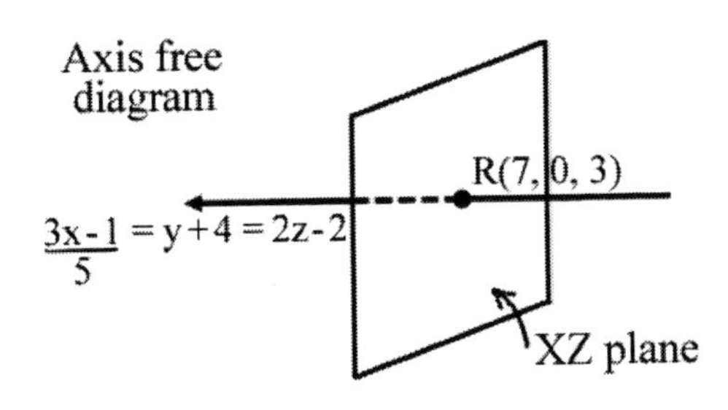

36. $\frac{x-1}{2} = y = 3z + 6$ is the equation of line R. If $\overleftarrow{R}$ meets plane P with equation 2x – y + 3z = 8 at K. Find the value of K.

Answer:
Given $\overleftarrow{R}$: $\frac{x-1}{2} = y = 3z + 6$
Convert to parametric equation:
$x - 1 = 2t => x = 1 + 2t$
$y = t$
$3z + 6 = t => z = -2 + \frac{t}{3}$ ...(1)
Also given, P: $2x - y + 3z = 8$ ...(2)
Substitute (1) into (2):
$$2(1+2t) - t + 3\left(-2 + \frac{t}{3}\right) = 8$$
$2 + 4t - t - 6 + t = 8$
$4t - 4 = 8$
$4t = 8 + 4 = 12$
$\therefore t = 3$
Substitute t = 3 into (1):
$x = 1 + 2(3)$
$= 1 + 6$
$= 7$
$y = 3$
$z = -2 + \frac{3}{3}$
$= -2 + 1$
$= -1$
$\therefore$ Line R meets plane P at K (7, 3, –1). //

37. Find the equation of the line that is normal to the plane x – 2y + 5z = 18 and passes through A(3, –2, 0).

Answer:
Given plane: $x - 2y + 5z = 18$
=> Normal vector, $\underset{\sim}{n} = [1, -2, 5]$
Also given point, A = (3, –2, 0)
Given, line is normal to plane:
=> Normal vector, $\underset{\sim}{n}$ = direction vector of line
=> Direction vector, $\underset{\sim}{u} = [1, -2, 5]$
Thus, symmetric equation of line with direction vector, $\underset{\sim}{u}$ and passes through A:
$$\frac{x - x_1}{a} = \frac{y - y_1}{b} = \frac{z - z_1}{c}$$
$$\frac{x-3}{1} = \frac{y+2}{-2} = \frac{z-0}{5}$$
$$x - 3 = \frac{y+2}{-2} = \frac{z}{5}$$ //

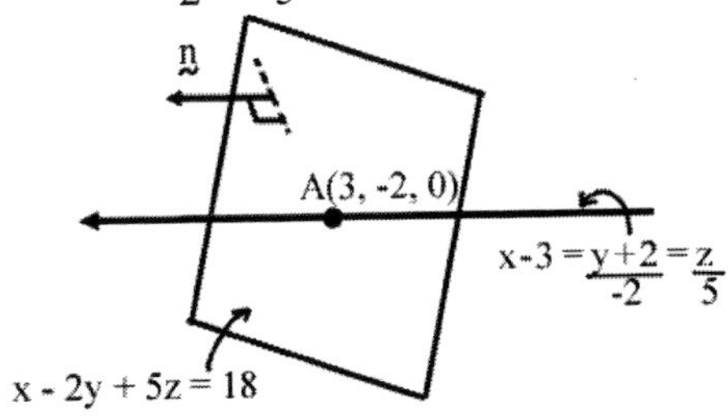

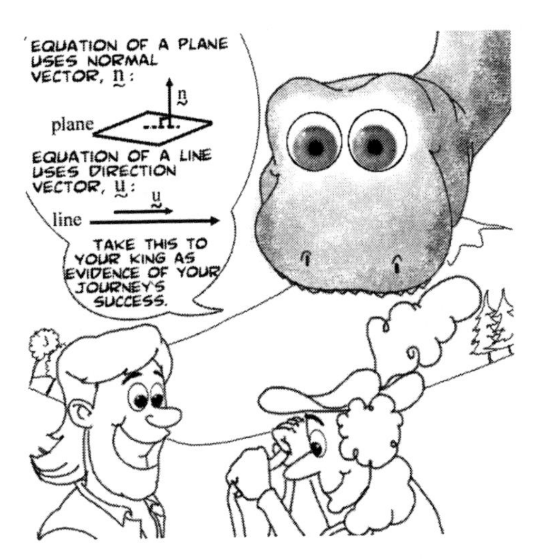

38. Line R passes through A(–2, 7, –1) and is perpendicular to the plane P: 2x + 2y – 3z = 12. Find the symmetric equation of line R.

Answer:
Given point, A = (−2, 7, −1)
Plane, P: 2x + 2y − 3z = 12
=> Normal vector, $\underset{\sim}{n}$ of P = [2, 2, −3]
Since $\overleftrightarrow{R}$ perpendicular to P
=> Normal vector, $\underset{\sim}{n}$ = direction vector, $\underset{\sim}{u}$ of $\overleftrightarrow{R}$
=> Direction vector, $\underset{\sim}{u}$ = [2, 2, −3]
Symmetric equation of line R with direction vector, $\underset{\sim}{u}$ and passes through A:

$$\frac{x-x_1}{a}=\frac{y-y_1}{b}=\frac{z-z_1}{c}$$

$$\frac{x+2}{2}=\frac{y-7}{2}=\frac{z+1}{-3}$$

$\therefore$ $\overleftrightarrow{R}$ is $\dfrac{x+2}{2}=\dfrac{y-7}{2}=\dfrac{z+1}{-3}$ //

39. Find the foot of the normal from A(3, −1, 5) to the plane, P: 3x − y − 2z = 14. Hence find the shortest distance from A to P.

Answers:
Given A = (3, −1, 5)
Plane, P: 3x − y − 2z = 14 ...(1)
=> a = 3, b = −1, c = −2, d = −14
=> Normal vector, $\underset{\sim}{n}$ of P = [3, −1, −2]
Let N = foot of the normal
Thus, parametric equation of $\overline{NA}$:
$x = x_1 + at$
$y = y_1 + bt$
$z = z_1 + ct$
=> x = 3 + 3t
y = −1 − t
z = 5 − 2t ...(2)
Substitute (2) into (1):
3(3 + 3t) − (−1 − t) − 2(5 − 2t) = 14
9 + 9t + 1 + t − 10 + 4t = 14
14t = 14
$\therefore$ t = 1
Substitute t =1 into (2):
x = 3 + 3(1) = 6
y = − 1 − 1 = −2
z = 5 − 2(1) = 5 − 2 = 3
$\therefore$ N (6, −2, 3)

$\therefore$ Foot of the normal is N(6, −2, 3). //
Shortest distance, $\overline{NA}$:

$$= \sqrt{(6-3)^2+(-2+1)^2+(3-5)^2}$$
$$= \sqrt{3^2+(-1)^2+(-2)^2}$$
$$= \sqrt{9+1+4}$$
$$= \sqrt{14} = 3.7417 \text{ units} //$$

Alternatively:
Using distance from point to plane formula:

$$= \frac{|ax_1+by_1+cz_1+d|}{\sqrt{a^2+b^2+c^2}}$$

$$= \frac{|3(3)+(-1)(-1)+(-2)(5)+(-14)|}{\sqrt{3^2+(-1)^2+(-2)^2}}$$

$$= \frac{|9+1-10-14|}{\sqrt{9+1+4}} = \frac{|-14|}{\sqrt{14}}$$ ⇦ absolute values are always positive

$$= \frac{14}{\sqrt{14}} = 3.7417 \text{ units} //$$

40. Point Q(1, 3, −2) floats above plane 2x + y − 3z = 39. Find the mirror image of Q on the plane.

Answer:
Given point, Q = (1, 3, −2)
Given plane: 2x + y − 3z = 39 ...(1)
=> normal vector, $\underset{\sim}{n}$ = [2, 1, −3]
Let N = image of Q on the plane
Parametric equation of $\overline{NQ}$:

$x = 1 + 2t$
$y = 3 + t$
$z = -2 - 3t \qquad \ldots(2)$
Substitute (2) into (1):
$2(1 + 2t) + (3 + t) - 3(-2 - 3t) = 39$
$2 + 4t + 3 + t + 6 + 9t = 39$
$11 + 14t = 39$
$14t = 28$
$t = 2$
Substitute $t = 2$ into (2):
$x = 1 + 2(2) = 1 + 4 = 5$
$y = 3 + 2 = 5$
$z = -2 - 3(2) = -2 - 6 = -8$
$\therefore$ Image of Q on the plane is N(5, 5, −8). //

Q(1, 3, -2)
2x + y - 3z = 39
N(5, 5, -8)

41. Given planes A and B are parallel. Show that if plane C intersects plane A at dihedral angle = 60°, then plane C must also intersects plane B at dihedral angle = 60°.

Answer:
Given conditions:
a) Planes A and B are parallel
b) Plane C intersects plane A at dihedral angle = 60°

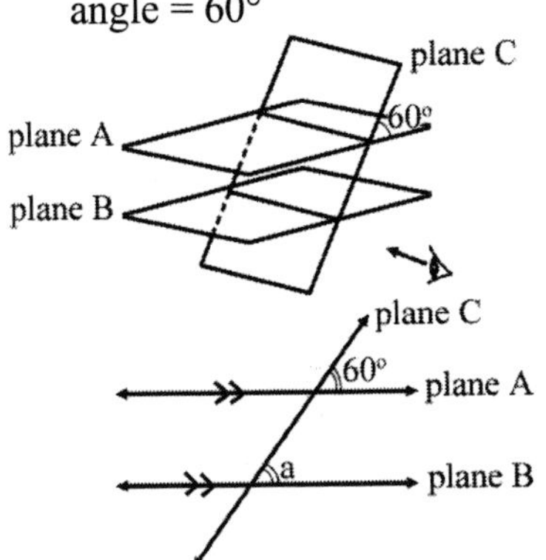

Let a = angle between plane C and plane B (i.e. dihedral angle between planes C and B).
Since plane A and plane B are parallel planes, corresponding angles are congruent.
=> $a \cong 60°$
$\therefore$ We have shown that when planes A and B are parallel, plane C that cuts plane A at dihedral angle = 60° must also cut plane B at dihedral angle = 60°. //

42. Planes A and B are two intersecting planes in space. If plane A: $2x + y + 2z = 3$ and plane B: $x - 2y + 2z = 4$, find the dihedral angle.

Answer:
Given plane A: $2x + y + 2z = 3$
=> normal vector of plane A, $\underset{\sim}{n}_1 = [2, 1, 2]$
Given plane B: $x - 2y + 2z = 4$
=> normal vector of plane B, $\underset{\sim}{n}_2 = [1, -2, 2]$
Let θ = dihedral angle between planes A and B

$$\cos\theta = \frac{a_1a_2 + b_1b_2 + c_1c_2}{\sqrt{a_1^2 + b_1^2 + c_1^2} \times \sqrt{a_2^2 + b_2^2 + c_2^2}}$$

$$= \frac{2(1) + 1(-2) + 2(2)}{\sqrt{2^2 + 1^2 + 2^2} \times \sqrt{1^2 + (-2)^2 + 2^2}}$$

$$= \frac{2 - 2 + 4}{\sqrt{4 + 1 + 4} \times \sqrt{1 + 4 + 4}}$$

$$= \frac{4}{\sqrt{9} \times \sqrt{9}} = \frac{4}{3 \times 3}$$

$$= \frac{4}{9}$$

$$\theta = \cos^{-1}\frac{4}{9} = 63.61°$$

$\therefore$ Dihedral angle between plane A and plane B is 63.61°. //

43. Plane A and plane B are two intersecting planes. If plane A is $x + 3y - z = 3$ and plane B is $x + 2y + z = 4$, show that the dihedral angle between the two planes is approximately 42°.

Answer:
Given plane A: $x + 3y - z = 3$
=> Normal vector, $\underset{\sim}{n}_A = [1, 3, -1]$
Given plane B: $x + 2y + z = 4$
=> Normal vector, $\underset{\sim}{n}_B = [1, 2, 1]$
Let θ = dihedral angle between planes A and B

$$\cos\theta = \frac{a_1a_2 + b_1b_2 + c_1c_2}{\sqrt{a_1^2 + b_1^2 + c_1^2} \times \sqrt{a_2^2 + b_2^2 + c_2^2}}$$
$$= \frac{1(1) + (3)(2) + (-1)(1)}{\sqrt{1^2 + 3^2 + (-1)^2} \times \sqrt{1^2 + 2^2 + 1^2}}$$
$$= \frac{1 + 6 - 1}{\sqrt{1 + 9 + 1} \times \sqrt{1 + 4 + 1}}$$
$$= \frac{6}{\sqrt{11} \times \sqrt{6}}$$
$$= 0.7385$$
$$\theta = \cos^{-1} 0.7385$$
$$= 42.39° \approx 42°$$
∴ We have shown that the dihedral angle between plane A and plane B is approximately 42°. //

44. P(2, 6, 3), Q(1, 5, 7), R(4, 1, a) and S(b, c, 9) are 4 points in space. If $\overline{PQ}$ and $\overline{RS}$ are parallel lines, find the values of a, b, and c.

Answers:
Given P = (2, 6, 3)
Q = (1, 5, 7)
Direction vector of $\overrightarrow{PQ}$: {in vector form}
$= [x_Q - x_P, y_Q - y_P, z_Q - z_P]$
$= [1 - 2, 5 - 6, 7 - 3]$
$= [-1, -1, 4]$
Also given R = (4, 1, a)
S = (b, c, 9)
Direction vector of $\overrightarrow{RS}$:
$= [x_S - x_R, y_S - y_R, z_S - z_R]$
$= [b - 4, c - 1, 9 - a]$
Since $\overline{PQ} \parallel \overline{RS}$
=> Direction vectors of $\overrightarrow{PQ}$ and $\overrightarrow{RS}$ are equal
=> $[-1, -1, 4] = [b - 4, c - 1, 9 - a]$
Comparing equal parts:
$-1 = b - 4$
∴ $b = -1 + 4 = 3$ //
$-1 = c - 1$
∴ $c = -1 + 1 = 0$ //
$4 = 9 - a$
∴ $a = 9 - 4 = 5$ //

45. ABCD is a parallelogram. A(1, 6, 5), B(2, 8, 7) and C(5, 3, 9). Determine the fourth coordinate, D. Subsequently find the midpoint of the parallelogram's diagonals.

Answers:
Given A = (1, 6, 5)
B = (2, 8, 7)
C = (5, 3, 9)
Parallelogram has 2 pairs of parallel lines
=> vector of $\overrightarrow{BA}$ = vector of $\overrightarrow{CD}$ ...(*)
Direction vector of $\overrightarrow{BA}$: {in vector form}
$= [x_A - x_B, y_A - y_B, z_A - z_B]$
$= [1 - 2, 6 - 8, 5 - 7]$
$= [-1, -2, -2]$
Let $D = (x_D, y_D, z_D)$
Direction vector of $\overrightarrow{CD}$: {in vector form}
$= [x_D - 5, y_D - 3, z_D - 9]$

Thus,
From (*):
Direction vector: $\overrightarrow{BA} = \overrightarrow{CD}$
$[-1, -2, -2] = [x_D - 5, y_D - 3, z_D - 9]$
Comparing equal parts:
$-1 = x_D - 5$
$\therefore x_D = -1 + 5 = 4$
$-2 = y_D - 3$
$\therefore y_D = -2 + 3 = 1$
$-2 = z_D - 9$
$\therefore z_D = -2 + 9 = 7$
$\therefore D(4, 1, 7)$ //

Midpoint of diagonal, $\overline{AC}$:
$= \left(\frac{x_1 + x_2}{2}, \frac{y_1 + y_2}{2}, \frac{z_1 + z_2}{2}\right)$
$= \left(\frac{1+5}{2}, \frac{6+3}{2}, \frac{5+9}{2}\right)$
$= \left(\frac{6}{2}, \frac{9}{2}, \frac{14}{2}\right)$
$= \left(3, \frac{9}{2}, 7\right)$

Midpoint of diagonal, $\overline{BD}$:
$= \left(\frac{x_1 + x_2}{2}, \frac{y_1 + y_2}{2}, \frac{z_1 + z_2}{2}\right)$
$= \left(\frac{2+4}{2}, \frac{8+1}{2}, \frac{7+7}{2}\right)$
$= \left(\frac{6}{2}, \frac{9}{2}, \frac{14}{2}\right)$
$= \left(3, \frac{9}{2}, 7\right)$
$\therefore$ Midpoint of the diagonals in parallelogram ABCD is $\left(3, \frac{9}{2}, 7\right)$. //

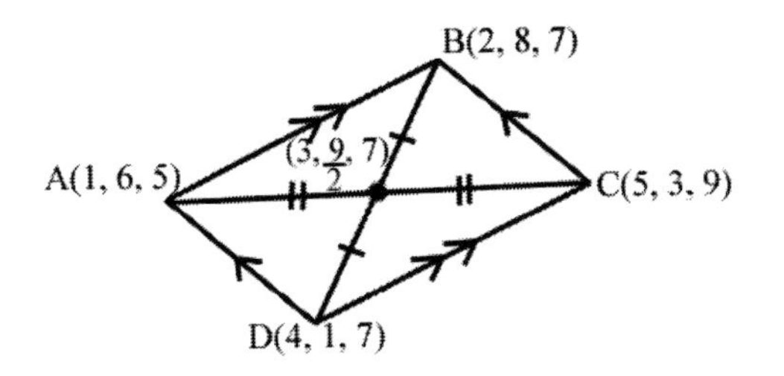

46. Show that A(2, −3, 1), B(1, 4, 3) and C(0, 11, 5) are collinear.

Answer:
Given A = (2, −3, 1)
B = (1, 4, 3)
C = (0, 11, 5)
Direction vector of $\overrightarrow{AB}$:
$= [x_B - x_A, y_B - y_A, z_B - z_A]$
$= [1 - 2, 4 - (-3), 3 - 1]$
$= [-1, 4 + 3, 2]$
$= [-1, 7, 2]$
Direction vector of $\overrightarrow{BC}$:
$= [x_C - x_B, y_C - y_B, z_C - z_B]$
$= [0 - 1, 11 - 4, 5 - 3]$
$= [-1, 7, 2]$
$\therefore$ Since points that are collinear are on the same straight line and therefore have the same direction vector. $\overrightarrow{AB}$ and $\overrightarrow{BC}$ have the same direction vector [−1, 7, 2]. B is common to both $\overrightarrow{AB}$ and $\overrightarrow{BC}$, hence points A, B, and C must be on the same line. Therefore we have shown that A, B, and C are collinear. //

47. Show that P(−1, 6, 3), Q(3, 4, −2) and R(11, 0, −12) are collinear points. Hence find the ratio which Q divides $\overline{PR}$.

Answers:
Given P = (−1, 6, 3)

Given Q = (3, 4, –2)
R = (11, 0, –12)

Since collinear points are on the same straight line, their direction vector must be the same.

Direction vector of $\overrightarrow{PQ}$:

$= [x_Q - x_P, y_Q - y_P, z_Q - z_P]$
$= [3 - (-1), 4 - 6, -2 - 3]$
$= [3 + 1, -2, -5]$
$= [4, -2, -5]$

Direction vector of $\overrightarrow{PR}$:

$= [x_R - x_P, y_R - y_P, z_R - z_P]$
$= [11 - (-1), 0 - 6, -12 - 3]$
$= [11 + 1, -6, -15]$
$= [12, -6, -15]$
$= 3[4, -2, -5]$
$= 3\overrightarrow{PQ}$ ...(*)

$\therefore$ $\overline{PQ}$ and $\overline{PR}$ have the same direction vector [4, –2, –5]. P is common to both $\overrightarrow{PQ}$ and $\overrightarrow{PR}$. Therefore P, Q, and R have been shown to be collinear. //

From (*):
$\overrightarrow{PR} = 3\overrightarrow{PQ}$

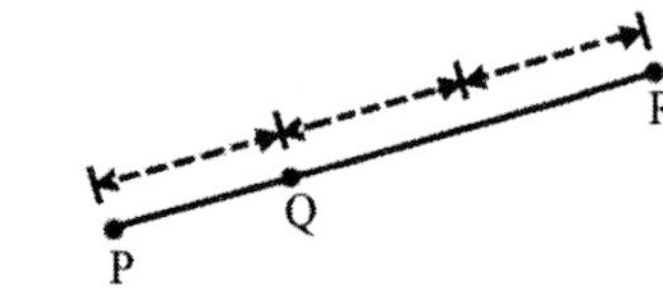

=> 2PQ = QR

$\frac{PQ}{QR} = \frac{1}{2}$

$PQ : QR = 1 : 2$

$\therefore$ Q divides $\overline{PR}$ in the ratio, $1:2$. //

48. A, B and C are collinear points, where A(–1, 3, 4), B(5, 5, 7) and C(23, a, b). If B divides $\overline{AC}$ in the ratio $1:3$, determine the values of a and b.

Answers:

Given A = (–1, 3, 4)
B = (5, 5, 7)
C = (23, a, b)

Also given A, B, and C are collinear

Direction vector of $\overrightarrow{AB}$:

$= [x_B - x_A, y_B - y_A, z_B - z_A]$
$= [5 - (-1), 5 - 3, 7 - 4]$
$= [5 + 1, 2, 3]$
$= [6, 2, 3]$

Direction vector of $\overrightarrow{BC}$:

$= [x_C - x_B, y_C - y_B, z_C - z_B]$
$= [23 - 5, a - 5, b - 7]$
$= [18, a - 5, b - 7]$

Also given $3\overline{AB} = \overline{BC}$

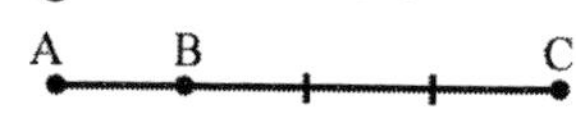

Thus,
$3\overrightarrow{AB} = \overrightarrow{BC}$
$3[6, 2, 3] = [18, a - 5, b - 7]$
$[18, 6, 9] = [18, a - 5, b - 7]$

Comparing equal parts:

$=> 6 = a - 5$
$\therefore a = 6 + 5 = 11$ //
$=> 9 = b - 7$
$\therefore b = 9 + 7 = 16$ //

49. Determine the equation of the sphere centered at (2, 3, 5) whose radius is 4 units.

Answer:

Given center point = (2, 3, 5)
Radius, r = 4

Equation of sphere:

$(x - x_1)^2 + (y - y_1)^2 + (z - z_1)^2 = r^2$
$(x - 2)^2 + (y - 3)^2 + (z - 5)^2 = 4^2$
$(x - 2)^2 + (y - 3)^2 + (z - 5)^2 = 16$ //

50. Find the equation of a sphere, center point (2, 1, 0) and radius $\sqrt{12}$ units.

Answer:
Given center point = (2, 1, 0)
Radius, $r = \sqrt{12}$
Equation of sphere:
$(x - x_1)^2 + (y - y_1)^2 + (z - z_1)^2 = r^2$
$(x - 2)^2 + (y - 1)^2 + (z - 0)^2 = \sqrt{12}^2$
$(x - 2)^2 + (y - 1)^2 + z^2 = 12$ //

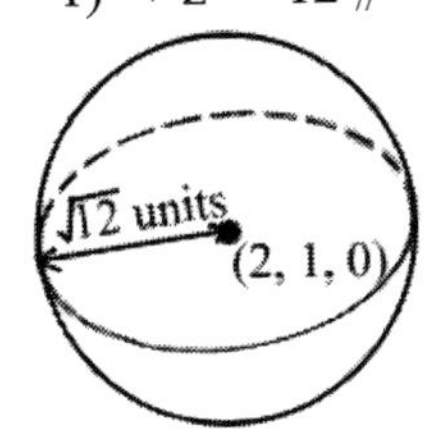

51. Find the equation of a sphere whose center point is (3, –4, 1) and diameter, 6.

Answer:
Given center point = (3, –4, 1)
Diameter = 6
Radius, r = diameter ÷ 2
= 6 ÷ 2
= 3
Equation of sphere:
$(x - x_1)^2 + (y - y_1)^2 + (z - z_1)^2 = r^2$
$(x - 3)^2 + (y - (-4))^2 + (z - 1)^2 = 3^2$
$(x - 3)^2 + (y + 4)^2 + (z - 1)^2 = 9$ //

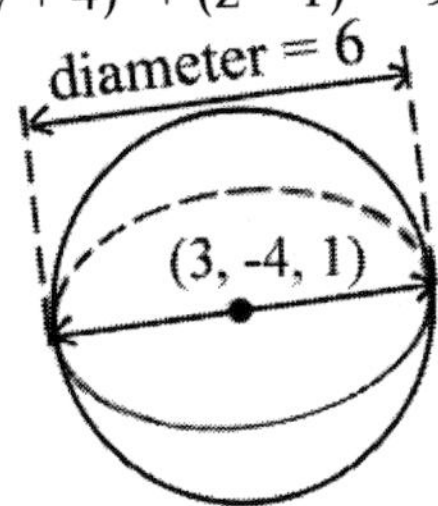

52. Plane A is parallel to XY plane and is 1 unit below it. Plane A cuts through a sphere with center point (2, 3, 4) and radius 6. Find the equation of the circle on plane A.

Answer:
Given center point of sphere = (2, 3, 4)
Radius of sphere, r = 6

Thus,
Equation of sphere:
$(x - x_1)^2 + (y - y_1)^2 + (z - z_1)^2 = r^2$
$(x - 2)^2 + (y - 3)^2 + (z - 4)^2 = 6^2$
$(x - 2)^2 + (y - 3)^2 + (z - 4)^2 = 36$ ...(1)

Also given plane A is parallel to XY plane and 1 unit below it.
=> All points on plane A have z = –1

Substitute z = –1 into (1):
$(x - 2)^2 + (y - 3)^2 + (-1 - 4)^2 = 36$
$(x - 2)^2 + (y - 3)^2 + (-5)^2 = 36$
$(x - 2)^2 + (y - 3)^2 + 25 = 36$
$(x - 2)^2 + (y - 3)^2 = 36 - 25$
$(x - 2)^2 + (y - 3)^2 = 11$

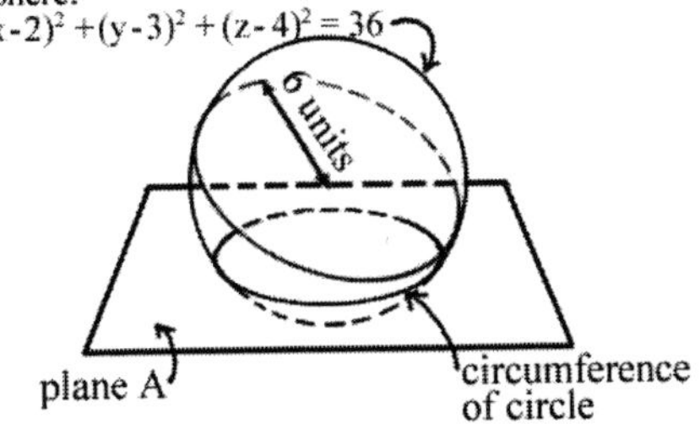

∴ Equation of the circle on plane A is:
$(x - 2)^2 + (y - 3)^2 = 11$ //

53. Given a sphere centered at origin and radius $\sqrt{11}$ units. Further given line R passes through A(1, 4, 2) and B(1, 6, 4). If $\overleftrightarrow{R}$ passes through the sphere, find the points where $\overleftrightarrow{R}$ meets the sphere.

Answer:
Given sphere:
Center point = origin = (0, 0, 0)
Radius, $r = \sqrt{11}$
Equation of sphere:
$(x - x_1)^2 + (y - y_1)^2 + (z - z_1)^2 = r^2$
$(x - 0)^2 + (y - 0)^2 + (z - 0)^2 = \sqrt{11}^2$
$x^2 + y^2 + z^2 = 11$ ...(1)

Also given line R passes through:
A = (1, 4, 2)
B = (1, 6, 4)
Direction vector of $\overleftrightarrow{R}$:
$= [x_B - x_A, y_B - y_A, z_B - z_A]$

$= [1 - 1, 6 - 4, 4 - 2]$
$= [0, 2, 2]$
Parametric equation of $\vec{R}$:
Using point A: {choose either A or B}
$x = 1 + 0t = 1$
$y = 4 + 2t$
$z = 2 + 2t$ $\quad$ t = real value ...(2)

To find the points where $\vec{R}$ cuts the sphere:
Substitute (2) into (1):
$(1)^2 + (4 + 2t)^2 + (2 + 2t)^2 = 11$
$1 + (16 + 16t + 4t^2) + (4 + 8t + 4t^2) = 11$
$1 + 16 + 16t + 4t^2 + 4 + 8t + 4t^2 = 11$
$21 + 24t + 8t^2 = 11$
$8t^2 + 24t + 21 - 11 = 0$
$8t^2 + 24t + 10 = 0$
$2(4t^2 + 12t + 5) = 0$
$2(2t + 1)(2t + 5) = 0$
$\therefore t = -\frac{1}{2}$ or $-\frac{5}{2}$

When $t = -\frac{1}{2}$:

Substitute $t = -\frac{1}{2}$ into (2):
$x = 1$
$y = 4 + 2\times\left(-\frac{1}{2}\right)$
$= 4 + (-1)$
$= 4 - 1$
$= 3$
$z = 2 + 2\times\left(-\frac{1}{2}\right)$
$= 2 + (-1)$
$= 1$
$\therefore$ Point is (1, 3, 1)

When $t = -\frac{5}{2}$:

Substitute $t = -\frac{5}{2}$ into (2):
$x = 1$
$y = 4 + 2\times\left(-\frac{5}{2}\right)$
$= 4 + (-5)$
$= 4 - 5$
$= -1$
$z = 2 + 2\times\left(-\frac{5}{2}\right)$
$= 2 + (-5)$
$= -3$
$\therefore$ Point is (1, −1, −3)
$\therefore$ $\vec{R}$ cuts the sphere at (1, 3, 1) and (1, −1, −3). //

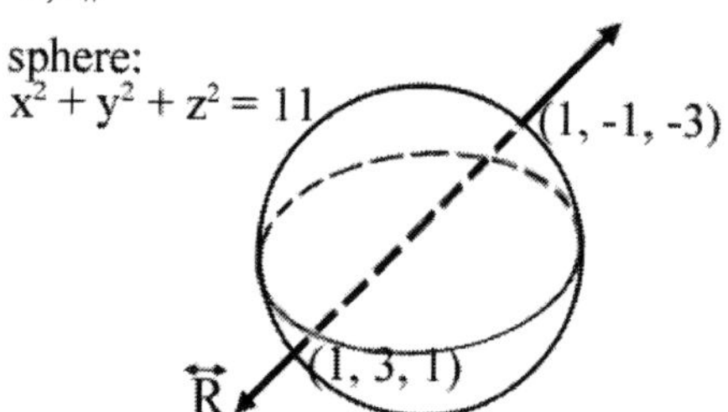

54. A straight line L in space passes through (2, 3, 2) and is perpendicular to plane P whose equation is $2x + y + 2z = 5$. If line L pierces a sphere whose center point is (1, 1, 4) and radius $\sqrt{18}$ units, find the points where line L cuts the sphere.

Answer:
Given L = straight line
=> L passes through (2, 3, 2)
=> L is perpendicular to plane P
Given plane, P: $2x + y + 2z = 5$
=> Normal vector, $\underset{\sim}{n} = [2, 1, 2]$

Since $\vec{L}$ is perpendicular to P, therefore the direction vector of straight line L must equal to the normal vector, $\underset{\sim}{n}$ of plane P.

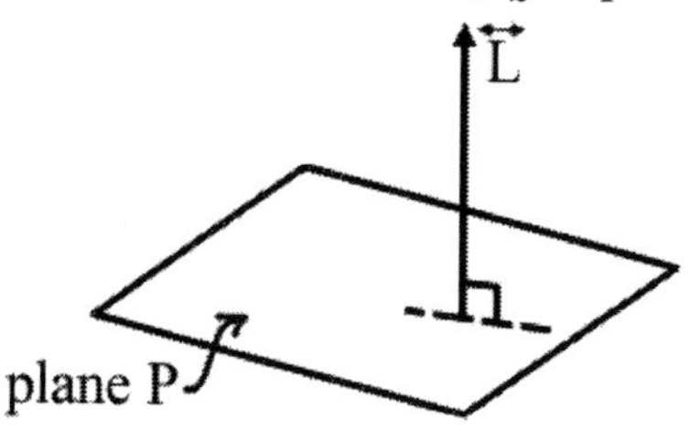

Parametric equation of $\vec{L}$:
$x = 2 + 2t$
$y = 3 + t$
$z = 2 + 2t$ $\quad$ where t = real number ...(1)

Also given sphere:
Center point = (1, 1, 4)
Radius, $r = \sqrt{18}$

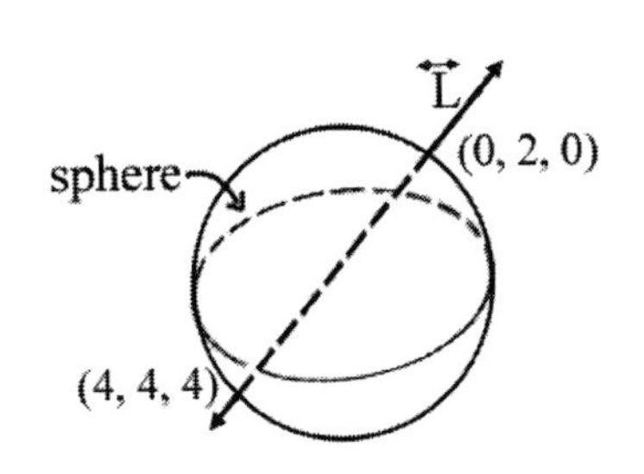

Equation of sphere:
$(x - x_1)^2 + (y - y_1)^2 + (z - z_1)^2 = r^2$
$(x - 1)^2 + (y - 1)^2 + (z - 4)^2 = \sqrt{18}^2$
$(x - 1)^2 + (y - 1)^2 + (z - 4)^2 = 18 \quad ...(2)$

To find where $\overleftrightarrow{L}$ cuts sphere:
Substitute (1) into (2):
$\Rightarrow (2 + 2t - 1)^2 + (3 + t - 1)^2 + (2 + 2t - 4)^2 = 18$
$(1 + 2t)^2 + (2 + t)^2 + (2t - 2)^2 = 18$
$(1 + 2t + 2t + 4t^2) + (4 + 2t + 2t + t^2) + (4t^2 - 4t - 4t + 4) = 18$
$1 + 4t + 4t^2 + 4 + 4t + t^2 + 4t^2 - 8t + 4 = 18$
$9 + 0t + 9t^2 = 18$
$9t^2 - 9 = 0$
$9(t^2 - 1) = 0$
$9(t - 1)(t + 1) = 0$
$\therefore t = 1, \text{ or } -1$

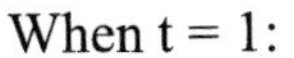

When t = 1:
Substitute t = 1 into (1):
$x = 2 + 2(1)$
$= 2 + 2$
$= 4$
$y = 3 + 1$
$= 4$
$z = 2 + 2(1)$
$= 2 + 2$
$= 4$
$\therefore$ Point is (4, 4, 4)

When t = −1:
Substitute t = −1 into (1):
$x = 2 + 2(-1)$
$= 2 - 2$
$= 0$
$y = 3 + (-1)$
$= 3 - 1$
$= 2$
$z = 2 + 2(-1)$
$= 2 - 2$
$= 0$
$\therefore$ Point is (0, 2, 0)

$\therefore$ $\overleftrightarrow{L}$ cuts the sphere at (4, 4, 4) and (0, 2, 0). //

Index

Index